The ASHRAE Guide for Buildings in Hot and Humid Climates

Expanded 2nd Edition

Lewis G. Harriman III
Joseph W. Lstiburek, Ph.D, P.Eng.

Supervising Technical Committee

Technical Committee 1.12
Moisture Management in Buildings
Chair - Rodney Lewis, P.E.
Houston, TX

Project Management

Lew Harriman
Mason-Grant Consulting
P.O. Box 6547
Portsmouth, NH 03802 USA
www.MasonGrant.com

Project Monitoring Subcommittee

Chair:	**Raymond Patenaude, P.E.**
	The Holmes Agency, St. Petersburg, FL
Region XIII	Suei Keong Chea
	Carrier Corp. Sarawak, Malaysia
TC 1.12, 4.4	Larry Elkin, P.E.
	Applied Building Sciences, Charleston, SC
TC 8.12	Douglas Kosar
	Gas Technology Institute - Chicago
TC 1.12	John Murphy
	Trane Commercial Systems, LaCrosse, WI
TC 4.4	William Rose, R.A.
	U. of Illinois - Building Research Council
TC 9.12	Lynn Werman, P.E.
	Werman Consulting, Omaha, NE
Co-Sponsor:	**U.S. Department of Energy** - Drury Crawley
Co-Sponsor:	**Munters Commercial Systems** - Michael Hayes
Co-Sponsor:	**Southern Company** - Pradeep Vitta
Co-Sponsor:	**Venmar CES** - Paul Pieper

Supplemental Reviewers

Region XII:	Ronald Bailey, P.E.
	Bailey Engineering, Jupiter, FL
TC 1.12	Carl Lawson
	Hanson Services, West Palm Beach, FL
TC 4.4	Sean O'Brien, P.E.
	Simpson, Gumpertz & Heger, NY, NY
Region XIII:	Chee Sheng Ow, Ph.D.
	Universiti Teknologi Mara, Malaysia
Region XIII:	Chandra Sekhar, Ph.D, C.Eng.
	National University of Singapore
SSPC 55:	Stephen Turner
	CTG Energetics, Providence, RI

ISBN 978-1-933742-43-4
American Society of Heating, Refrigerating and Air-Conditioning Engineers, Inc.
1791 Tullie Circle, N.E., Atlanta, GA 30329
(404) 636-8400 www.ashrae.org

Library of Congress Cataloging-in-Publication Data

The ASHRAE guide for buildings in hot and humid climates / Lewis G. Harriman III ... [et al.]. -- 2nd ed.

p. cm.

Summary: "Focuses on needs of owners, architects and engineers who build and manage buildings in hot and humid climates; includes info on building enclosures, dehumidification, sustainability, mold avoidance, energy reduction, moisture management and techniques for reducing energy consumption in hot and humid climates, based on real-world field experience and ASHRAE research"--Provided by publisher.

Includes bibliographical references.

ISBN 978-1-933742-43-4 (hardcover)

1. Air conditioning. 2. Building--Tropical conditions. 3. Dampness in buildings--Prevention. 4. Humidity--Control. I. Harriman, Lewis G., 1949- II. Title: Guide for buildings in hot and humid climates.

TH7687.A785 2009

697.9'3--dc22

2008049708

The ASHRAE Guide for Buildings in Hot and Humid Climates - Second Edition

ISBN 978-1-933742-43-4

1791 Tullie Circle, NE
Atlanta, GA 30329
www.ashrae.org

Printed in the United States of America
Printed using soy-based inks.

Disclaimer

ASHRAE has compiled this publication with care, but ASHRAE has not investigated, and ASHRAE expressly disclaims any duty to investigate, any product, service, process, procedure, design, or the like that may be described herein. The appearance of any technical data or editorial material in this publication does not constitute endorsement, warranty, or guaranty by ASHRAE of any product, service, process, procedure, design, or the like. ASHRAE does not warrant that the information in the publication is free of errors, and ASHRAE does not necessarily agree with any statement or opinion in this publication. The entire risk of the use of any information in this publication is assumed by the user.

Copyright

The ASHRAE Guide for Buildings in Hot and Humid Climates

Major Sections & Chapter Titles

Preface To The 2nd Edition

The first edition of this book contained a subset of the information we provide here. It dealt with the broad, cross-cutting issues of thermal comfort, ventilation air, energy consumption and mold. In this 2nd edition, the book has expanded from 100 to over 300 pages.

As the book expanded, it became apparent that although experts often agree about general principles, digging into the details sometimes generates passionate debate. Strongly-held opinions based on decades of the different experiences of our expert advisors made writing this second edition quite a challenge.

So it's useful to keep in mind that the suggestions presented in this book include a broad range of opinions and judgements. It is quite possible—even probable—that there will be different opinions between experts about any single suggestion. But the authors trust and expect that taken as a whole, the information provided here will be helpful when making the key decisions about design and operation of buildings in hot and humid climates.

Above all, what we have tried to achieve is a clear and engaging presentation of the critical issues. Most experts will probably agree that as long as the key issues are given some attention, more often than not the building will be quite successful. It's when the decision makers are simply *not aware* of the issues that the real problems occur.

For example, it's not obvious to most architectural designers that the design of a building's glazing will govern the comfort of the occupants, the cost of its HVAC system and the building's energy use for all time. Nor is it obvious to HVAC designers that sealing up the connections in exhaust duct work will greatly reduce the risk of mold. But when the *entire team is aware* of the importance of glass design, the importance of overhanging the roof and importance of sealed duct connections, the decisions the team makes on behalf of the owner are likely to be better. Then we will have achieved the purpose of this book: to improve buildings in hot and humid climates for the benefit of their owners, for their occupants and for society as a whole.

Acknowledgments

This book was prompted by the long-standing sustainability concerns of Terry Townsend, P.E., President of ASHRAE during 2006 and 2007. Based on his concern that without ASHRAE guidance, hot and humid climate design practices may not be as sustainable as what will be needed by future generations, President Townsend asked the ASHRAE Board to approve this special project.

This expanded second edition has been made possible by the technical and financial support of: the Office of Building Technologies of the U.S. Department of Energy; the Commercial Systems Division of Munters Corporation in San Antonio, Texas; the Services Division of the Southern Companies in Birmingham, Alabama and Venmar CES Inc. in St-Léonard-d'Aston, Quebec. On behalf of the Project Committee and of the future readership of this book, we express our great appreciation for the support of these generous sponsors, without whom this second edition could not exist.

Dedication

We also appreciate the support of the many donors of the technical material, photos, diagrams and field experiences which enrich and enliven this book. General principles, while useful, are much easier to understand, to remember and to apply when their relevance is made clear through real-world experience and examples. We are very grateful for those experiences, and for the enormous amount of time volunteered by our reviewers and by our Project Monitoring Committee to help improve the text. Wherever the book is clear, accurate and useful, it is largely because of the contributions and the oversight of these generous experts. To them, we dedicate this second edition.

Lew Harriman
Portsmouth, NH
January, 2009

Common Issues In Hot & Humid Climates

The Building Enclosure

HVAC Design

HVAC Design

Construction

Chapter 1

Introduction

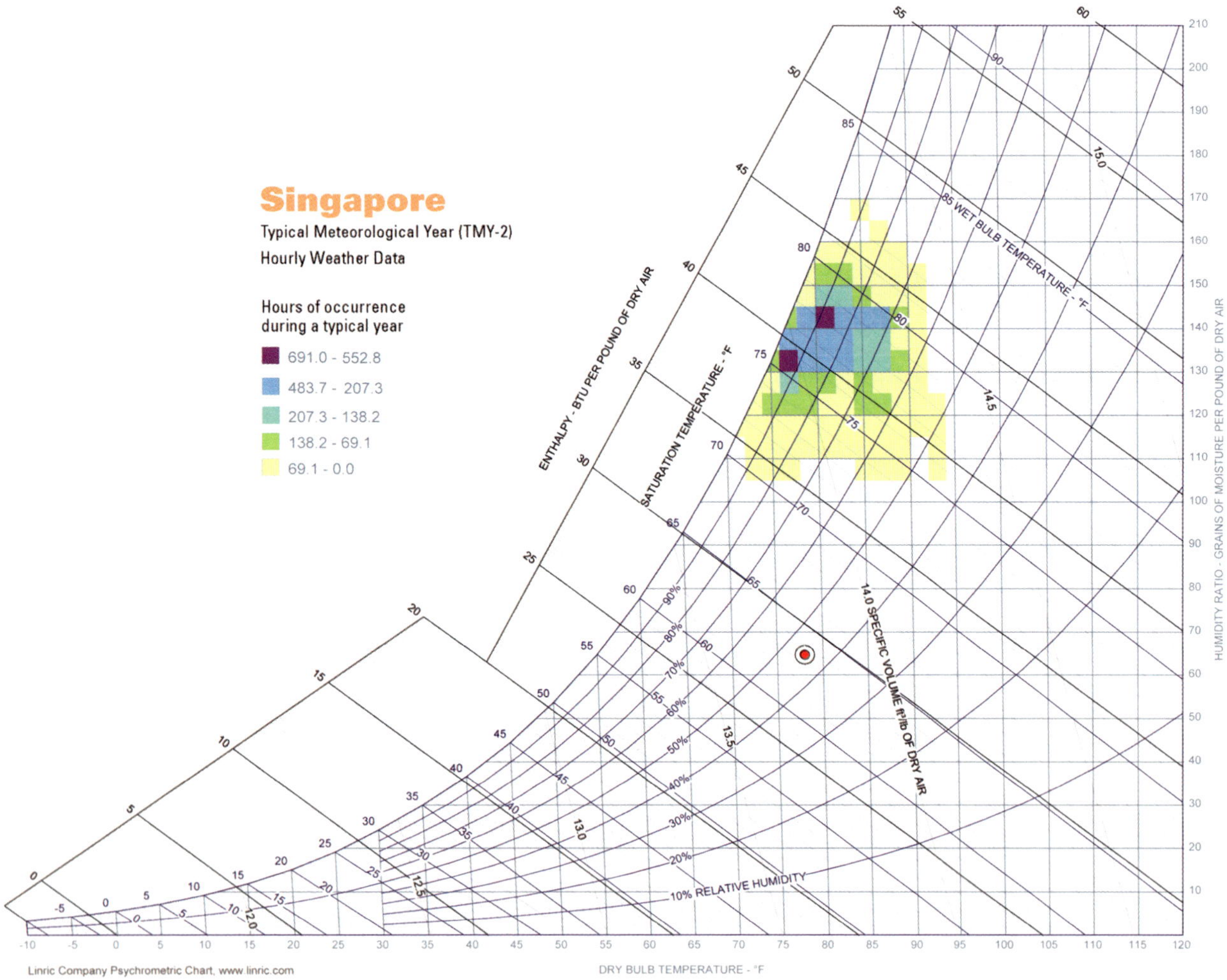

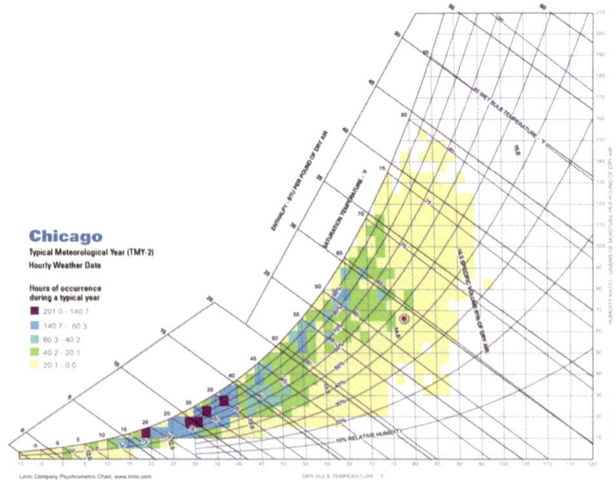

Fig. 1.1 Hot & Humid Climates

In mixed climates like Chicago's, there are certainly many hours each summer when the weather is both hot and humid. But in Singapore, ***all*** *the hours are* ***very*** *humid, and most of the hours are also hotter than the indoor temperature.*

The adjectives "hot" and "humid" suggest the principal challenges addressed by this book. Namely, how to keep heat and humidity ***out*** *of a building.*

Background and Purpose

In October of 2006, after decades of professional practice in the hot and humid regions of North America and the Caribbean, and after trips to support ASHRAE's efforts in South Asia and the Middle East, Terry Townsend P.E., ASHRAE's President during 2006 and 2007, proposed that ASHRAE should answer three questions, and answer them quickly:

1. What should owners, Architects, HVAC designers, contractors and building operators all be thinking about when they build and operate air conditioned buildings—in a sustainable way—in hot and humid climates worldwide?
2. What are the few really critical issues for achieving excellence and long-term sustainability *in these regions*, as opposed to the thousands of critical-but-common issues for achieving excellence in any climate?
3. Most importantly, what sort of simple and practical suggestions can ASHRAE provide—which are focused clearly on hot and humid climates—to help busy and overworked professionals make better decisions about their buildings?

Because of the worldwide acceleration of construction in hot and humid climates, the ASHRAE Board was impressed with the importance and relevance of the questions, and also concerned that the answers were not immediately obvious. That concern led to the publication you hold in your hands.

Readership, Scope & Limitations

The purpose of this book is to help technical professionals design, build and operate commercial and institutional and multi-unit residential buildings in hot and humid climates. If you are not a technical professional, or are interested principally in single-unit residential buildings, this book may meet fewer of your needs.

Overview rather than all the details

We write primarily for the professionals who have to make the overall decisions with respect to buildings, as opposed to the working-level designers or maintenance technicians who need to know exact and detailed specifics which allow "my job to get done by Friday." Specifics will need to come from other sources.

Also, the worldwide range of equipment is far too broad to include here. And critical equipment details become obsolete as manufacturers change their designs. Finally, enough detail to be useful would violate the commercial impartiality required for ASHRAE publications.

The information in this book should help you define the most essential aspects of the building's enclosure, and to specify the general performance requirements of its HVAC systems. It will also discuss some useful design considerations for operation and maintenance. But to complete a set of plans or specifications, you will need to have more extensive discussions with equipment suppliers. This book is focused on the "big picture" decisions.

This book is frequently redundant

The Authors believe that most readers will first read the particular sections which are most related to their immediate projects. So we have tried to make each chapter as complete as possible in itself. Consequently some of the text and graphics appear in more than one chapter. Repetition reduces the number of annoying references to other chapters, but it will be redundant for those intrepid souls who read the book cover-to-cover.

Wide range of topics, narrower range of climates

This book discusses the design of the building and its mechanical systems. And it also includes many issues related to construction and installation, as well as some aspects of building operations and maintenance. So the scope of this book is very broad in its range of topics.

1.2 Rain, heat and condensation

The suggestions contained in this book focus on ways to avoid the potential problems which can arise when a cool, air conditioned building is exposed to these heavy loads, over decades.

On the other hand, the book is narrowly focused on the most important aspects of those topics as they relate to buildings in hot and humid climates. For example, there is no discussion of heating systems, even though heating is often necessary in some buildings in hot and humid climates. And while insulation is important in all climates, solar heat gain through windows and humidity loads from ventilation are especially important in hot and humid climates. So windows and humidity are discussed in more detail than insulation.

Why the broad range of topics has become important

During the first 100 years of the Society's existence, ASHRAE guidance was primarily focused on the design of HVAC & refrigeration systems and equipment. The majority of the membership had less immediate interest in the surrounding issues of building design, construction and installation, or in the operation and maintenance of the building and its systems.

However, ASHRAE's range of concern has expanded greatly with the establishment of Standards 90.1 (Energy Standard for Buildings Except Low-rise Residential Buildings) and Standard 62.1 (Ventilation for Acceptable Indoor Air Quality). In its second century, the Society has become a cognizant authority in these two areas. So in addition to HVAC design, it is important to understand and to improve the decisions made during architectural design and building operations which influence the building's energy use and its indoor air quality.

Scope of this book vs. responsibilities of HVAC designers

As ASHRAE broadens its range of concern to include building design and operational issues, HVAC designers—the core constituency of the Society for its first 100 years—are sometimes uncomfortable.

In the U.S., litigation related to failures or perceived failures in buildings has been popular, at least among attorneys. So HVAC designers are sometimes concerned that if an ASHRAE publication discusses a given topic in depth, the HVAC designer for the building might somehow be held responsible for success or failure in that area.

But to reduce energy use in buildings and to ensure acceptable indoor air quality, issues other than HVAC system design must be addressed—by somebody. The Society serves the general public and it serves technical professionals of all kinds, not only HVAC designers. This book provides *suggestions—not regulations or standards—* for many aspects of building and HVAC design, construction and operation. The book does not assign responsibility for any aspect of building design or performance to anybody in particular. That is a matter best left to contract documents and to building codes.

Reader Input

Finally, we assume that the material contained here is simply a good beginning. It can certainly be improved as readers use the information and reflect upon it in light of their own experiences. We encourage you to contribute those experiences to the ongoing improvement of this book. We welcome all constructive comments, additions and suggestions to improve any aspect of this book's text, graphics or photos. Please address your remarks to:

Lew Harriman
Mason-Grant Consulting
P.O. Box 6547
Portsmouth, NH 03802 USA
LewHarriman@MasonGrant.com
Tel: (603) 431-0635
Fax: (603) 427-0015

Chapter 2

Improving Thermal Comfort

By Lew Harriman

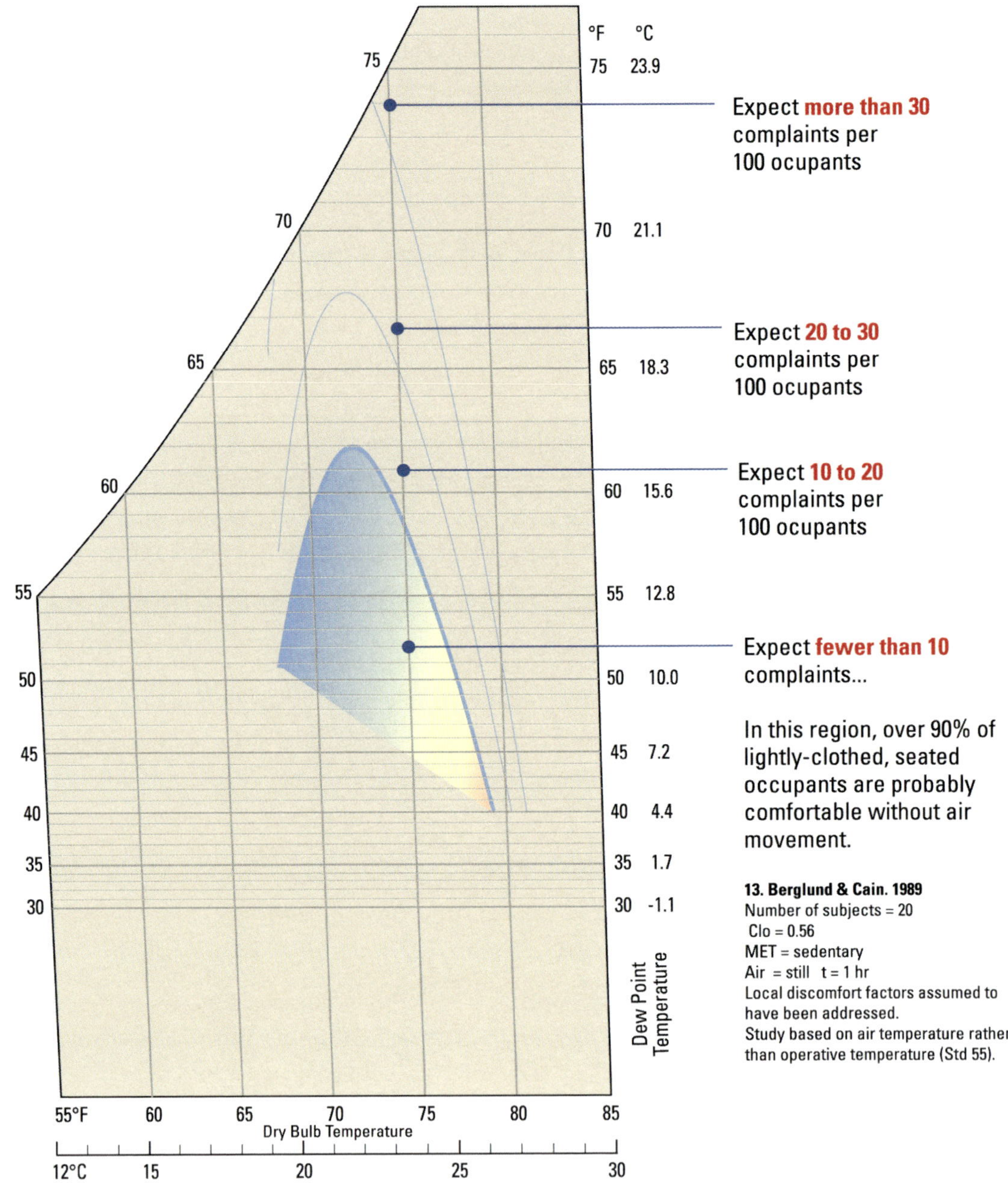

Figure 2.1 Reducing thermal comfort complaints

Unlike structural and electrical engineering in which codes demand that design capacity be over 100% of expected loads, standard HVAC design practice assumes that only 80% of occupants will be thermally satisfied.

Improving beyond 80% probable satisfaction requires a better-than-minumum building enclosure and a more-than-usually-effective HVAC system. Interestingly, these improvements also reduce operating costs and help owners meet energy reduction targets.

Key Points

Without the need for thermal comfort, there would be no need for buildings. In hot and humid climates, protective cages with roofs would serve most other purposes.

So it's useful to keep in mind that the need for thermal comfort is the often unseen and neglected *foundation* for most of the decisions made by a building owner and its Architect, and nearly all of the decisions made by the HVAC designer and the HVAC operating staff.

To increase the probable number of occupants who will find the building to be comfortable, consider implementing these suggestions:

1. Design and construct the enclosure so it is very well-insulated, and so that it keeps solar heat and glare out of the building through exterior shading for all windows. In addition to exterior shading, reduce the size of all windows to a minimum, especially any windows which face west.
2. Keep the indoor temperature above 74°F and below 79°F, while also keeping the indoor dew point below 55°F. [Above 23.3°C and below 26.1°C and below a 12.8°C dew point].
3. Use *more* rather than fewer air handling systems, for a closer match to the different and dynamically-changing internal heat loads in different zones. This improves comfort, and also makes each system simpler, less costly to operate and more reliable.

Thermal Comfort - A Moving Target

Given the fundamental importance of thermal comfort, one might expect that, after several thousand years of designing buildings, the subject would be well-understood. And that if nothing else, the public could safely assume that in the 21st century, Architects and Engineers could ensure that all buildings will be thermally comfortable.

But the public is often thermally discontented. To most Engineers and Architects, it comes as an unwelcome surprise to learn that surveys of occupant satisfaction consistently show major shortcomings in thermal comfort.

For example, surveys performed every year for the U.S. General Services Administration's Public Building Service indicate that one of every three occupants is dissatisfied with the indoor temperature.[1] No other aspect of building function even comes close to the occupants' dissatisfaction with environmental control. It seems clear that thermal comfort deserves more attention from building professionals than it has received in the past.

A good first step is understanding that perception of thermal comfort is heavily influenced by the social and cultural context of the occupants, because those factors govern the occupants' expectations and their responses to thermal stimuli. Perception of comfort by an office worker in Hong Kong is different than the perceptions of an elementary student in Hawaii or of a hotel guest in New Orleans. Different also are the socially-acceptable responses of these people to perceived discomfort. Social context and cultural differences are just as influential over the perception of comfort as the easier-to-quantify variables such as air temperature, humidity and velocity.

This is the first and most important fact to keep in mind when designing for thermal comfort in buildings: thermal comfort is a complex and dynamically-changing mixture of a large number of variables, many of which cannot be calculated and controlled by the Architect or the HVAC designer or the building operator. Those are some of the many reasons why, at any given moment, some percentage of any group of people located in the same space will not be comfortable.

The reasons for this fact are explained in great detail in Chapter 8 of the 2005 ASHRAE Handbook—Fundamentals (Thermal comfort).[2] A more tightly-compressed discussion is presented in ASHRAE Standard 55 (Thermal Environmental Conditions for Human Occupancy).[3] Also, the *logic* behind the current provisions of Std 55 (2004) is discussed and explained by an article written by members of the Std 55 committee, and published in the ASHRAE Journal.[4] These documents are very useful for understanding the full range of variables for human comfort, in all climates.

The purpose of this chapter is to compress the information still further, by focusing on aspects of thermal comfort which are especially relevant for air conditioned buildings in hot and humid climates. Based on those factors, this chapter also provides specific suggestions for what an owner, Architect and HVAC designer can do to achieve a higher percentage of comfortable people in their buildings.

Success in thermal comfort is governed by expectations

The certainty of thermal discomfort for some percentage of the building occupants, some of the time, is not what an owner, Architect, Engineer or occupant wants to hear. Most of us have become accustomed to near-miracles of technology in many parts of our lives. So building occupants as well as building professionals have high expectations for thermal comfort.

But universal and continuous thermal comfort is so difficult to achieve that ASHRAE standards don't even suggest that as a goal. Structural codes might require a building frame with enough strength to meet 160% of the expected stress. Electrical codes might require wiring with current-carrying capacity for 125% of the design load. But for thermal comfort, the current ASHRAE goal is that only 80% of occupants should expect to be satisfied.

Budgeting for building enclosures and HVAC systems could probably be improved if building owners and occupants understood that the *standard* practices of architectural and HVAC designers are only expected to satisfy 80% of the occupants.

In other words, the owner's expectation of thermal comfort should be balanced by his understanding that, for the cost of the typical buildings of the past, he should expect that 20 out of 100 occupants may want to complain of thermal discomfort.

Conversely, since *standard* practices only aim for 80% satisfaction, the owner might wish to consider suggestions which can be an improvement over those traditional practices.

Interestingly, the less-common practices which reduce the number of comfort complaints will also reduce energy and operational costs. Done with reasonable care, improving comfort also makes it easier to meet energy reduction targets established by commercial imperatives or by government regulations.

Specific suggestions in this chapter will include measures for reducing hot and humid air infiltration, reducing solar loads from windows, keeping the dew point under control and providing a more stable thermal environment.

Improving the percentage of comfortable people

Even though we cannot expect to satisfy 100% of occupants 100% of the time, it seems possible for Architects and Engineers to do a better job of providing thermal comfort than has been typical of many buildings. The GSA survey described earlier found that 80% of *all complaints* about buildings relate to thermal comfort.[1] One suspects that it should be possible for clever designers to improve comfort at least a little bit. Perhaps in the future, 80% of complaints about buildings could be about parking, or elevators, or lighting, or the bathrooms... or some mixture of other building characteristics.

Further, if comfort is achieved for more occupants, there are direct cash benefits in addition to fewer complaints. When a building fails to provide thermal comfort, occupants take matters into their own hands, usually by increasing energy consumption. When a traveler cannot speak the local language, he often resorts to speaking his own language ***loudly.*** In a similar way, when thermal comfort is not forthcoming from the building and its HVAC system, occupants often "shout louder" at the HVAC system by twisting the thermostat to crank-up the AC system and overcool the building.

Whenever one feels cold in a building when it's hot outdoors, energy costs are high. When occupants say they are too cold in a hot climate, the Architect and the HVAC designer could have done a better job for both comfort and energy use—provided that the owner gave them the budget to do so.

Dynamic & social nature of comfort perception

Thermal comfort perception is more complex than most other technical problems because comfort is different for different people, even when everything is at steady-state. And outside of the research lab, people are never at steady-state.

Comfort perception changes as people add or subtract clothes and as they increase or reduce their physical activity. Also, as people enter and leave a building, their recent thermal history influences their perception of comfort and their current expectations. For example, in Bangkok the preferred temperature for *transitional* areas (lobbies, entry ways and foyers) was measured to be 80°F [26.7°C]. And the *lower* limit of thermal acceptability for those transitional spaces was found to be 78°F [25.5°C].[5]

Also, the occupants' visual perception of the indoor environment changes their comfort perceptions and expectations. For example, one would expect to be cold in a refrigerated meat locker which has shiny, white aluminum walls and is a small, confined space. So if a room has a normally-comfortable air temperature—but also has the look and feel of a meat locker—the occupants are likely to feel cooler than they really are. This effect was quantified by research at Kansas State University.[6] A "meat locker-style" test room was perceived to be more than 2.7°F [1.5°C] cooler than that same room after it was paneled with wood and carpeted, even though the air temperature and humidity levels during both tests were identical.

Fig. 2.2 Social factors

The social context is a heavy influence on clothing choices. It is unlikely that both of these occupants will be thermally satisfied at the same combination of temperature, dew point and air velocity. If the lady is comfortable, the gentleman is probably going to be too warm. But given the social circumstances of an otherwise pleasant and expensive dinner, neither person is likely to complain, no matter how uncomfortable they may become.

In addition to the dynamic changes and the effect of visual differences, the social situation and cultural differences will increase or reduce the amount of *attention* focused on thermal comfort, which either increases or reduces its importance to the occupant at any specific moment.

For example, a crowd of teenagers at a high school dance pays little attention to thermal comfort. Their clothing choices are governed by their social impressions and romantic concerns rather than by any concern about thermal comfort.

For an example of the influence of culture on comfort complaints, consider a study of local office workers done by researchers from Hong Kong Polytechnic University.[7] The researchers noted that thermal preference responses were skewed by the traditional upbringing and business culture of Chinese office workers.

Hard work and no complaining are basic assumptions of Chinese middle class life. So these subjects were reluctant to express dissatisfaction with any working condition provided by their supervisors. Questionnaires used culturally-adapted terms, slightly different from ASHRAE English, so the occupants could express thermal dissatisfaction without implying criticism of their office buildings or their companies.

Also, temperature preferences were skewed by the cultural need to appear respectfully formal by wearing traditional Northern European/North American business clothing. Suit-wearing Hong Kong office workers preferred slightly cooler temperatures than what was preferred in a similar study of office workers in Bangkok, where the style of formal business clothing was more closely adapted to the hot and humid climate.

Avoiding cold buildings in hot climates

Especially in hot climates, most occupants do not want cold buildings. "Comfortable" is different from "cold."

Searching for optimal temperature levels, one study performed in Sacramento, California looked at the relationship between comfort complaints and adjusting temperature set points to reduce energy costs.[8] The research showed that during hot weather, cold complaints could be expected when the indoor temperature fell to 73°F [22.8°C]. In the same environment, hot complaints would not be expected until the indoor temperature rose above 77°F [25°C]. Interestingly, overall energy costs were also minimized when air temperature stayed between those limits.

Of course, local preferences and specific types of occupancies may call for higher temperatures, such as the preference for 80°F [26.7°C] in transitional spaces in Bangkok mentioned earlier.

The importance of uniformity and stability

Some time after entering a building, occupants become adapted to the new environment. After that point, which varies, they become much more sensitive to fast changes in temperatures, and more sensitive to drafts and to temperature differences within the same space. Conversely, when the temperature stays very uniform around the occupants, the building is perceived to be more comfortable, even if the temperature is slightly above or below the otherwise ideal range. This effect has significant implications for architectural and HVAC designers who want to improve comfort.

Research shows that given stable and uniform conditions, an *additional 10%* of occupants are likely to be satisfied at any given temperature.[2] In other words, one could expect the number of satisfied occupants to rise from 80% to 90%, as long as air temperature, radiant temperatures and air velocities stay uniform around the occupants, and provided that fast temperature *changes* are avoided. (A fast change can be defined as temperature falling by more than 4°F [2.2°C] in less than an hour.)

Stable, uniform temperatures with low air velocities are consequences of buildings which are well-insulated from solar loads. Without a large and highly-variable heat load coming through the enclosure, well-insulated buildings can have simpler and smaller HVAC systems. These can remove the relatively small internal loads more smoothly, so that indoor temperatures stay stable.

For an example, consider a poorly-insulated manufactured building of the type formerly called a "trailer home." With the usual oversized cooling unit, the home should (in theory) be comfortable even with its poor insulation. But the combination of excess cooling capacity and high solar heat loads create very unstable conditions—temperatures and humidities which switch rapidly between overcooling and overheating. Air temperatures near the cooling unit are too cold, while temperatures near the sun-facing wall are too high. Air temperatures are not comfortably uniform in the same space, and the air temperature swings rapidly, leading to discomfort.

Fig. 2.3
Cold buildings are not comfortable
The absurdity of cold buildings in hot climates is obvious to all. Occupants are uncomfortable and energy use is excessive.

Keeping the dew point low helps avoid the need to overcool the space, providing comfort for a wider variety of occupants. This effect can be seen in results from the field research displayed graphically by figure 2.10

Compare that thermally-chaotic trailer home to a well-built house with excellent solar shading over all windows, combined with continuous, sprayed-foam insulation in all the walls, and more sprayed foam insulation applied to the underside of the roof. In that home, the solar load on most days is almost negligible. So the cooling equipment can be much smaller, which means it will not rapidly overcool the space, nor does it need to. Temperatures stay more stable, so comfort is enhanced.

Architecture - The Foundation of Comfort

The baseline heat load is governed by the owners' functional and aesthetic decisions about building orientation, solar shading of the windows and their total glazing area. After those decisions have been made, the architectural designer controls the percent of those baseline loads which enter the building. Then the HVAC designer figures out how to remove the remaining loads as smoothly as possible.

Fig. 2.4
Solar shading improves comfort
This courthouse building in Puerto Rico, built more than 50 years ago, is an excellent example of good and poor practices with respect to thermal comfort. The tower at left has unshaded windows, so special glass will be needed to keep solar loads out. The older main building has loggias and roof overhangs. These effectively eliminate solar loads for the building's windows for most of the day, while providing visual interest for the occupants and the general public.

1. Shaded windows improve comfort & reduce glare

If a person stands between a wall of ice and an open furnace, and if the average of the two surface temperatures is 76°F (24°C) mathematically, he should be comfortable. But of course he won't be. One side will be too cold because it's losing too much heat to the ice by radiation, and the other side will be too hot because the furnace is radiating too much heat.

In a hot and humid climate, the sun shining through windows can feel like an open furnace, and the interior flooded by cold air can feel like ice. To improve thermal comfort, reduce the heat load and glare entering though the windows by shading them on the outside.

To reduce the load still further reduce the percentage of glazing and use low-emissivity insulating glass. This admits visible light, but provides better insulation against convective and conductive heat gain, and it excludes thermal infrared and high-energy ultraviolet energy. Reducing solar loads improves thermal comfort in three ways.

- The windows pass less radiant heat to the occupants, so their "hot-side" is not as hot. The sides of their bodies which face the windows are more comfortable.
- With reduced solar gain through the windows, there is less glare from both the windows and from reflective surfaces inside the room. Reducing eyestrain and facial muscle tension also improves the perception of comfort.
- Because less radiant heat enters the room, the peak cooling load is reduced, allowing the HVAC designer to use smaller systems.

Of course, the HVAC designer has the tools to remove any cooling load, and to keep the temperature of the air the same at the thermostat regardless of load changes—provided the owner has enough money, and provided he has allocated that money to the mechanical system.

But high loads require larger and more expensive equipment, better air mixing and more energy. And keeping the air temperature constant *at the thermostat* does not guarantee comfort. In theory, any heat load can be overcome by adding more money to the mechanical system. But in practice, even with an unlimited budget, the mechanical designer will not usually be able to fit enough duct work and enough supply air diffusers into the ceiling space to provide smooth, even temperatures throughout an overheated space.

By shading the windows, the owner and Architect make an immediate and significant improvement in thermal comfort, as well as reducing the size, construction budget and operational cost for the mechanical system.

Conversely, when the owner and Architect choose not to use solar shading, or at least reduce the percentage of glazing or use spectrally-selective insulating glass, they should expect more comfort complaints from occupants, as well as larger and more expensive mechanical equipment and higher operational costs.

2. Less glass on east and west faces = better comfort

In colder and mixed climates, designers know that glass on the north face means higher heating loads and glass on the south face means higher cooling loads.

But closer to the equator in hot and humid climates, the sun is more nearly overhead. So over a full day, more heat comes through the roof, and through any east- or west-facing windows.

The negative effect of glass is greatest on the west face. Direct solar loads accumulate all afternoon, after the entire building has been heated up during the morning. So the west-facing glass loads accumulate on top of the already high load from earlier in the day, driving the peak heat load for the building to very high levels.

Anything the owner and Architectural designer can do to minimize glass area on the east and especially the west face will improve thermal comfort (and significantly reduce the size, complexity and cost of operating the cooling system.) When windows must face east or west, then the designer and owner should definitely plan for sun-shading and low-emissivity insulating glass to reduce the glare and heat load as much as possible.

3. Comfort is unlikely in glass boxes

Recent architectural aesthetic preferences have sometimes favored large, visually-dramatic glass-walled, light-filled atria and entry lobbies. There is a currently-popular misconception that glass technology has become so advanced that unshaded glass walls will reduce cooling loads by nearly as much as solid, multi-layer insulated

Fig. 2.5 Reducing the high solar load from the west face

Constructed in the mid-1960's, this federal building was designed to minimize the heat loads from the west face, where the solar loads are highest.

The west face is narrow, and it has very little glass. Also, the glass is shaded by vertical aluminum louvers. The stone facade has an air space behind, to limit heat transmission, and to provide a drainage plane for any water intrusion. 40 years later, those architectural design elements perform as well as they did when they were first installed, keeping comfort high and operating costs low.

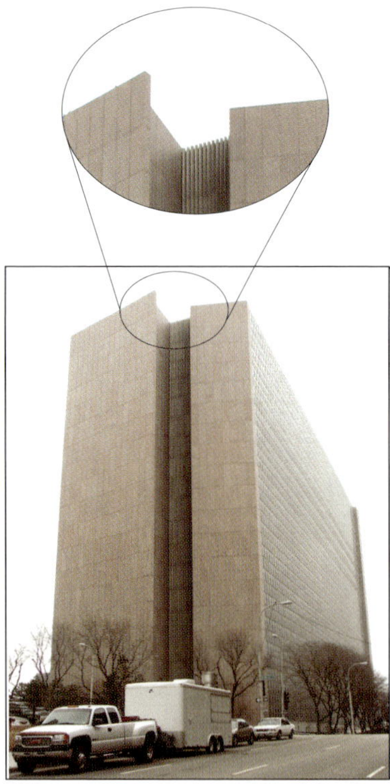

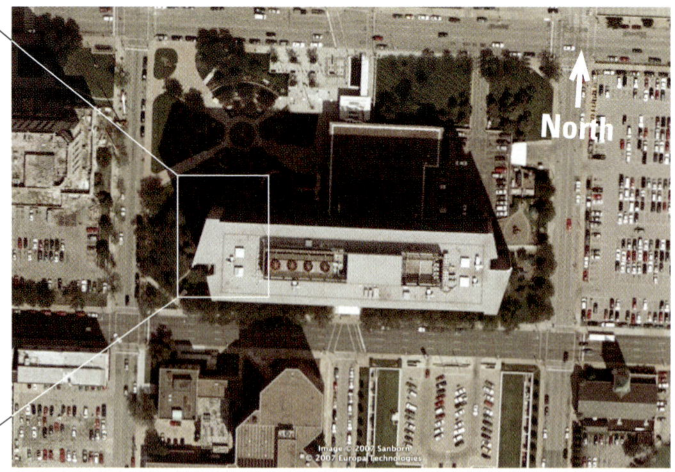

Fig. 2.6
Don't expect comfort inside glass boxes in hot climates
Located in a hot climate, this signature building was notoriously uncomfortable in spite of a very large, expensive and maintenance-intensive mechanical system. High heat loads through the glass made it necessary to retrofit another cooling system under the floor in the lobby, to reduce heat stress for guards stationed at the metal detectors. Operating costs for this extra "comfort band-aid" were estimated at $1,000 per guard, per year. These comfort issues did not escape the attention of occupants and of the press and of the local TV news reporters. Interestingly however, the building's budget, comfort, maintenance and energy issues were not apparently considered to be important by judges of architectural design competitions.[9a,b,c,d,e]

walls. This is not the case. Modern glass may be thermally slightly better than *older* glass—but it still conducts *over five times more* solar heat than a well-insulated solid wall. Also, a solid wall does not transmit *any* heat from solar radiation, while even very costly glass and its framing transmits between 30 and 70% of solar radiation into the building to become a heat load.

Owners and architectural designers planning new projects may not be aware, until the building is complete, of the negative effects that unshaded glass boxes have on thermal comfort. Unshaded glass in hot and humid climates eats up the mechanical budget at a horrifying rate. And large glass surfaces make it very difficult for any HVAC system—no matter how creative, expensive, complex and costly to operate—to keep occupants thermally comfortable.

The popular press makes the public keenly aware of the comfort shortcomings of such buildings, even if architectural and engineering magazines tend to be silent on the subject, especially when the poorly-performing building has been favored by architectural critics with design awards.[9]

Given the silence of professional journals on this topic, HVAC designers can perform a service to all concerned, especially occupants, by keeping owners and architectural designers aware of this relationship at the earliest stages of conceptual design.

Namely: more unshaded glass facing the sun = higher probability of comfort complaints + higher costs for mechanical system construction, operation and maintenance.

4. Tight, well-insulated exterior walls avoid sharp changes

The lower the heat load through the building enclosure, the smaller the cooling system can be, and the easier it is for that system to remove loads as they change through the day. With smaller loads come greater internal temperature stability and uniformity, and less potential for comfort complaints.

In practice, this means the architectural designer should insulate all the walls very well. Also, detail all the joints so they do not have big cracks and holes.

Most designers are surprised to learn that typical low-rise buildings leak a great deal of air. And the leaks are mostly though big holes and cracks, avoidable by better detailing by the architectural designer and by contractors who follow those better instructions.

For example, the leakage rate in 70 low-rise commercial and institutional buildings was measured by the Florida Solar Energy Center to be between 0.5 and 3.0 air changes per hour.[10] Just for a moment, consider those numbers—with light pressure for testing, one or two complete air changes every *hour*, leaking through cracks and holes in the building walls, even when the HVAC systems are turned off.

For comfort, the concern is not so much about insulation that is not thick enough, or for walls which are not hermetically sealed. The more important concern is to make sure that a moderate thickness of insulation is actually in place, and that it is *continuous* and *without* holes and gaps.

Regarding air tightness, the concern is at the joints and seams. Just be sure that there are no big holes or cracks around *penetrations* such as those made for pipes, windows, through-wall AC units and electrical cables. Cracks under window sills, where they are not visually obvious, are at the root of many comfort and mold problems.

Also, make sure that *long* joints are really sealed well, especially where the exterior walls connect to the roofing assembly. Soffits under overhung roofs are notorious for large open seams that leak vast amounts of air, especially in school and restaurant construction.[11]

In summary, at the very least, one should not to be able to see daylight through cracks or holes, when looking from inside the building or from inside the attic.

5. High ceilings and personal fans allow low-cost comfort

The heat removal rate from bare skin depends strongly on the air temperature, but also its flow rate over the skin. Until the indoor dew point gets quite high, one can obtain a comfortable heat removal rate at very low cost by increasing air velocity rather than reducing air temperature. That's one reason why historically, ceiling fans have been a popular means of achieving comfort in hot and humid climates.

The construction cost and energy consumption of a slowly-rotating ceiling fan are very low compared to chilling and forcing air through ducts and diffusers. So ceiling fans are often used in residential buildings, because they save energy by reducing the number of hours that the cooling system must operate to provide comfort. In energy-conscious construction and in developing countries, where both electrical power and construction budgets are severely limited, ceiling fans provide a very favorable ratio of cost to comfort.

Assuming the owner is content with this low-cost, low-energy strategy for improving thermal comfort, the architectural designer rather than the HVAC designer has to take the first step towards implementation. Low ceilings and circulating fans are not a good combination, and the architectural designer controls the ceiling height. Also lighting needs some thought when ceiling fans are used.

Ceiling fans mounted underneath lights will "chop" the illumination, creating an unpleasant flashing effect for the occupants. The effect is especially pronounced with the traditional downward-directed fluorescent lighting seen in office buildings. This problem can be avoided by using indirect illumination for the general ambient (such as daylighting or lighting reflected off the ceiling) combined with task lighting at the work surface.

Ceiling fan manufacturers have specific advice about mounting height. The consensus appears to be that for safety, the fan should be mounted so the blades are at least 7 ft. [2.13 m] above the floor. For best comfort, a blade height of 9 to 10 ft. [2.74 to 3.0 m] is an improvement, and the fan should not be so tightly-mounted against the ceiling that air flow is obstructed. Taken together, manufacturer recommendations appear to suggest an ideal ceiling height of 12 ft. or more, with a fan blade height of 10 ft. [3.65 m and 3.0 m]. As a minimum, manufacturers recommend a ceiling height of 8 ft. with at least a 7 ft. blade height [2.44 and 2.13 m].

Both ceiling and personal fans can provide comfort during periods of low cooling loads, without the need to operate the cooling equipment. In particular, some fan arrangements can allow occupants to *partially control their own environment*. Individual control of air velocity across the skin allows adjustment for different body types and activity levels, increasing thermal comfort at very low cost.

6. One fan room per floor = better comfort + simpler systems

With the traditional centralized, all-air cooling systems in large or tall buildings, owners and Architects are often reluctant to allow the HVAC designer enough fan rooms to ensure comfort and to allow simple-to-operate mechanical systems.

The reluctance is understandable. Air handlers take up a lot of space. Floor space is expensive, and there is *never* enough space on each floor or in each wing of the building for all the functions and people that the owner needs to have co-located.

Fig. 2.7 Holes, gaps and seams
Outdoor air that is pulled into the building accidentally makes it nearly impossible to keep occupants thermally comfortable, no matter how big and expensive the AC system might be. The concern is not for a 100% hermetic seal. Rather, the architectural designer and builder should focus on sealing gaps and closing holes. Also, close up any long seams, such as those around through-wall AC units and those where the roof assembly meets the exterior walls.

So eliminating fan rooms from each floor or each wing, and centralizing the air handling equipment so it serves 4, 6 or even 12 floors or four wings from a single mechanical room is one of the first ideas that occurs to the Architect and owner when the budget becomes a problem. The thinking is usually along the lines of: "Since the total cooling tonnage is still the same, I can increase the size of duct work (which is comparatively inexpensive) in return for regaining the more expensive and much-needed floor space, and reducing both the number of pieces of equipment and the number of locations where maintenance must be performed."

Unfortunately, serving several floors or several wings with a single, distant cooling system forces the HVAC designer into a no-win choice. He can increase the complexity of the system with more controls, more fan horsepower, undersized noisy duct work and more dampers—or sacrifice thermal comfort. Often, the result is a system which is unreliable and difficult to control because of its size and complexity, and is therefore uncomfortable for occupants. When occupants complain of being too cold during hot weather, the problem can often be traced to cost-cutting decisions that forced the HVAC designer to over-centralize the systems without the budget to make those larger systems responsive to the large number of distant zones they must serve.

Fig. 2.8
Clothing in hot & humid climates
As suggested by this 1994 meeting of the Heads of State of New Zealand, Papua, Australia, Malaysia, China, and Chile, clothing preferences in hot, humid climates are often adapted to the ***outdoor*** *environment—even for formal indoor gatherings.*

The crossed arms of several of these powerful people could be a reflection of political circumstances or cultural differences. On the other hand, arms crossed over the chest rather than behind the back, combined with a preference for hot beverages, such as the tea in the hands of the man at left, are typical of overcooled occupants who are trying to keep their bodies warm in an overcooled building.

A classic example is a large, multipurpose government building. A few operations like law enforcement operate at all times, while the rest of the building is vacant after normal working hours. If the building is only served by two or four air handling systems in a 12-story building, an entire large system must operate, clumsily, to provide any air conditioning at all. Similarly, a single-story but very large school might need air conditioning for only a few classrooms for adult education in the evenings.

Large central cooling systems are most effective in providing comfort when they serve zones on the same floor or the same wing of a building which have nearly identical occupancy schedules and similar heat loads. When systems are forced to serve distant floors with different occupancies and different heat loads, comfort suffers and complexity rises, along with fan energy use.

Instead, when comfort is a concern, plan for smaller systems and more of them, to allow a closer match between system operation and local loads. This approach improves comfort and also reduces the energy cost of moving *large amounts of air* to provide comfort in a small percentage of an otherwise unoccupied building.

HVAC Suggestions for Better Comfort

Now we'll move from the architectural influences on comfort to what the HVAC designer can do to improve occupant satisfaction.

1. Design HVAC systems for real clothing preferences

In cold and mixed climates, HVAC designers logically assume that clothing levels will vary, but will tend towards greater clothing coverage, especially in the social context of business occupancies. But this is not an appropriate assumption for most occupancies in hot and humid climates, and often not an appropriate assumption for business situations.

For example, in the hot and humid climate zones of the US, where the HVAC design may be based on nationwide layouts and equipment sizing, the thermally-aware observer will sometimes notice conference attendees slipping outdoors to "warm-up" from excessively cold temperatures in a meeting room or conference center.

In most occupancies, clothing preferences of people in hot and humid climates are for much less coverage than is common in cold and mixed climates. And this is particularly true in Asia, Africa and the Caribbean, where part of the business culture includes clothing that is functional in the outdoor weather as well being suitable for cold air conditioned buildings.

Three suggestions become apparent based on this common preference for light clothing in hot and humid climates:

- For most occupancies, consider an indoor design temperature closer to 79°F than to 75°F [26°C rather than 24°C], and keep the dew point below 55°F [12.8°C].
- Consider several stages of cooling capacity, including modulation. This helps avoid the extra-cold temperatures which come from equipment that cannot shed enough of its capacity to avoid overcooling under everyday loads.
- Direct supply air outlets so they do not blow cold air directly onto occupants. This advice applies especially to the unitary equipment (packaged, single-room cooling units) often used in crowded occupancies, such as schools, or in small rooms such as hotel guest rooms, eldercare resident rooms and hospital patient rooms. Occupants don't like it when these units blow cold air on them, especially when they don't have the choice of controlling their location with respect to the noise and chilling effect of the supply air stream. Chapter 33 of the 2005 ASHRAE Handbook—Fundamentals is titled "Space Air Diffusion", and it contains extensive and useful guidance to avoid these comfort problems.

2. Dry ventilation air helps avoid temperature swings

Drying the ventilation air helps the cooling system avoid a common source of sharp temperature swings that annoy occupants.

If the ventilation air enters the cooling system in its raw state, the main cooling system must chill all or part of the supply air deeply to remove the humidity load. In some cooling systems, this may require reheating the supply air, because the ventilation air humidity load is almost constantly high, even when the outdoor air temperature might be neutral.

To improve comfort, dry the ventilation air with a dedicated system. Then the cooling equipment only needs to respond when internal sensible heat loads rise, avoiding the need (during most operating hours) to overcool the supply air to dry it.

Perhaps because of the problems in the US with mold in buildings, a dedicated outdoor air dehumidification system has become a favored alternative in recent years. For example, for federal buildings in the US, dedicated ventilation dehumidification equipment was made a requirement in 2003, not only as a response to the occupant dissatisfaction with the indoor environment[1], but also to limit the risk of mold and bacterial growth.[12]

3. Constantly-cold coils can also dry air effectively

If the ventilation air is not pre-dried, the humidity load it carries must be removed in the main cooling system to ensure comfort. This can be done using a cooling coil or desiccant dehumidifier which responds to a humidistat rather than to a thermostat. The key is to keep the coil cold constantly, so that its surface is cold enough to really condense and remove the full humidity load whenever the outdoor dew point is above the target indoor dew point, which is most of the operating hours in a hot and humid climate.

A variable air volume (VAV) system keeps the supply air temperature constantly low. To reduce cooling capacity as loads fall, a VAV

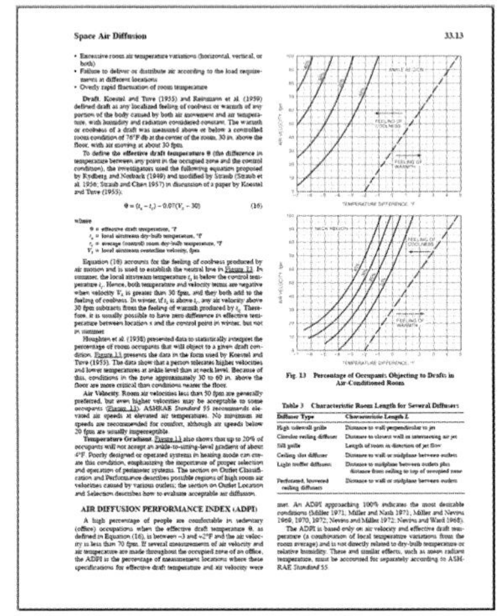
Space Air Diffusion 33.13

Fig. 2.9 Avoiding cold air drafts

Chapter 33 of the 2005 ASHRAE Handbook—Fundamentals is titled "Space Air Diffusion." It provides practical guidance for HVAC designers who want to reduce the risk of cold air drafts from cooling equipment and systems.

system reduces the air flow rather than changing its temperature, saving fan energy—the component that takes the most energy to operate. With its constantly-cold cooling coil, a VAV system can provide dehumidification as well as energy-efficient cooling, provided that:

- The supply air dew point is indeed low enough to provide adequate dehumidification.
- The coil *stays* cold, any time the indoor dew point is above the specified indoor dew point. In other words, don't re-set the supply air temperature higher to "save energy" if the indoor dew point is above the desired set point. Otherwise the building will feel cold and clammy.
- The energy to re-heat any overcooled supply air comes from waste heat, such as that from refrigeration condensers or a heat recovery wheel. Otherwise the system may not meet ASHRAE's energy consumption guidelines and some states' building codes.

A key point about VAV systems in hot and humid climates is that reheat hours can be greatly reduced and in some cases eliminated if the minimum air flow settings are low enough. Usually, 30% of full flow is still high enough to ensure adequate ventilation. But setting the minimum flow at 50%, as one might do to avoid air mixing problems in heating climates, usually results in unnecessary hours of reheat in hot and humid climates.

Another strategy is to place a separate dehumidifier or constantly-cold cooling coil in a bypass. In that arrangement, most of the air goes through the main system, which is optimized for cooling alone. At the same time, a smaller portion of the supply air goes through the bypass, where it is dried deeply by a dehumidifier or constantly-cold dehumidification coil. Then that dry air is blended back into the supply air before it is delivered to the space.

Again, it is best to use waste heat to provide the needed reheat or desiccant reactivation in order to meet energy codes and to avoid high energy costs. The incoming ventilation air dew point will be above the indoor target dew point nearly all year long in hot and humid climates. So unless waste heat is being used, adequate dehumidification for comfort and for building protection can appear to be very expensive.

4. Drier air expands the comfort range in mixed uses

Uncomfortable building occupants are often heard to say: "It's not the heat... it's the humidity." In the past, the usual HVAC design practices have sometimes ignored this common observation, under the assumption that if the air is cold enough, high humidity does not matter very much in the comfort equation.

This is true enough, based on research tests of the one-person, uniform-clothing, single-activity-level laboratory situation. But high humidity is more problematic when *many different* metabolic rates, clothing levels and different body types must be accommodated in the *same space*.

Consider an assisted living facility occupied by older, frail, sedentary residents and also by much younger, heavier, and very hardworking staff. When humidity is high, the staff is extremely uncomfortable at the warmer temperatures preferred by residents. However, residents would be uncomfortably cold at the low temperatures preferred by the staff.

Dropping the dew point in the building allows the active staff to release more heat by evaporation, while still allowing the temperature to stay warm enough for residents' comfort.

A similar metabolic and body mass mismatch is common between teachers and elementary school students, and between restaurant servers and their customers. In all of these occupancies, comfort can be achieved for a wider variety of body types, clothing preferences and activity levels by dropping the dew point and increasing air movement.[13, 14] These measures increase the effectiveness of evaporation, rather than relying on the brute force method of making the air colder and using more of it.

Illustrating this point, consider figure 2.10, which shows research results from 10 schools in Georgia.[15] Five of the schools were equipped with older-style cooling systems, which lacked the ability to dry the air on demand. The other five schools were equipped with dedicated ventilation dehumidification systems which could dry the incoming air, keeping the indoor dew point below a specified level.

In all 10 buildings, teachers controlled the thermostats. In the buildings with random humidity and lower ventilation rates, teachers set the temperature lower by as much as 6°F (3.3°C) compared to the temperatures set by teachers in the drier, more highly-ventilated buildings. In other words, when given the choice of colder and more humid vs. warmer and drier, teachers preferred the warmer temperatures. The low dew point allowed this preference to become clear. Apart from better comfort, higher temperatures probably cost less to maintain in hot and humid climates. The research described here modeled the energy net cost reduction of the warmer, drier buildings at 18 to 23% less than the colder buildings, when given the same ventilation air supply rates.

5. Capacity modulation avoids sharp changes

Sharp temperature changes are at the root of many complaints about thermal comfort in buildings. The occupants are too hot, so they ask for cooler temperatures. The systems respond, and then overshoot the desired condition on the cold side.

This problem is very common when the heat load is moderate, or very low. During those hours, the system has far too much capacity. High capacity at low load makes any equipment inherently unstable and difficult to control. Its like using an airplane for a trip to the corner convenience store—overshooting the desired destination is difficult to avoid.

For better comfort in smaller systems and smaller spaces, the HVAC designer can provide smoother modulation of capacity at low loads by following one or more of these suggestions. The lowest-cost suggestions are first:

- Stop oversizing the cooling equipment. This is one reason why occupants add clothing layers indoors in hot and humid climates. Cooling equipment is sized for peak loads—which automatically means it is larger than what will be needed for 99% of operating hours. Then the

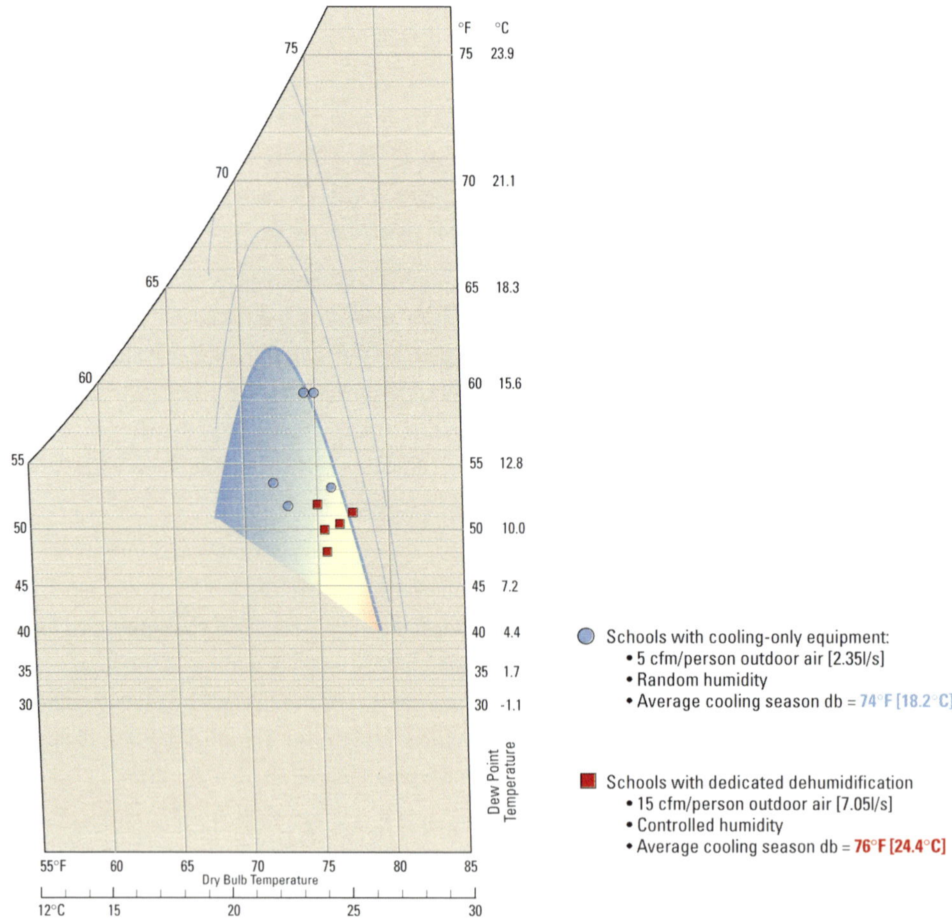

Fig. 2.10 Drier air widens the comfort zone

Field measurements of 10 schools in Georgia show that when the dew point is controlled, teachers preferred warmer temperatures. The addition of dedicated dehumidification equipment allowed better comfort, at much higher ventilation rates and lower energy cost.[13]

designer might add 10 to 20% on top of the peak cooling estimate, "just in case." This guarantees that the system will be larger than needed for 100% of its operational hours, making it very difficult to avoid overshooting the desired temperature. The usual result is discomfort from rapid switching between too-cold and too-hot. Instead, size cooling equipment at—or just under—the peak load as estimated through calculations. This reduces costs as well as increasing comfort.

- Specify variable-speed motors or multiple stages of capacity for the cooling equipment, so overall capacity can be modulated smoothly instead of switching on and off in large increments. This greatly improves comfort, reduces energy use and avoids more expensive solutions.
- Split the cooling load between several pieces of equipment rather than using one large one. Bring the capacity on in discreet stages, as loads rise. This is usually the most expensive suggestion, because it means more equipment and more mechanical room space. But it also provides even better comfort, much less energy use, less maintenance cost and better reliability.
- Make sure there is a separate component someplace in the system that will respond to a humidistat, keeping the indoor air dew point below 55°F [12.8°C] regardless of what is happening on the cooling side of the system. Without a dedicated dehumidification component, indoor humidity can become uncomfortably high when the system's cooling capacity is reduced to avoid overcooling.

For better comfort in larger buildings and larger spaces, the same suggestions apply. But the larger budget will also allow the use of more sophisticated controls that allow the equipment to begin reducing capacity slightly in advance of falling loads, rather than long after loads have reduced and occupants are already uncomfortable. The increased cost and maintenance complexity of sophisticated controls is often justified in large buildings, not only for reasons of comfort, but also for reductions in energy use, which can offset the maintenance expense.

6. Higher velocity VAV diffusers avoid "cold air dumping"

"Loud diffusers"—diffusers selected at high velocity—generate noise complaints in sleeping areas. So noise from diffusers is often viewed as a negative feature. But for variable air volume systems in non-sleeping areas, slightly elevated sound levels from supply air diffusers have both comfort and productivity benefits.

If the diffusers in a VAV system are silent at the peak design air flow, it's a warning sign that, when the air flow is reduced at low load conditions, cold air will "dump out" of the diffuser. In other words, it will fall as a cold column of air onto the occupants instead of mixing uniformly into the room air. Higher supply air velocities at peak design flow help avoid this common reason for cold air complaints.

Also, in offices, schools and many other occupancies, there is a social benefit to low-level "white noise" generated by a diffuser. If an office is too quiet, any small noise or conversation between people becomes very audible, and therefore annoying to others in the space. Complete silence is not beneficial to the working or social environment. Selecting supply air diffusers for VAV systems at the top of their flow range helps prevent cold air dumping at part load, and also provides a more acoustically-neutral environment.

Detailed Study of Thermal Comfort

ASHRAE Standard 55, and Chapter 8 of the 2005 ASHRAE Handbook—Fundamentals are very helpful in gaining an understanding of the many interacting factors which govern the perception of thermal comfort. In addition to temperature, much has been published by ASHRAE on the subject of humidity and its particular influence on thermal comfort in the ASHRAE Humidity Control Design Guide.[15]

These three publications reflect the complexity of their subject material. For the student of thermal comfort, and for those who need to understand the details in *all* their interacting complexity, the publications listed below will be very helpful.

References

1. Smealle, Peter 2003. "Building occupant and customer satisfaction survey results for 2001" *Proceedings of the General Services Administration, Public Building Service Workshop on Building Occupant and Customer Satisfaction.* Published by the Office of the Chief Architect, GSA, Washington, DC. (Mr. Smealle's full presentation is recorded as a video supported by slides, on this DVD report. The survey involved all GSA buildings nationwide, leased as well as owned. Total individual responses = 81,337.)
2. Chapter 8 (Thermal Comfort) *ASHRAE Handbook—Fundamentals, 2005.* ASHRAE, Atlanta, GA www.ashrae.org
3. ASHRAE Standard 55 (Thermal Environmental Conditions for Human Occupancy) ASHRAE, Atlanta, GA www.ashrae.org
4. Olesen, Bjarne and Brager, Gail. "A better way to predict thermal comfort" *ASHRAE Journal,* August 2004, pp:20-28.
5. Jitkhajornwanich, Kitchai et al. 1998. "Thermal comfort in transitional spaces in the cool season of Bangkok" ASHRAE Transactions, Volume 104, Part 1.
6. Rohles, Frederick, Ph.D., Fellow and Life Member, ASHRAE. "Temperature and temperament - A Psychologist looks at comfort" *ASHRAE Journal,* February 2007, pp:14-22.
7. Chan, Daniel, et al. "A large-scale survey of thermal comfort in offices in Hong Kong" ASHRAE Transactions, Vol 104, Part 1.
8. Federspiel, C.C., R. Martin and H. Yan 2003. *Thermal comfort and "call-out" (complaint) frequencies.* Final report, ASHRAE research project RP-1129.
9. Architectural designers do not always recognize the high probability of thermal discomfort in glass buildings in hot climates. Nor, apparently do their peers assign much importance to thermal comfort when bestowing design awards. But the general public often feels differently, as indicated by these press clippings about a large Federal Courthouse in Phoenix, Arizona. The design is basically a very large glass box, in a hot climate, completed in 2000.

   a. Pitzl, Mary Jo. *The Arizona Republic*, September 8th, 2001. "Phoenix Federal Building Has'm Sweating - Courthouse Hothouse" "...Thomas Zlaket, Chief Justice of Arizona, got the only laugh of the event when he joked that the building must have been designed by someone who had never lived in Phoenix during the summer. The steamy situation "Seemed ripe for a lawsuit", he joked."

   b. Kamman, John. *The Arizona Republic*, May 6th, 2002. "Atrium's Dual Identity: Blunder—Shining Symbol" "...after atrium temperatures fluctuated in the courthouse's first year between the low 40's and the high 90's, GSA paid $56,000 to install conventional heating and cooling through a false door to give relief to staffers at the metal detectors. The additional system will consume an estimated $6,700 a year in energy or around $1,000 for each person stationed there."

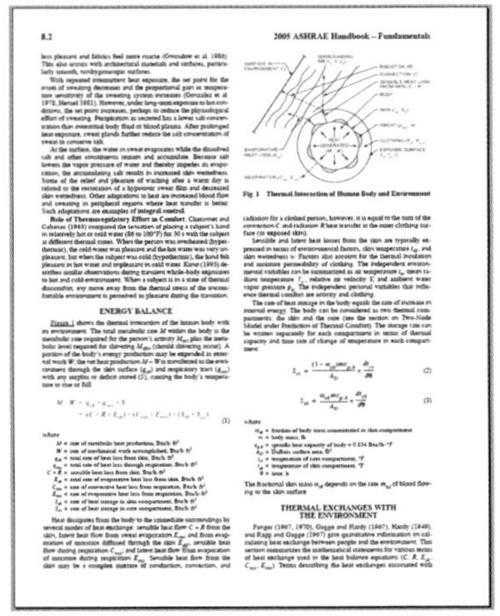
8.2 2005 ASHRAE Handbook—Fundamentals

ENERGY BALANCE

THERMAL EXCHANGES WITH THE ENVIRONMENT

Fig. 2.11 Theoretical foundation

Chapter 8 of the 2005 ASHRAE Handbook—Fundamentals is titled "Thermal Comfort." It provides the detailed theory, along with the hygrothermal and metabolic calculations which support ASHRAE's current understanding of thermal comfort in both hot and cold environments.

c. Kamman, John. *The Arizona Republic*, May 13th, 2002. "New Woes Emerge at Federal Courthouse: Windows Cracking..."

"...Meanwhile, 13 windows on the west side of the building recently cracked, reportedly because the wrong type of coating was applied to reduce sunshine and heat coming through them. The problems came to light after the *Arizona Republic* reported last week that construction of the $127 million Courthouse exceeded the budget by at least $16 million and was finished 17 months late."

d. *Architectural Record Magazine* - "2002 Honor Award Winner: Phoenix United States Courthouse". American Institute of Architects, Washington, DC.

e. *USA Weekend.com* - Special Report, September 1st, 2002. "Breaking New Ground. Inspirational. Amazing. These structures set the pace for American Architecture in the 21st Century." (Phoenix Federal Courthouse). "...Over the past year, USA Weekend and the American Institute of Architects collaborated to come up with this list of the great architectural works of the 21st century. The AIA provided five of its most esteemed members to take part as expert judges. They are: ..."

10. Cummings, James B., Withers, C. R. Withers, N. Moyers et al. 1996. *Uncontrolled air flow in non-residential buildings.* Final report. FSEC-CR-878-96. April 15th, 1996. Florida Solar Energy Center, Cocoa, FL

11. Ask, Andrew. "Ventilation and air leakage." *ASHRAE Journal*, Nov. 2003, pp.28-34 ASHRAE, Atlanta, GA www.ashrae.org

12. Chapter 5 - Mechanical Systems. *Facilities Standards for the Public Buildings Service (P100 - 2003/2005)* Office of the Chief Architect, U.S. General Services Administration, Washington, DC.

13. Berglund, L.G, and W.S. Cain. 1989. "Perceived air quality and the thermal environment." *The Human Equation: Health & Comfort*. Proceedings of the ASHRAE/SOEH Conference, IAQ '89, Atlanta, GA. pp.93-99. ASHRAE, Atlanta, GA www.ashrae.org

14. Fischer, J.C., and C.W. Bayer,. 2003. "Failing grade for most schools: Report card on humidity control" *ASHRAE Journal*, May, 2003, pp.30-39 ASHRAE, Atlanta, GA www.ashrae.org

15. Harriman, Lewis. G. III, G. Brundrett and R. Kittler. *ASHRAE Humidity Control Design Guide for Commercial and Institutional Buildings.* 2001/2006 ISBN 1-883413-98-2 ASHRAE, Atlanta, GA. www.ashrae.org

Image Credits

Fig. 2.3 ©Rhymes With Orange, Hillary B. Price. Reprinted with permission of King Features Syndicate

Fig. 2.4 Courtesy of the U.S. General Services Administration, Public Buldings Service

Fig. 2.6 ©Arizona Republic. Reprinted with permission

Fig. 2.7 ©CDH Energy, Cazenovia, NY. Reprinted with permission

Fig. 2.8 Courtesy of the National Archives of Australia: A8746, KN22/11/94/62

Chapter 3

Managing Ventilation Air

By Lew Harriman

Fig. 3.1
Ventilation dilutes indoor pollution
The purpose of ventilation is to dilute the concentration of pollutants generated indoors by people, the building and its furnishings.

Ventilation improves indoor air quality—as long as the incoming air is both cleaned and dried.

Key Points

Before air conditioning was commonplace, owners and designers of buildings did not need to be especially concerned about managing ventilation air. Ventilation management used to be easy—just open the windows. When more people occupied the building, the windows could be opened wider.

But with air conditioning, everything changed. Buildings are now closed-up, because they must contain and preserve their expensive cool air. Consequently, air conditioned buildings need mechanical ventilation. And that turns out to be more complicated than just cooling down some outdoor air.

Over the last 20 years, cooling costs, indoor mold problems[1], and the health hazards from small-particle outdoor air pollution[2] have all focused a great deal of attention on ventilation. What we now understand about ventilation is that in hot and humid climates, just bringing outdoor air into an AC system does not automatically improve indoor air quality. Outdoor air must be cleaned and dried, not just cooled. Some observations and suggestions for minimizing the cost and maximizing the effectiveness of ventilation air include:

- Clean, dry ventilation air is a precious commodity. Measure it carefully and control its volume and location, so that it is not wasted by the way you produce it and use it. Also, don't produce any more of it than you need for the actual number of people occupying the building.
- Humid ventilation air has often ruined a building's walls and furnishings by contributing to mold growth. Humidity aids mold and bacterial growth, and these make the indoor air quality worse—not better. Therefore, make sure that *all ventilation air is dried, at all times*.
- "Exhaust ventilation" systems which create building suction contribute to major mold growth problems. To avoid suction and humid air infiltration, balance the sum of all the exhaust air flows with a slightly greater amount of dried make-up air. But also, be sure to seal up *all* plenums, *all* exhaust ducts and *all* duct connections so they are airtight, using mastic or similarly durable adhesive sealants. Otherwise, leaking exhaust air duct connection will create suction behind walls and above ceilings, leading to mold growth in those locations.
- Ventilation dehumidification and cooling loads per cfm or per l/s are much higher and more variable than the loads in the return air. Therefore the maintenance and operating personnel need to devote more time, more attention and more of their budget to the ventilation system's air flow controls, its filters and its dehumidification components.

Measuring and conserving ventilation air

To most building owners and HVAC designers, it is obvious that clean, dry ventilation air costs a lot to produce. What's not as obvious is that most HVAC systems produce *far* too much and *far* too little ventilation for the occupant's real needs. There is a real opportunity to save operating costs by reducing the amount of ventilation air when the building is lightly occupied. And there's an equally big opportunity to greatly improve indoor air quality by delivering enough ventilation air to match the true occupancy.

To most owners, it comes as an unwelcome surprise to hear that "standard" HVAC systems don't actually measure and control the amount of ventilation air that is produced, nor do they vary the amount of ventilation air to each space in proportion to it's actual occupancy. In most systems, ventilation air volumes are set early in the design process—usually based on a series of highly questionable assumptions about occupancy and about the air volume that flows through dampers set at certain positions. Also, there are even more error-prone assumptions about the amount of ventilation air which actually reaches a given occupied space after mixing into the larger supply air flow.

Measured Ventilation Air Flow - 510 Office Buildings

ASHRAE Standard Requirement (At the time of measurement)
Excess ventilation

	Per Person cfm/person [l/s/person]	Per Unit Floor Area cfm/ft2 [l/s.m2]	Air Changes Per Hour
Mean (Average)	**117** [55]	0.4 [2.04]	2.00
Median (Most Common)	**63** [30]	0.2 [1.03]	0.98
Standard Deviation	**158** [74]	0.5 [2.54]	2.45
Std 62.1-1989	**20 [9]**	**0.07 [0.36]**	**0.34**

Fig. 3.2

Past practices for ventilation management have been wasteful

Measuring the actual ventilation in 510 office buildings, NIST research showed the gross over-ventilation provided by traditional budget-starved HVAC design practices.[3] To avoid this waste, the occupancy must be measured or sensed, and then the ventilation air must be measured and directed to the spaces which need it—not the ones which don't. This requires a more deliberate approach to ventilation system design and a larger budget for it's construction, but it greatly reduces energy waste and operating expense.

This assumption-based approach may be a long-time standard, but it's ventilation effectiveness is very poor, nearly always. When ventilation air flows are actually measured, it becomes obvious that buildings are *greatly* over-ventilated *and* under-ventilated—usually at the same time in different parts of the building.

Consider the graph shown in figure 3.2. The graph shows the field test results from 510 office buildings throughout the United States. These site measurements were taken by the staff of the National Institute of Standards and Technology (NIST).[3] The measured values show that, compared to the then-standard requirement of 20 cfm/person, the average ventilation rate was 117 cfm/person, and the most typical rate was 63 cfm/person. Indeed, in more than 20% of the buildings, ventilation rates were even higher than 275 cfm/person. Any way you look at it, it's dismally wasteful performance compared to the actual need of only 20 cfm/person.

Given this degree of sloppiness in typical ventilation control strategies, it's no wonder that ventilation air gets a reputation for being an expensive luxury in hot and humid climates. Make no mistake; it is indeed expensive to produce clean, dry air when starting with highly-humid and relatively dirty outdoor air. But with a more deliberate approach to ventilation control, under typical operating conditions it's usually possible to use less than half of the ventilation air traditionally used by most HVAC systems, without compromising the indoor air quality.

The more careful approach *measures* the ventilation effectiveness continuously, and then *raises or lowers* the ventilation air flow according to the distinct and constantly-changing needs of each occupied space.

To see why this would be useful, consider the graph shown in figure 3.3. It shows the concentration of carbon dioxide (CO_2) in each of many rooms served by a single system, along with the CO_2 concentration in the common return air from all of those rooms.[4] Carbon dioxide is a product of human metabolism. Since most people in the

Fig. 3.3

CO_2 concentration is an excellent indicator of human occupancy

The datalog shows that while the average ventilation seems adequate (CO_2 in the return air), two rooms are actually grossly under-ventilated while others are over-ventilated. An independent ventilation system can avoid such wasted energy, while ensuring adequate ventilation to the more densely-occupied spaces.[4]

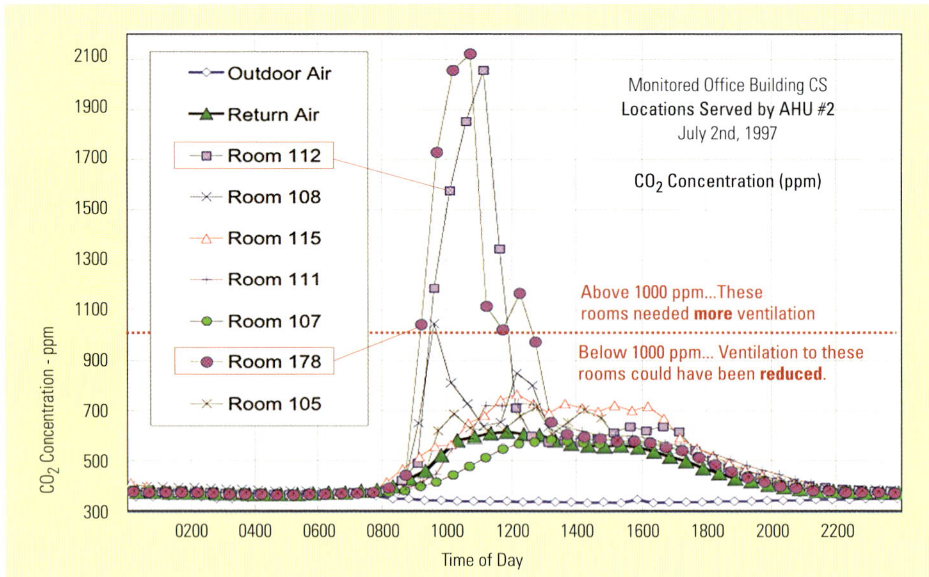

same space have similar activity levels, the concentration of CO_2 is a good indicator of the number of people in the space, and therefore a good indicator of the amount of ventilation air needed to dilute the contaminants generated by those occupants. Outdoors, the baseline CO_2 concentration is usually between 300 and 450 ppm. Indoors, if the system is providing enough ventilation to dilute occupant-generated contaminants, the CO_2 concentration will not rise more than about 700 ppm above the outdoor concentration. That's why an indoor air CO_2 concentration below 1000 ppm is a general indication of good indoor air quality. Below 1000 ppm, the amount of ventilation air can probably be reduced. But as the CO_2 concentration rises too far above that level, there's some evidence that human productivity may slowly decline, as measured by tests of academic performance.[5,6]

Returning to figure 3.3, note that the average CO_2 concentration in the combined return air stream never rises above 600 ppm. This would indicate rather good indoor air quality, on average. Indeed, such a low concentration would suggest that the amount of ventilation air could be *reduced* by about 50% before the average concentration would rise above the 1,000 ppm level—an opportunity for savings. On the other hand, it's also clear from the graph that the indoor air quality in rooms 112 and 178 is comparatively poor. In those rooms, the CO_2 concentration suggests they needed more than twice as much ventilation air as they received between the hours of 0830 and 1330.

Based on the patterns seen in figure 3.2 and 3.3, it's apparent that for both economy and better indoor air quality, the better ventilation system will *vary the ventilation air volume to each space*, rather than providing a constant volume to all spaces.

Ideally, the ventilation system will be a separate, dedicated variable air volume system, independent of the cooling system, with dedicated duct work for both supply and return air. That's the approach mandated for federal buildings by the P-100 Federal Facility Standard.[7] A dedicated system allows individual control for each space, based on either a schedule in the building automation system or based on the CO_2 concentration in each space. And it allows for easy energy recovery from the returning exhaust air. This approach saves operating costs by reducing both the amount of ventilation air and its cost of production, without risk to the indoor air quality.

Of course, such complete, real-time modulation of ventilation air to every space may not always be affordable in the construction budget, no matter how much it might reduce the operating budget. So in many occupancies, it's possible to simplify the sensors and air flow controls, reducing installed costs without *excessive* risk to indoor air quality, especially for spaces where the occupancy is relatively predictable.

For example, compared to a CO_2 sensor, an occupancy sensor is a very simple and low-cost device. So the designer could place occupancy sensors in small conference rooms or classrooms instead of CO_2 sensors. The ventilation air sent to those spaces could be controlled by a simple two-position damper. If the room is occupied, it gets the *maximum* amount of air needed for its rated occupancy. If it's not occupied, it gets the *minimum* amount of air needed to dilute building-generated contaminants.

That approach will still mean that sometimes, those spaces will be under or over-ventilated. But it saves money in construction and makes for a simpler system to maintain. And for *most* of the operating hours, occupancy sensors will help avoid the *extreme* under or over-ventilation shown in figure 3.2.

What probably makes the most economic sense in most buildings is a mixture of these approaches. Simple occupancy sensors can be used for small spaces with smaller and more predictable numbers of occupants. CO_2 sensors would be more helpful for large spaces with a much larger and less predictable number of potential occupants, and for occupancies which occur at less predictable times and durations.

For example, classrooms in K-12 schools are not especially large. And chances are good that when they are *occupied*, their occupants would benefit from full ventilation.[5] In contrast, a gymnasium is a large space with highly variable occupancy. And its large interior volume contains so much un-breathed air that a gym may not need very much ventilation air when its occupancy is light, or when it is only occupied for a short period. So a gymnasium would be a good candidate for a CO_2-controlled ventilation air flow, while the classroom ventilation air flow could be switched between minimum and maximum flow by the lower-cost and simpler occupancy sensors.

But here's the main point. It saves operating costs to *only produce and deliver the amount of dry, clean ventilation air needed by the number of people that really occupy each space.* This approach produces better indoor air quality at much lower costs than relying on the owner's and designer's early guesses and assumptions about occupancy. It's costly and wasteful to let early *assumptions* rather than real-time *measurements* determine the ventilation air flows for the life of the building.

Drying ventilation air—all the time

Most owners are aware that excessive humidity leads to mold growth and indoor air quality problems. So another unwelcome surprise to many owners is that ventilation air is often *not* completely dried before it enters the occupied spaces.

Sometimes this is because the HVAC budget is too tight. Other times it's because the HVAC designer may not be aware of the issue and just chooses to cool the air slightly instead of drying it deeply enough to remove internal loads. And sometimes the system might not be operating the way the designer intended. But no matter how or why it happens, humid ventilation air creates problems. Excessive humidity carried by ventilation air must be removed.

Figure 3.4 shows the dimension of the potential problem, for an office building. In most commercial and institutional buildings, the ventilation air carries *by far* the largest humidity load of any of the humidity load elements. For humidity loads in other types of buildings, see figure 11.1 in Chapter 11—Estimating Dehumidification Loads. In all of them, ventilation air is the largest component of the dehumidification load.

To understand the importance of the ventilation load in relation to the total load from an actual building, see figure 3.5. That figure shows field-measured loads from a school in Georgia, during two periods when the school was occupied, but when neither the outdoor temperature nor the humidity was at it's peak.[8] In other words, at times when the running loads were more typical than the extremes used for design.

Note that during these two periods, the ventilation load accounted for 45% and 41% of the entire combined cooling and dehumidification loads for the building.

That example shows why HVAC designers in hot and humid climates often prefer to remove the dehumidification and cooling loads from the ventilation air before it mixes with the return air.

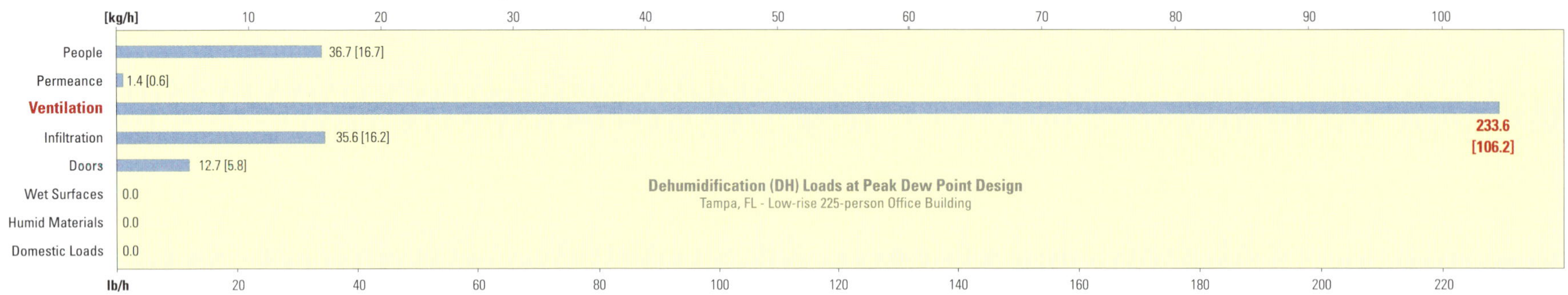

Fig. 3.4 - Ventilation is the largest dehumidification load

The graph shows humidity loads for a 3-story, 225-person office building located in Tampa, FL. Note that ventilation accounts for more than 73% of the total dehumidification load. That's why it's so important to dry the ventilation air in a hot and humid climate.

Fig. 3.5 - Ventilation loads are high any time the building is occupied

*Ventilation loads are very high per cfm [l/s]. Also, the ventilation load is mostly dehumidification rather than cooling. That's why it's important, and usually more energy-efficient, to remove the ventilation humidity **before** it can enter the occupied space, rather than trying to remove it from the return air stream.*

After ventilation air mixes with return air, it's more complicated to remove excess humidity without affecting the room air temperatures. That difficulty is one reason that many buildings are overcooled in hot and humid climates. Cooling systems are usually optimized for cooling, not for dehumidification. To actually dry the air when cooling loads are *not* at their peak, it's difficult for some systems (especially the common, low-budget, constant-volume cooling systems which lack an independent dehumidification capability) to avoid overchilling some parts of the building.

Interestingly, it's not especially difficult to avoid high humidity and overcooled rooms. If the ventilation air is dried down to a 55°F dew point, it no longer adds humidity to the building. And if the ventilation air is dried still further, to perhaps a 50°F or even a 45°F dew point, it will absorb some of the internal humidity loads. Then, the cooling components can be reset to produce air which is not quite so cold, which saves energy and helps avoid overchilling the occupants.

There are several ways to dry the ventilation air. Among the more popular are dedicated outdoor air systems (DOAS) which use some combination of desiccant and cooling technologies, and variable air volume systems which are optimized for hot and humid climates. Both of these types of systems can be designed to dry the supply air deeply, any time the outdoor air dew point is above the desired indoor dew point.

But the most important characteristic of an *effective* system is that it will dry the ventilation air *all* the time, not just when a thermostat calls for cooling. Independent control of humidity (independent of the need for cooling) is a feature the building owner should insist on, and that the HVAC designer should keep in mind as a fundamental requirement for AC systems in hot and humid climates. To summarize this point, the elements of effective dehumidification of ventilation air include:

- Drying ventilation air *any time* it is above a 55°F dew point—not just when cooling is required.
- Measuring, displaying and controlling the *indoor dew point*, rather than the relative humidity. The rh changes constantly, and is a less reliable indicator of mold risk and building damage than is the dew point.

Avoiding building suction and infiltration

To most building professionals in hot and humid climates, it's obvious that a building must avoid suction and infiltration of humid air. But often, the problems lead to suction and infiltration are not as obvious. Most designers believe that if the ventilation air volume exceeds the exhaust air volume, the building will not suck in humid outdoor air. Unfortunately, designing for a slight excess of ventilation air is not enough. It's just the first step. Leaky air connections and dirty filters are often the reasons that buildings "go negative."

Certainly, begin by making sure the sum of the ventilation air exceeds the sum of the exhaust air volumes. But after that first step, sealed duct connections need attention from the HVAC designer. Then, clean outdoor air filters must be addressed by both the designer and the maintenance staff.

Average positive pressure alone does not prevent infiltration

Providing more make-up air to the building than the sum of the exhaust air flows is an essential first step. But it's not enough.

Counter to intuition, excess ventilation air does not provide positive pressure *everywhere* on the building's periphery. Pressure can be positive in rooms, but negative in the building cavities which separate those rooms. Then, where building cavities connect to the exterior wall (as they all do, sooner or later) humid air can be pulled into the building, even while the rooms themselves show a net positive pressure with respect to the outdoors.

Leaky return air duct connections, leaky exhaust air connections and leaky air handler cabinets are usually responsible for the slight suction in building cavities which leads to humid air infiltration. When an exhaust or return duct connection is not air tight, it pulls in air from the building cavities it passes through. That provides the suction which leads to infiltration from the weather.

Another common example is a wall-mounted AC unit with a leaky cabinet or a return air inlet which is not sealed to the finished wall surface, as shown in figure 3.6. Gaps and unsealed seams allow the fan to pull in air from the wall cavities as well as from the room. In fact, the air pulled from the wall cavities is often what *provides* the positive pressure in the room which seems so reassuring! While the room itself is positive with respect to the outdoors—the wall cavity is still negative. So outdoor humidity is pulled into that cool wall cavity, where it provides the moisture which supports mold growth.[9]

That's why all duct connections, all air handler cabinets, and all return air plenums must be sealed-up, air tight *in addition* to providing a net excess of ventilation air for the building as a whole.

Clean ventilation air filters are both a maintenance and a design responsibility

When the outdoor air filters are clogged, the exhaust air flows may exceed the design ventilation air flows. Clogged outdoor air filters reduce make-up air flow, which allows exhaust fans to create building suction and therefore air infiltration through joints and doors.

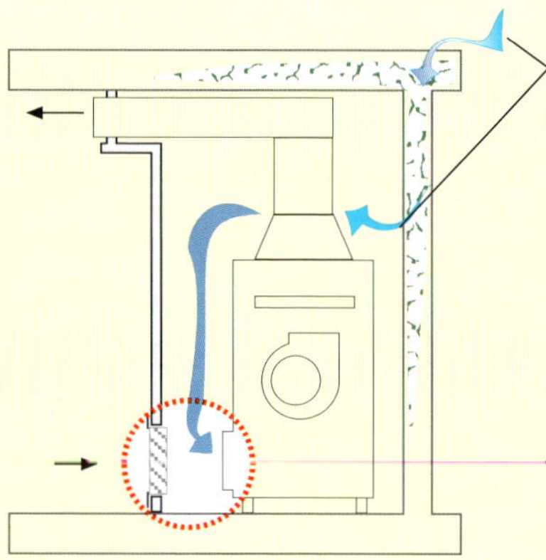

Beware of Accidental Suction ...
All walls leak air. The rate of leakage depends on the pressure difference across the wall. Leaks on the return (suction) side of the HVAC equipment create negative pressure in building cavities. This increases infiltration, which causes condensation and building damage.

... Caused by Incomplete Return Air Connections
This unit pulls air from the room...but also pulls air from building cavities, increasing infiltration from outside. Hard-connecting and sealing the return grill to the unit's inlet with ductwork avoids wall suction, reducing condensation and mold growth in the walls.

Fig. 3.6 Wall-mounted HVAC Units
If the return air inlet is not sealed to the interior finish, the unit can pull air from building cavities instead of just from the room. The fan suction will pull in humid outdoor air, supporting the fungus seen in the photo.

Fig. 3.7

Operaters must sometimes mitigate architectural design deficiencies

The owner saved money on floor space, but made it nearly impossible for the designer to fit the equipment into the mechanical room. The filter access is on the back side, over the wooden bridge built by the operating staff to allow them to change the filters. ASHRAE Standard 62.1 (Ventilation for acceptable indoor air quality) now requires adequate maintenance access in the HVAC design. This is an example of how NOT to design a building for adequate maintenance of indoor air quality.

At first, this seems like a pure maintenance issue: change the outdoor air filters regularly—at least once a month in most areas, and more often in highly-polluted urban areas and areas near highways. But if the designer has not provided practical *access* to those outdoor air filters, then one cannot blame the maintenance staff for their failure to change them monthly.

Figure 3.7 shows an example of the problem. The maintenance staff built the wooden structure to be able to get to the side of the unit where the filter access was located—against the far wall of the grossly-undersized mechanical room. Filters for this unit have to be custom-ordered in narrow widths. Standard filters are too wide to slide into the casing, because the side access is inadequate.

To be fair, the owner of this historic building decided how big this mechanical room would be—not the HVAC designer. To help avoid such cramped spaces in new construction, the astute HVAC designer can point to the requirements outlined in ASHRAE Standard 62.1 - Ventilation for acceptable indoor air quality.[10] That document calls for adequate space for maintenance of ventilation air systems. The space shown in figure 3.7 is certainly not adequate, and therefore does not comply with ASHRAE Standard 62.1.

A low-cost tool for improving filter and other maintenance access is the "Andy Stick." This simple device, shown in figure 3.8, has proven to be very cost-effective in improving indoor air quality.[11] Maintenance professionals in particular have found they can make their access requirements more clear (and sometimes more emphatic) to owners and HVAC designers when they bring a variety of Andy Sticks to design conferences where floor space is being allocated.

Greater O & M attention and budget for ventilation

The temperature, humidity and particulate loading of the ventilation air varies widely—much more widely than the temperature, humidity and particulate loading of the return air. With wider variation and higher loads per cfm [l/s], things go out of adjustment more rapidly and more frequently.

We have just discussed the need for monthly changes of outdoor air filters. When outdoor air filters are keeping all that ventilation particulate out of the building, the return air filters will not need such frequent changes.

Next is the adjustment of the dampers which control the ventilation air volume. In outdoor air, bearings and linkages corrode and seize-up far more often than the dampers which control the return air. So frequent lubrication and adjustment are more necessary for ventilation air dampers.

Cooling coils which dry ventilation air will be condensing moisture *constantly*. That means that the condensate must have a clear path to it's drain so it can flow freely and not collect to grow bugs, mold and rot in the pan. For the designer, this means the condensate pan must be sloped in two directions towards the drain connection, and it must have a proper trap—one which is deep enough to hold enough water to resist the fan pressure, and one which can be easily clean out with a brush when it collects dirt, twigs, feathers and all the other stuff which somehow manages to bypasses the filter and drain down the face of the coil along with the condensate. Figure 3.9 shows such a well-designed, cleanable trap.

Fig. 3.8 The "Andy Stick"
An effective tool for quality assurance during on-site inspections, an Andy Stick[11] *quickly measures compliance with the aisle width specified for maintenance access. Adequate access is now a requirement of ASHRAE Std 62.1.*

For the maintenance staff, all that dirt and condensate means the drain pan, the drain trap and its condensate line will need frequent brushing and cleaning.

Sensors located in the ventilation air are another frequent maintenance item. Air pressure sensors clog with dirt. Temperature sensors corrode and both humidity sensors and CO_2 sensors go out of calibration because of regular condensation and near-condensation in ventilation air.

For all of these reasons, components which measure, control, clean and dry the ventilation air will need a disproportionately large share of the available maintenance time and budget. But it's worth it. Without that maintenance attention, all that humidity and particulate can create havoc in the rest of the system—raising overall maintenance costs and fouling up both comfort and energy budgets.

Summary

Ventilation air is essential for diluting indoor air contaminants generated by occupants and by the materials and furnishings of the building itself. But frequently, adding more ventilation air creates more problems than it solves, especially in hot and humid climates. To avoid the problems, clean and dry the ventilation air, and then measure and provide that expensive high-quality air to each space in proportion to its actual occupancy.

For more discussion of the logic of ventilation, along with more detailed requirements for ventilation design, see Chapter 15 - Designing Ventilation Systems. For a quick summary of the ventilation air flow requirements of ASHRAE Standard 62.1-2007, see figure 11.28 at the end of Chapter 11 - Estimating Dehumidification Loads.

References

1. Mudari, David and Fisk, William J.; "Public health and economic impact of dampness and mold." *Indoor Air,* June 2007. Volume 17, Issue 3. pp 226-235. Journal of the International Society of Indoor Air Quality and Climate, Blackwell Publishing, www.blackwellpublishing.com

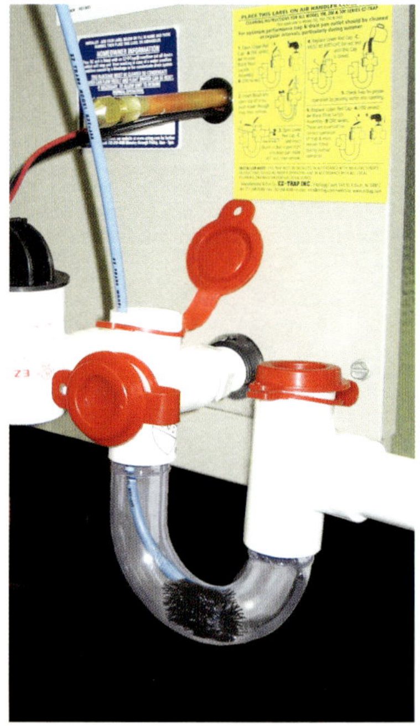

Fig. 3.9 Cleanable drain trap
Any clog is visible to the operator, and a brush can be easily inserted to clear the debris. Such features help reduce operating problems with ventilation components, which require extra maintenance atention because ventilation loads are so high and so variable.

2. United States Environmental Protection Agency "The Particle Pollution Report - Current Understanding of Air Quality and Emissions Through 2003." December, 2004. EPA 454-R-04-002 U.S. EPA Office of Air Quality Planning & Standards, Emissions, Monitoring & Analysis Division, Research Triangle Park, NC www.EPA.gov/air

3. Persily, Andrew; Gorfain, Josh; Brinner, Gregory. "Ventilation Design and Performance in U.S. Office Buildings." *ASHRAE Journal*, April 2005, pp.30-35 ASHRAE, Atlanta, GA www.ashrae.org

4. Bearg, David, 2007. "CO_2- vs. CFM-based ventilation assessements—Advantages and disadvantages of two monitoring options." *HPAC Engineering*, August 2007. pp.36-41. Penton Media, Cleveland, OH. www.HPACEngineering.com

5. Fisk, William. "A review of health and productivity gains from better indoor air quality." 2000. LBNL-48218 Lawrence Berkeley National Laboratory, Berkeley, CA. http:repositories.cdlib.org/lbnl/LBNL-48218

6. National Academy of Science "Emergency and continuous exposure guidance levels for selected submarine contaminants" 2007. The National Academies Press, Washington, DC. Text is online at no cost at http://www.nap.edu/catalog/11170.html Printed edition: ISBN 0-309-10661-3

7. Chapter 5 - Mechanical Systems. *Facilities Standards for the Public Buildings Service (P100 - 2005)* Office of the Chief Architect, U.S. General Services Administration, Washington, DC.

8. Fischer, John; Mescher, Kirk; Elkin, Ben; McCune, Stephen and Gresham, Jack. 2007. "High-performance schools—High marks for energy-efficiency, humidity control, indoor air quality and first cost." *ASHRAE Journal*, May 2007, pp.30-46. ASHRAE, Atlanta, GA.

9. Chapter 16 - Managing Building Pressures. Harriman, Brundrett & Kittler, 2008. *ASHRAE Humidity Control Design Guide*, ISBN 1-883413-98-2 ASHRAE, Atlanta, GA

10. ASHRAE Standard 62.1-2007 (Ventilation for Acceptable Indoor Air Quality) ASHRAE, Atlanta, GA www.ashrae.org

    Also: The 62.1 User's Manual ASHRAE/ANSI Standard 62.1 (Ventilation for Acceptable Indoor Air Quality) 2005 ASHRAE, Atlanta, GA www.ashrae.org ISBN 1-93862-80-X

11. The "Andy Stick" This tool is named for Andrew Äsk P.E., the innovative young engineer who first designed, manufactured and used it. The Andy Stick can be used as a motivational device during design conferences with Architects and owners. But it is primarily used, in different lengths, as a measurement tool for testing whether adequate space has been provided to allow access to equipment for adustments and for maintenance in mechanical rooms and above ceilings.

Image Credits

3.3 David Bearg, Life-Energy Associates, Concord, MA

3.5 John Fischer, SEMCO, Inc. Atlanta, GA

3.6 Joseph Lstiburek, Building Science Inc. Westford, MA

3.7 Mason-Grant Consulting, Portsmouth, NH

3.8 Mason-Grant Consulting, Portsmouth, NH

3.9 E-Z Trap. Inc. Edison, NJ

Chapter 4

Reducing Energy Consumption

By Lew Harriman

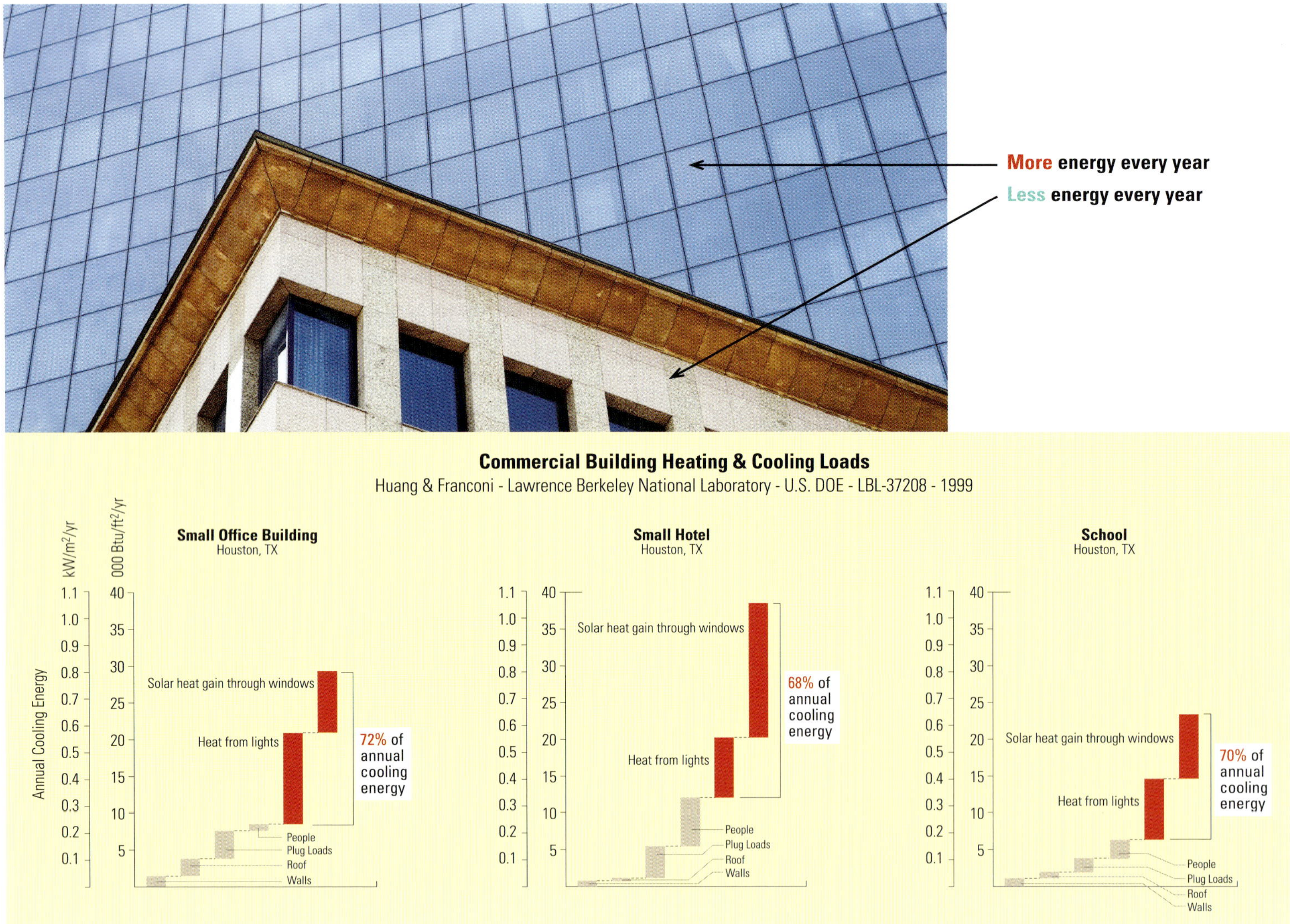

Fig. 4.1

Windows and energy consumption [1]

In most commercial and institutional buildings, windows dominate the cooling loads in two ways: the amount of solar heat they let into the building, and the amount of usable daylight they provide.

Solar heat is a direct and obvious cost. But their usable light can reduce the lighting power load, which in turn reduces the heat shown here, which is generated by lights.

Consequently, the owners' preferences for windows and enclosure designs govern the energy consumption of most air conditioned buildings in hot and humid climates.

Key Points

It takes a great deal of energy to cool a building in a hot and humid climate. The amount of energy depends on how much heat and humidity accumulates inside the building. So the key to using less energy is keep the heat and humidity out of the building. Smaller loads allow less energy use—larger loads force the mechanical system to use more energy.

From these simple relationships, it follows that owners and architects are the primary decision makers with respect to energy use in buildings. They control the loads imposed or avoided by the building's enclosure.

Summarizing the suggestions in this chapter, to reduce energy use in hot and humid climates:

1. Minimize exterior glazing, install insulating low-e glass and shade it. If this first suggestion is not adopted, there is less use in reading the rest of this chapter. Energy consumption will be relatively high, even after all other efforts.
2. Avoid glazing which faces west, so that AC systems can cost less to buy and can use less energy for the life of the building.
3. Design the exterior enclosure and its glazing so that the sun provides daylighting at the perimeter of the building. And design the perimeter lighting so that it modulates, generating heat only when solar daylighting cannot provide adequate illumination.
4. Design and construct the exterior enclosure so it is air tight. In particular, seal up attics and crawl spaces instead of venting them, eliminate air leaks in parapet walls, and install flashing for all the joints around through-wall air conditioning units and other wall penetrations.
5. Commission new buildings and systems, so the operational reality of their energy consumption matches their owners' requirements and their designers' intent.
6. Seal up *all* the air duct connections and make *all* the supply and return air plenums air tight, using mastic.
7. Provide the budget needed, and ask the mechanical system designer to ensure that the HVAC system automatically reduces the amount of ventilation air so that it matches the building's occupancy, as that occupancy reduces during evenings, nights, weekends and holidays.
8. Capture and re-use the heat from the cooling systems' condensers, and the energy contained in exhaust air.
9. Keep the indoor dew point low enough that the temperature can be comfortably warm, avoiding the energy wasted by overcooling the building to provide comfort.
10. Invest in constant commissioning for existing buildings, to provide high and essentially risk-free returns on investment, in addition to both better comfort and reduced energy consumption.

To reduce energy consumption, reduce the loads

In hot and humid climates, reducing energy consumption requires reducing the cooling and dehumidification loads. After those loads are reduced, one can size and control the mechanical systems so they do not use more energy than what is really needed to remove the smaller loads, as they rise and fall over the course of the year.

The largest cooling load—by far—is the heat that comes through the glazing; windows, glass curtain walls and skylights. That load is kept out of the building by shading the windows, by making them small, and by using glazing which excludes most of the sun's radiant heat.

The next-largest cooling load is generated by the lights inside the building. Cooling loads from lighting can be reduced by making productive use of light from the sun for as many hours as possible, and by using lamps which produce less heat and more light.

So both of these very large cooling loads are governed by the design of the windows, and that design is controlled by the aesthetic preferences of the owner. If the windows are small, shaded and are effective in daylighting, the building will use very little energy. If the windows are large, unshaded and not effective for daylighting, the building will use a great deal of energy.

Moving on to humidity, there are usually only two loads of major significance: humidity in the ventilation air, and humidity in the air which leaks into the building through holes, cracks and open joints in the exterior walls.

The number of people occupying the building governs the size of the ventilation air load. And the number and size of the cracks, holes and joints in the walls determine how much humid outdoor air will be pushed into the building by wind, or pulled into the building by any leaks in the duct connections of the air distribution system.

Again, the two largest humidity load factors (and therefore energy needed to remove humidity) are controlled by the owner and the architect. High occupancy requires the mechanical system to bring in a great deal of ventilation air, so the *peak* humidity loads will be unavoidably high. However, occupancy changes over the course of a day. Also many buildings are almost empty at night and during weekends. Indeed, schools are nearly vacant for weeks. So the ventilation air flow and therefore the annual humidity load can be greatly reduced, *if* there is a budget for the necessary equipment and controls.

Regarding air infiltration, if the building is more nearly air tight (if it is built without lots of open joints and has an air barrier), then the humidity load from air infiltration will be low, provided the duct connections are also sealed up air tight, so they don't pull in humid air from outdoors.

Fig. 4.2 ASHRAE Standard 90.1

The guidance provided by ASHRAE Std 90.1-2004 is constantly evolving, helping owners, designers and regulatory authorities all over the world who wish to reduce energy consumption in buildings.

The suggestions in this chapter build on that foundation, but they go beyond the provisions of the current edition, to help decision-makers who wish to build for future improvements to minimum requirements. Through a decision by the ASHRAE Board in 2007, buildings built to the 2010 standards will be expected to be 30% more energy-efficient than those built to 2004 standards.

After the heat and humidity loads have been reduced to their minimums through the decisions made by the owner and the architect, the mechanical system designer has the ability to remove those loads using less energy or more, depending on how closely the HVAC system can reduce energy consumption as the loads are reduced, and depending on how much energy is re-used rather than wasted after a single use.

ASHRAE Std 90.1 - Detailed Requirements

This chapter provides a brief summary of measures which will reduce energy consumption. Beyond this summary, ASHRAE has several more comprehensive publications which provide the specific details needed for design, after the big-picture decisions are made.

The most important of these comprehensive publications is ASHRAE Standard 90.1, which is titled: Energy standard for buildings except for low-rise residential buildings.[2] It is a public consensus standard, approved by the American National Standards Institute (ANSI). It is sponsored jointly by ASHRAE and the Illuminating Engineering Society of North America (IESNA.org).

The purpose of Standard 90.1 is to provide *minimum* requirements for energy-efficient design of new buildings. The provisions of the standard are not mandatory in and of themselves, because ASHRAE has no regulatory authority. However, the energy consumption of cooled and heated buildings has risen to about 40% of global energy consumption—a significant concern to many regulatory organizations. Because Standard 90.1 is the product of a rigorous, public and international consensus process, regulatory authorities in many North American and international jurisdictions have adopted some or all of the provisions of the standard into building codes. In those locations, the relevant portions of Std 90.1 have essentially become minimum legal requirements for owners, architects and HVAC designers.

Standard 90.1 and its user guide

Standard 90.1 provides very detailed instructions on how to achieve the current minimum energy efficiency targets in buildings. It runs to 180 pages of text and tables in its 2004 edition. Although it is quite comprehensive, it is written in "code language" to make it easier for regulatory authorities to use its provisions in laws. Consequently, the standard itself is rather terse, emphasizing instructions rather than the explaining the logic behind those instructions. Reading the standard is like reading a building code.

To understand its logic and to see examples of how the provisions of Std 90.1 can be implemented, the interested reader is encouraged to obtain the 90.1 user's manual—a separate publication of over 300 pages which includes a CD with helpful software.[3] The user's manual helps the interested professional understand not only what the standard says, but why it says it, with examples of how the provisions can be implemented, and examples of potential trade-offs which can help achieve the energy targets.

Standard 90.1 requirements become more challenging

Std 90.1 is constantly changing as the Society provides guidance for ever-greater reductions in energy consumption. For example, the ASHRAE Board of Directors voted in 2007 to improve the standard so extensively that when a building complies with the 2010 edition, it will use 30% less energy than a similar building which only complies with the 2004 edition. The Board determined that based on recent experience in both North America and Europe, a goal of 30% annual energy reduction below the 2004 edition is well within the capability of designers, while still being economically practical for owners and without compromising the health or comfort of occupants.

That said, as of the publication date of this book, few believe that a 30% annual energy reduction over std 90-2004 is going to be simple or easy for all buildings in all locations. Many of the common (and most energy-wasteful) design practices for building enclosures, windows and HVAC systems will have to change. Buildings will have to become much more air-tight than they are at present. Windows will have to exclude more heat than they do now, while providing more useful daylight. Lights and HVAC systems will have to modulate more of their capacity, and modulate in more steps than has been typical practice. And there will have to be more effective conservation and re-use of wasted energy than has been common in the past.

These improvements can only be accommodated by a much earlier and much more interactive collaboration between owner, architect and HVAC designer than what has been typical of building projects in recent decades. *Specifically, the exterior walls and windows and the interior surface finishes will have to be designed so they reduce the size of the mechanical system and the amount of interior lighting.*

Owners and architects will need to ask more of their lighting and mechanical engineers *as the project is being conceived*—rather than after the enclosure, its fenestration and the interior design and finishes have been established. And those engineers will need to be willing to provide more detailed calculations and more alternatives at an earlier stage than in the past if they are to affect the changes in the building enclosure needed to meet the coming energy targets.

This will mean more time for design, and therefore in some cases a higher early-stage cost for professional fees. But these modest increases in early costs will often be immediately offset by lower construction costs, and *will always* provide far lower operating costs for the life of the building.

This chapter vs. ASHRAE Std 90.1

This chapter is not a substitute for the highly detailed guidance contained in Standard 90.1 and its user's manual. The information contained here is both more than the current standard, and less. More, in that this chapter suggests ways to reduce energy consumption which will anticipate future improvements to Standard 90. Less in that it would be pointless to duplicate the highly detailed and complex instructions of the current standard and its user's manual.

This chapter is for decision makers who seek the overview of the key principles of energy reduction as they apply to all buildings in hot and humid climates. This is the "birds eye view" of the most important issues rather than the details needed for construction documents of building enclosures and HVAC systems. For detailed design, the reader is encouraged to read the later chapters in this book, and also to read Standard 90.1 and its user's manual.

ASHRAE Advanced Energy Guides

Going beyond the 2001 and 2004 editions of Standard 90.1, ASHRAE has published a series of Advanced Energy Design Guides. As of the publication of this book, the series includes separate volumes for small offices[4] and for retail buildings.[5]

These guides show the path to a 30% reduction in energy consumption compared to Standard 90.1. To accomplish such a large reduction, each guide discusses a single class of buildings rather than only those requirements common to all types of buildings. Also, for the building type in question, they describe climate-specific recommendations for each of the eight climate zones defined by the US Department of Energy for the United States.

The guidance takes note of what is actually possible in an office or a retail store. For example, it's a fact of life that offices are less occupied during evenings and weekends than retail stores. Also, retail stores simply must use more interior lighting than offices to achieve their purpose. These differences call for different guidance for energy reduction. Similarly, buildings in cold climates need more heat, while buildings closer to the equator are dominated by the energy used for cooling. So the guidance for each structure type changes with climate as well as with the buildings' functions.

This chapter borrows heavily from these Advanced Energy Guides for basic principles, and for some of the appropriate target values for achieving large energy reductions in hot and humid climates.

For more extensive details for retail buildings and small offices, and for energy reduction design criteria in cooler or drier climates, the reader is encouraged to read these ASHRAE Advanced Energy Guides. The large amount of information they contain for specific building types cannot be duplicated here in a single chapter.

Suggestions For Reducing Energy Use

The suggestions which follow are presented roughly in order of their importance for new buildings in hot and humid climates, based on the assumption that the building will be air conditioned.

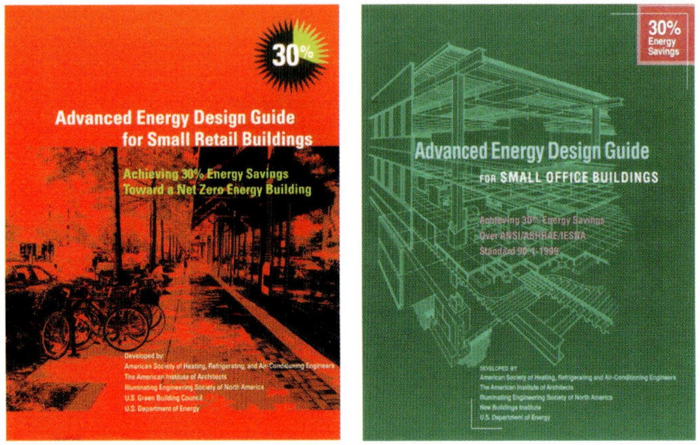

Fig. 4.3
Advanced Energy Guides
Pointing the way to improved performance beyond the 2004 edition of Std 90.1, the ASHRAE Advanced Energy Guides are focused on climate differences, and on specific types of buildings.

Several of the suggestions discussed in this chapter are based on design elements for hot & humid climates which are described in more detail by the Advanced Energy Guides for small office buildings[4] and for retail stores.[5]

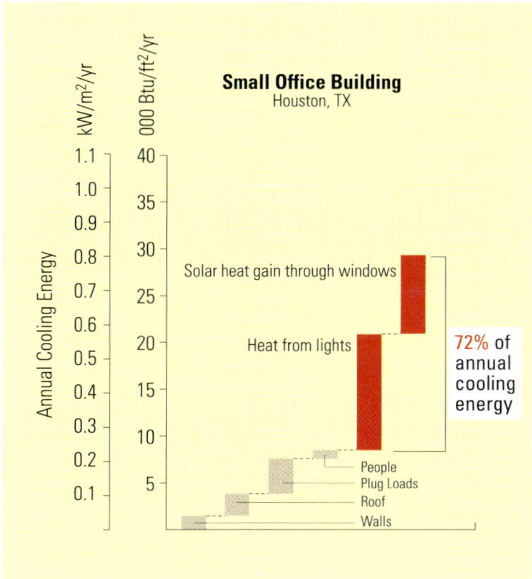

Fig. 4.4
Windows control solar heat & light
Solar heat and light strongly influence cooling loads. So annual energy use is essentially governed by windows.[1]

1. Reduce the cooling load from windows

Figure 4.4 shows why window design is the first and most important priority for those who seek to reduce energy consumption. The graph displays the cooling loads for a typical low-rise office building in Houston, TX.[1]

Note that 72% of the entire annual cooling load is generated by the heat coming though the windows combined with the heat generated by the indoor lighting. So the first priority will be reducing the amount of heat that comes though the windows. And the second priority will be making productive use of the light which comes through those redesigned windows, so that the interior lighting can generate less heat.

The easiest and least expensive way to achieve both goals is simply to use fewer windows on the building, and smaller ones. Design them as a narrow band of horizontal, low-e glazing high up, near the ceilings of each floor, so the reduced window area is still effective for daylighting. Using fewer, smaller windows—ones which are high, narrow, horizontal and designed specifically for daylighting—has many benefits for the building, including lower energy use and reduced construction costs.

Only one problem, really. Ever since humans decided to move out of caves, most commercial building occupants (with the possible exception of call center employees) have become accustomed to having windows at eye level which let them see the outdoors.

So, to reduce the window load while still providing visibility to the outdoors, design the fenestration with two stacked windows in each set. The lower window—at eye level—is for visibility. The upper window is for daylighting. Both are shaded from direct sunlight, and neither is located on the east or west faces of the building. Figure 4.6 shows an example of what such a design looks like from the outside, from the perspective of the public. Figure 4.5 is a photo from a different building, but it shows what daylighting looks like to the building occupant on a sunny day.

Here's why window shading is so important for reducing heat load and energy consumption. Figure 4.7 shows the air conditioning load from one square foot [or one

Fig. 4.5 Daylighting can provide pleasant illumination
Since heat loads peak during the day, reducing electric lighting when sunlight is available will reduce the cooling load generated by lights, which will save energy and also reduce the installed cost of cooling systems.[6]

Fig. 4.6 Exterior design enables daylighting, or makes it impractical
Wide windows near ceilings provide daylighting to reduce electric lighting during peak cooling periods. Buildings designed for effective daylighting look different. Note the exterior light shelf, which shades the lower view windows and also bounces more light into the building through the upper daylight windows.[6]

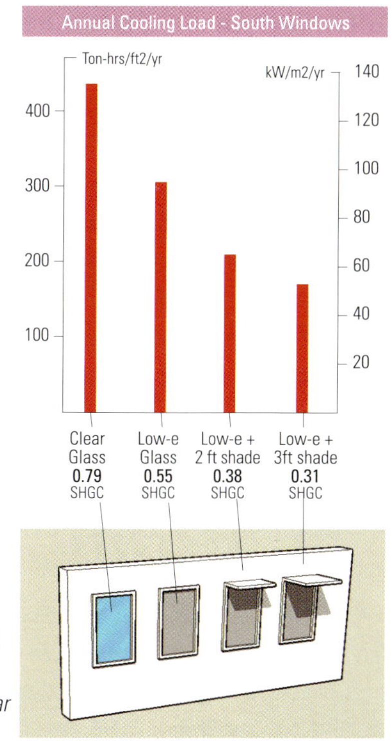

Fig. 4.7
Windows and cooling load
To reduce cooling load, use low-e glazing and exterior solar shading[7]

square meter] of a south-facing window on a hypothetical small office building located in Miami, Florida. [7]

The difference in annual heat load is due to the difference in the solar heat gain coefficients (SHGC) of the four alternative designs.

To calibrate the reader's expectations, note that single-pane clear glass usually has a SHGC of about 0.79. In other words, the single-pane clear glass only blocks 21% of the radiant heat loads from the sun.

In contrast, consider that to reach a 30% reduction in overall energy consumption compared to the 2004 edition of Std 90.1, the recommended SHGC for all office windows in hot and humid climates is 0.31. In other words, 69% of all solar radiant heat should be blocked by *all* windows—if the energy reduction target is to be achieved, and if the owner and architect want 20 to 40% of the wall surface to be windows. (If the SHGC of the windows is over 0.31, and if 40% of the walls are windows, that 30% energy reduction probably cannot be achieved, no matter how clever the HVAC designer might be, nor how much extra money he is given for exotic equipment. The heat loads will simply be too high, for too much of the year.)

Looking again at figure 4.7, note that using better glazing makes a significant improvement over single-pane clear glass. In that diagram, the second window uses double-pane, insulating low-e glass. The term "low-e" describes the emissivity of the glass surface. If the emissivity is high, the surface emits more heat. If emissivity is low, the surface may still be hot, but it *emits less of that heat* to the indoor environment by radiation. By placing the low-e coating on the inside of the exterior pane of glass, much less of the heat that warms that exterior pane is radiated to the interior pane. And the gas-filled space between the two panes helps keep the convective heat from the outer pane from reaching the inner pane.

The combination of that gas-filled gap and the low-e coating on the indoor surface of the exterior pane keeps 45% of the suns' radiant heat from moving into the building through the glass. In other words, the SHGC of that glazing is 0.55. That's a major improvement. But the reader will quickly note that it's not nearly good enough to meet the energy reduction target of 30%. For that, shading will be required. No amount of clever glass technology can reach the target maximum SHGC of 0.31 by itself without compromising visible light transmission. Overhangs are needed to shade the windows and to keep radiant solar heat off of the glass and therefore out of the building.

Given that some shading above the windows will be needed, how far off the wall surface must that shading project? The quick answer is that, if one assumes double-pane, insulating low-e glazing, a target SHGC of 0.31 can be reached if the horizontal shading projection is at least 40% of the vertical height of the window. As seen in figure 4.7, that much shading (a projection factor of 0.4) allows the baseline solar heat gain coefficient to reduced by an additional 33%. (The SHGC is multiplied by a factor of 0.67.)

For the engineer, this single example is a gross oversimplification of the possibilities. Only the ASHRAE Advanced Energy Guides for Small Office Buildings and some local energy codes currently require a SHGC as low as 0.31. And the exact value depends on the glazing surface area, which may be more or less than 20 to 40%. Also, it matters whether the windows are located on the south or the west faces of the building, as we shall see next.

But to the owner and the architect, such details are often tedious and of secondary concern. For decision-makers concerned with the big picture, the most useful messages include:

1. To make a significant reduction in energy consumption compared to past good practices (let alone compared to past poor practices), the *glazing and shading of its windows will become the architect's very first concern* as he or she is considering the building's look and feel.
2. If the building is to reach energy targets of ASHRAE standard 90.1 after 2010, the windows will almost certainly have to be shaded, as well as being made with at least double-pane, low-e glazing.

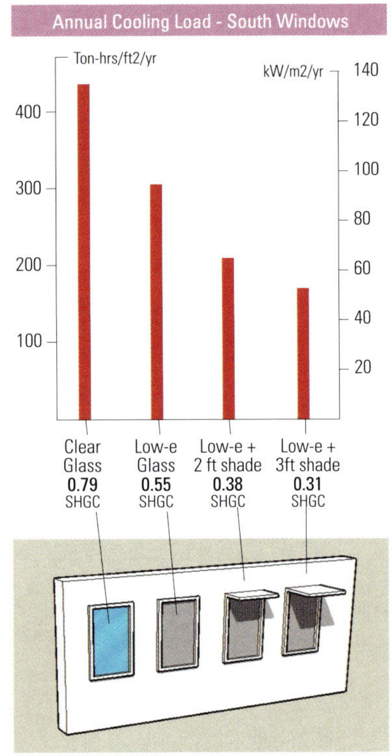

Fig. 4.7
Windows and cooling load
To reduce cooling load, use low-e glazing and exterior solar shading [7]

3. Even with such glazing and excellent shading, window surface areas over 20% of the wall surface make it more difficult to meet energy targets. Windows totalling over 40% will probably require heroic complexity and costly engineering budgets to keep energy consumption within ASHRAE and international energy code targets.

These messages are most useful to owners and architects at the earliest stages of a project because they highlight a key point. Namely, if the lighting and mechanical engineers are not deeply involved at a very early stage in project planning, the enclosure design may make the building incapable of meeting its energy targets.

Increasingly in the future, engineers should expect to be asked by owners and architects to provide quantitative advice before the details of design are established. This can be a sometimes unfamiliar and occasionally uncomfortable circumstance for some technical professionals, who may be accustomed to entering the process long after such key details are decided.

But if energy in buildings is to be reduced, owners and architects will have to seek early advice, and engineers will have to be prepared with quantitative alternatives at that early stage—before the look-and-feel of a building becomes part of its marketing to prospective owners and tenants, and before local review boards have passed judgement on the design. After regulatory bodies have approved the building's look-and-feel, any meaningful changes needed in the enclosure design to achieve energy targets may be a practical impossibility.

2. Avoid west-facing glass

In hot and humid climates, west-facing glass is especially wasteful of energy and wasteful of the HVAC budget. It also requires the owner to give up more floor space for mechanical rooms.

This is because any glass on the west face has a much greater impact on the size of the mechanical system than does glass on any other face of the building.

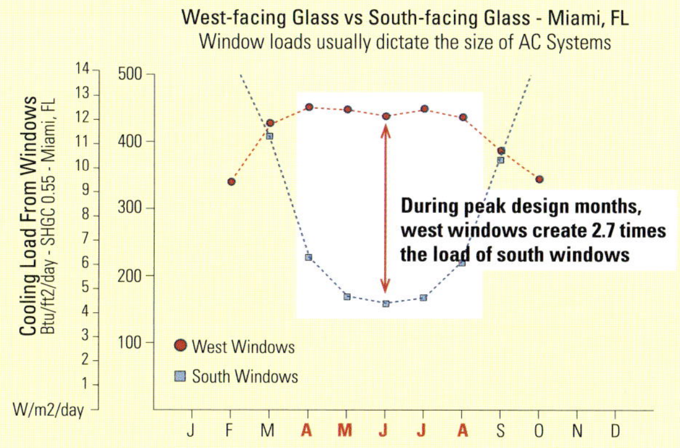

		Jan	Feb	Mar	Apr	May	Jun	Jul	Aug	Sep	Oct	Nov	Dec
South	Btu/ft2/day	609	537	404	236	173	159	174	221	380	543	616	604
	W/m2/day	17	15	11	6	5	4	5	6	10	15	17	16
West	Btu/ft2/day	271	339	424	465	458	433	453	432	392	345	276	237
	W/m2/day	7	9	12	13	12	12	12	12	11	9	8	6

Fig. 4.8 West-facing windows often govern HVAC costs

West-facing windows pass maximum heat into the building just at the time when all other cooling loads are also at their peak (afternoons in the summer).[7] So they often govern the size of the cooling system, and therefore its cost and its annual energy consumption. Reducing or eliminating west-facing glass reduces both energy consumption and installed costs.

Figure 4.8 illustrates why this is so. It shows calculations for the solar radiant heat gain through an unshaded, but double-glazed insulating window which uses low-e glass.[7] Looking at the summary table for the same size window located on either the south or west faces, the total annual load is actually smaller for the west-facing window. But it's the *peak load during the hottest months* which sizes the cooling system—not the net annual load.

The graphed cooling loads for April through August show the problem. In the summer months, the west-facing window lets in 2 to 2.7 times more heat than the same size window on the south face. What's even worse is that the west-facing window lets in all that heat in the afternoon—after the entire building has been heated up from the solar load, and heated up by loads generated internally by lights and by occupant activities.

Of course, the solar load is equally large on the east face of the building. But that east-face load comes earlier in the day, before the other heat loads have reached their peak. It's the west-facing glass load that really forces the decision between a larger and a smaller cooling system. The size of that system determines its cost, and determines the amount of mechanical room space it will need.

And a large system is inherently more difficult to modulate during periods of low loads. A system sized for large cooling loads from a great deal of west-facing glass sometimes overchills the whole building during the early morning. Large systems often force the building operators to choose between running that big system at it's lowest (but still much too large) capacity—effectively freezing the occupants—or not running the cooling at all, which makes the occupants too warm. The usual choice is to freeze the occupants rather than fry them, or to alternately freeze and fry them, as the system struggles to smoothly modulate its extra-large cooling capacity. Building operators often find that occupants can be rather intolerant of this effect.

Avoiding (or at least minimizing) west-facing glass reduces the peak cooling load for the building. The cooling system can be smaller, which saves energy and saves money in the construction budget. It is also simpler and less expensive to operate and is likely to be more responsive and therefore be more comfortable for occupants during 99.6% of building's life (the part-load hours).

3. Reduce the heat from lights, using daylighting

As seen in figures 4.1 and 4.7, the next largest cooling load after exterior glass is the load from interior lighting. For really significant reductions in the building's annual energy consumption, the design can reduce lighting power consumption when sunlight is available.

Daylighting—using less electrical power and more sunlight to light the interior during daytime hours—can make a significant reduction in annual energy consumption, especially in hot and humid climates, which are closer to the earth's equator than are cold climates. So sunlight is available for more of a building's occupied hours compared to cold or moderate climates, offering the potential for even larger energy savings.

Further, it's often the case that a buildings' peak electrical power *demand* comes during clear, hot afternoons during the summer, when cooling loads are at their peak. So it can be especially cost-effective to use daylighting to reduce that peak load. Peak cooling load determines the size and cost and complexity of the cooling systems, as discussed earlier. And on the economic side, the combined peak lighting and cooling load also determines the cost of electrical power for buildings located in places where power is limited. In those locations the annual energy cost is often determined by the maximum peak power *demand* rather than by the total Kw consumed.

Every Watt saved in lighting reduces the building's power consumption by at least 1.2 Watts, because the lighting load appears twice in the energy budget. First, it appears as lighting power. Then part of that power appears again as a heat load for the cooling system. The cooling system uses additional power to remove the heat generated by the lights. That additional power consumption will be between 20 and 30% of the power used by the lights, depending on the efficiency of the cooling system.

So the potential annual energy cost reduction from daylighting can be significant. On the other hand, a building which has effective daylighting often looks different from a building in which daylighting cannot be effective. And a building which uses daylight effectively will probably cost more to construct.

Before the first cost even becomes a factor, one common reason that buildings in hot and humid climates have sometimes neglected the potential of daylighting is because the critical enabling decisions are made very early in project planning. If the owner and architectural designer are not aware of the measures needed to take advantage of the climate's high daylighting potential until after the look and feel of the building's exterior are decided, it may be impractical to consider changes. Both regulatory approval and the marketing of the building may already be based on a structure which cannot use daylighting effectively.

Daylighting does not work properly (and does not save any energy) unless the owner understands and agrees that shape, look, and feel of the building's exterior *and* its interior are going to be part of the lighting system, and will therefore need to be designed accordingly. Also, the lighting budget will probably rise, because the building will still have to be lit after dark. So the number of fixtures will probably be no different from other buildings. And to actually save power, the lighting output will have to be modulated by a control system as more or less daylight is available.

More specifically, to achieve the most significant reduction in lighting power and cooling load, the building will need:

1. Wide daylighting windows placed high on the walls, near the ceiling of each floor.
2. Light shelves projecting outwards from the exterior wall underneath those windows.
3. Exterior sun shades above those windows.
4. Glazing in those windows which passes a significant portion of visible light - ideally more than 60%, while still keeping out most of the sun's radiant heat.
5. Light-colored ceiling and wall finishes which reflect incoming daylight evenly, without glare, into occupied spaces.
6. Automatic lighting controls which sense the current indoor lighting level near the occupants' reading surfaces, and which switch off or dim the lights, modulating energy use as daylighting rises and falls.

Fig. 4.9 Windows for daylighting

Conventional windows set at eye level do not transmit enough daylighting to allow a reduction in lighting power. The key exterior architectural elements for success are shown in this photo.[6] Also, interior surfaces need to reflect rather than to absorb the incoming light.

Excellent advice for detailed design of exterior enclosures and glazing to optimize daylighting can be found in references 6 and 8, are described at the end of this chapter. They provided the photos seen here as figures 4.5, 4.6 and 4.9, and some of the advice they contain is summarized briefly below.

Figure 4.9 shows what the exterior of a building looks like when it is designed for effective use of daylight, along with callouts indicating the key elements of such a design. Each of those elements plays an important role:

Wide windows, located near ceilings

The deeper the daylight penetrates into the building's interior, the more electrically-powered light it can displace. A small percentage of daylight penetrates very deep into the building, of course. But as a rule of thumb, enough daylight to be effective only penetrates to 1.5 times the height of the window.[8] In other words, if the tallest part of the window is 6 ft off the floor, adequate daylight will only penetrate as far as 9 ft. into the buildings interior [if the top of the window is 1.8m off the floor, adequate light may penetrate to 2.7m]. So the higher the top of the window, the deeper the light penetration and the more effective daylighting can be.

Also, the *wider* the window at that tall height, the more light will be able to penetrate into the building. It's largely a matter of the total window area at it's maximum height (its width at the top) which determines how much light can penetrate.

So the ideal fenestration for daylighting is a narrow band of windows circling the entire building near the ceiling of each floor.

Unfortunately, by itself this arrangement only allows the occupants to see the sky. That's why daylit buildings usually have two sets of windows stacked on top of one another. The lower window is set at eye level for a pleasant visual connection to the outdoors. The upper window is set close to the ceiling, to maximize the *depth* of light penetration and it is also wide, to provide a maximum *amount* of light.

Exterior light shelves

Light shelves which project outwards from the exterior wall just below the daylighting windows benefit the building in three important ways. First, they increase the amount of light coming through the daylighting windows, so that the penetration depth increases by about 30%—2.0 times the height of the window instead of only 1.5 times the maximum window height.[8]

Next, they act as sun shades for the view windows set below the daylighting windows. That shading usually reduces the solar heat gain though those larger windows enough to meet the strict energy reduction targets of the International Energy Code, and enough to meet the recommendations of the ASHRAE Advanced Energy Design Guides for hot and humid climates (a solar heat gain coefficient of less than 0.31).

Finally, if those projections are attached all along their inward edge, they can reduce the risk of mold and other microbial growth inside the building. An attached projection will force any rainwater that's flowing down the walls off and away from the windows underneath them. The joints around windows are the usual places where water gets into the walls to support microbial growth indoors. As one experienced building scientist has often observed: "If the window doesn't get wet—it can't leak."[9]

Glazing which passes visible light but keeps out heat

It is obviously important for daylighting windows to allow as much visible light as possible into the building. A minimum visible light transmission of 60% (VT= .60) is a useful rule of thumb, and more is better.[8] Clear glass transmits about 80% of visible light, but it's important for daylighting windows to have a low solar heat gain coefficient, because after all, they are still windows. Without a low SHGC the daylight windows would waste in cooling any energy savings they might gain from daylighting.

The difficulty is that glass with an extremely low SHGC can reduce the visible light transmission far below 50%. Tinted or highly reflective windows are an example, some having a solar heat gain factor as low as 0.20, but which also pass only 24% of visible light (VT = 0.24)

Fortunately, given the hundreds of glazing combinations currently available, with modern glass it's possible to strike a reasonable compromise between keeping heat out while allowing visible light in. Double pane, "spectrally-selective" low-e glazing can be obtained with a VT of 0.71 and a SHGC of only 0.38 . Such windows would be quite effective for daylighting while still excluding a great deal of heat.[8]

Sun shades above daylighting windows

On the other hand, even with solar heat gain coefficients as low as 0.38, the daylighting windows will not meet the target values for solar heat exclusion which are suggested by the ASHRAE Advanced Energy Guides. So if the owner wants to meet the target of 30% less energy than the 2004 edition of ASHRAE Std 90.1 the SHGC of all windows will need to be less than 0.31. That means the daylighting windows will also need sun shades.

Light-colored, diffusively-reflective ceilings and walls

If the interior surfaces are dark, they will simply absorb the incoming daylight and re-radiate that energy in the form of heat, eliminating the benefit of the daylighting windows.

To avoid wasting that daylight by turning it into heat before it can be useful to occupants, the interior finish on the ceiling must be light and highly reflective—and also diffusive. Mirrors on the ceilings would reflect nearly all the incoming light, but the glare would be visually unbearable. A matte-finish, white ceiling tile is ideal.

Also, the walls should also be white or at least extremely pale, so they don't absorb that daylight, either. The look-and-feel of an antique Mediterranean whitewashed interior is an excellent model. Those buildings use daylight very efficiently. Light comes through small windows and then bounces around the interior, reflecting off of whitewashed interior walls and providing considerable illumination for occupants during the daytime. That interior design evolved over centuries, before electric lighting became cheap and readily available.

It would be unfortunate if the interior designer decided on a dark-cave-sophisticated-nightclub look-and-feel for ceilings and walls. Dark colors quickly eliminate—with a single coat of paint—most of the benefits the owner was expecting from his investments in daylighting windows, light shelves and spectrally-selective glass.

Automatic lighting controls

Automatic lighting shutoff for unoccupied spaces is a baseline requirement of ASHRAE Std 90.1-2004, regardless of whether or not the architect and owner decide to take advantage of daylighting. But with daylighting, it makes sense to modulate the interior lighting, or at least to bring it on in stages, rather than simply turning lights on and off with a time clock.

Sunlight varies in intensity as the cloud cover changes, and of course as the day turns into night. So in most commercial occupancies, successful daylighting includes automatic controls to brighten or to dim the electrical lights as the daylighting levels rise and fall, keeping those changes imperceptible to occupants, or at least irrelevant to their activities.

In residential occupancies, there may be little need for automatic controls. The residents will either feel the need to turn lights on, or they won't. But in commercial buildings, where responsibility and authority for controlling lights is not always given to the occupants, automatic controls will be needed to really achieve the desired energy savings.

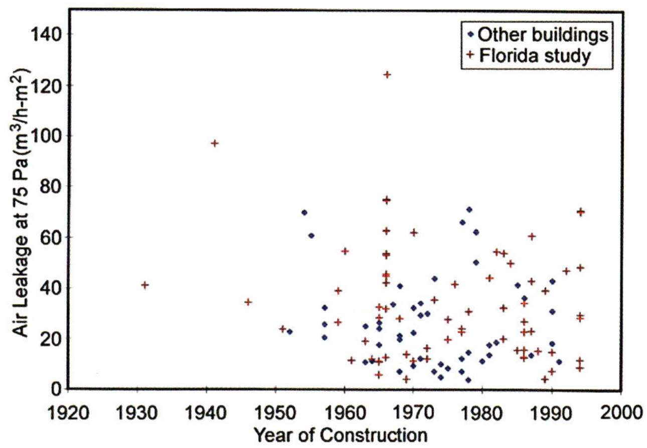

Fig. 4.10 So far, newer buildings are not actually tighter

Air leakage drives loads, which drive energy consumption. Field studies show that most US buildings have not yet shown improvements.[11]

4. Build an air tight exterior enclosure

When hot and humid air leaks into the building, the mechanical systems will use energy to remove those loads. So, to reduce the building's annual energy consumption, keep the hot and humid air out of the building. Otherwise, a leaky building enclosure is like propping open the door to an EnergyStar™ refrigerator, and then expecting that refrigerator to still meet EnergyStar standards for energy use.

That seems like obvious advice. It also sounds irrelevant, because most professionals who live in North America have heard so much about tightening up buildings over the last 20 years that they assume that buildings are now so tight that air leakage could not really be a problem. In some thoughtfully-designed and carefully-constructed single-family residences, this perception may indeed be justified.[10]

Also, in other countries, building codes have for some time required both air tightness and the post-construction testing which

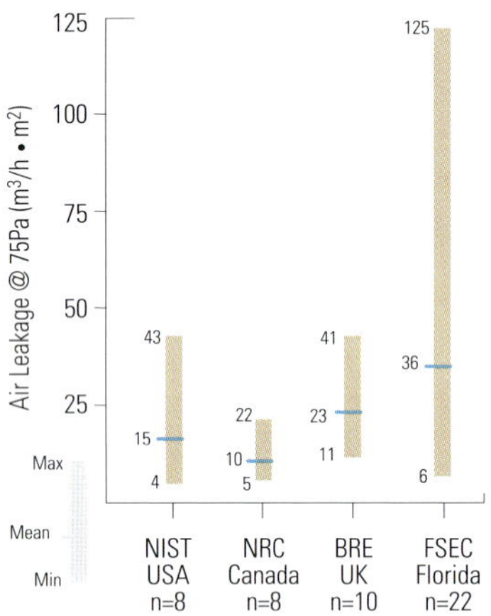

Fig. 4.11
Measured air leakage[11]

assures it. However, in the United States most housing and most light commercial and small institutional buildings are very, very leaky. And there has been no indication of significant improvement since 1990.

Figure 4.10 shows the air leakage rates for 117 buildings constructed between 1932 and 1998, as studied by the U.S. National Institute of Standards and Technology (NIST).[11] The data show that there is no correlation between year of construction and air tightness.

There is even some evidence that commercial buildings in hot and humid parts of the US might even be more leaky than the norm. Figure 4.11 compares four sets of building leakage tests done in the Washington, DC area, in Canada, in the United Kingdom and in Florida. In the Florida building selected for this particular test, there was a very wide range of leakage rates. The tightest of the 22 Florida buildings was similar to those in other climates. But the leakiest buildings were more than twice as leaky as those tested in other climates.

This amount of infiltration can add up to a lot of hot and humid air. Figure 4.12 shows a study of 70 low-rise commercial and institutional buildings performed by building scientists from the Florida Solar Energy Center. The figure shows the results of field measurements of air leakage under passive ventilation conditions. In other words, the buildings were unoccupied, and the mechanical systems were off, so the only driving force was the light blower-door pressurization designed to overcome breezes and stack effect during the tests. The vertical axis shows the number of buildings in a particular air leakage category. The horizontal axis shows the different air leakage ranges which apply to the buildings tested.

Note that there were many buildings which only leaked air at a rate of 0 to 0.2 air changes per hour. On the other hand, most of those 70 buildings leaked air at rates between .05 and 2.0 complete air changes *every hour*.

To reduce this infiltration load, it's useful to design and construct the building to be more air tight than in the past. In some northern code jurisdictions, notably Canada and the State of Massachusetts as of the publication of this book, buildings must be built with a continuous air barrier, by code. And that air barrier must be called out in the plans, and it must be tested during construction.

To date, in hot and humid climates in North America an air barrier is not yet a code requirement. But such measures will save energy, as they do in colder climates. Another NIST study suggests that a 6% annual cooling savings can be expected from a tight building.[12]

In fact, the real concern is not so much achieving a hermetic air seal as it is simply not building the enclosure with wide gaps between

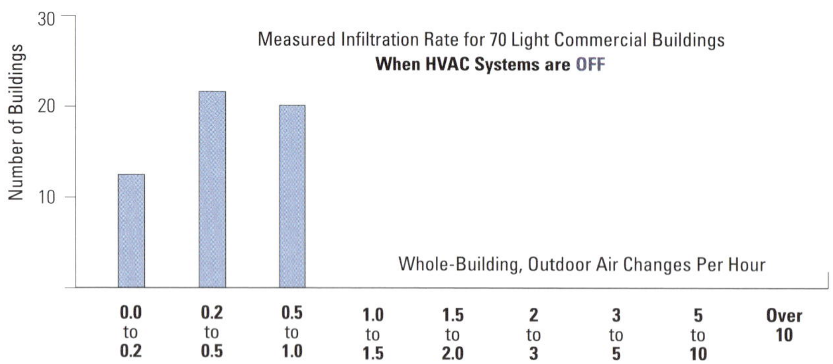

Fig. 4.12
Whole-building air leakage
Passive air leakage (when systems are off) is substantial in light commercial construction in the US, as measured in these 70 commercial and institutional buildings in Florida.[13]

Sealing up holes, gaps and joints in construction is the best way for the architect to help reduce energy consumption, after the overall building enclosure has been designed to keep out solar heat.

the roof and walls, or accepting leaky roof parapets or allowing ventilated soffits which let humid air into attics above dropped ceilings.

For the architect, the basic guidance is really rather simple: seal up the building enclosure so that air cannot flow through the exterior walls though big holes and cracks.

More specifically, the architectural designer can use the measures which are used to keep fire from spreading between floors and across walls. Don't allow wide gaps between occupied floors and attics, or around penetrations in the exterior walls, and especially not between the outdoors and the attic, if there is one. Design with very narrow gaps, when gaps are necessary for penetrations. Then seal those gaps tight with expansive fire sealants, as shown in figure 4.13.

Fig. 4.13
Sealing up holes and air gaps
The techniques and materials used for containing fire and noise are also effective in preventing the air leaks which lead to high cooling loads and energy waste in hot and humid climates.

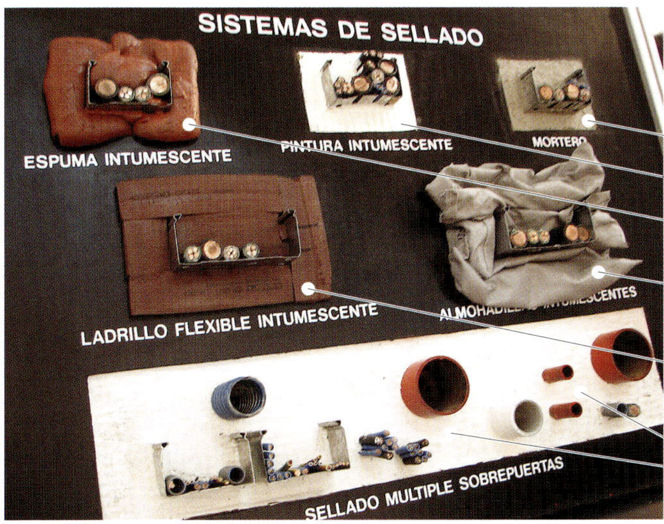

5. Commission new buildings and mechanical systems

Moving on to energy-saving suggestions for mechanical systems, the most useful measure is to make sure that the mechanical systems are designed so they can be operated and maintained easily, and that they are actually installed and functioning the way their designers intended. That's commissioning.

In large building construction, commissioning, testing, adjusting and balancing are often given a great deal of attention. But smaller buildings represent the majority of commercial and multi-unit residential construction. In those buildings the mechanical system is sometimes designed and installed at lower cost and with less deliberation, which can be a problem from an energy perspective.

A common public impression is that energy efficiency is largely a matter of which piece of equipment you install. Many assume that after deciding to use "energy-efficient equipment," as long as the surrounding system is designed and installed by professionals, all that needs to be done is to turn it on, and it will work properly and be energy-efficient. That's not a good assumption. No two buildings are exactly identical—every building is really a new prototype. So big and small problems will add up to a great deal of energy waste, overcoming any benefit from good equipment choices.

To visualize the integration and installation problems, imagine if cars were designed and manufactured like buildings. Several different teams of professionals, located remotely, would design separate aspects of each unique car. The laws governing the construction of each car would be different for every car made, in each different location. And the automobile assembly line would be erected outdoors in the weather, only once for each car, with different employees every time the assembly line was put in place. Of course, each component of each car would be bid separately. So sometimes a given car would be designed to have a gasoline engine, but the assembly managers might get a better price for a steam engine, and install that sort of power plant instead, assuming the controls and the interface with other drive components would work the same. Component suppliers would have all sorts of ways to improve (and cheapen) the car at the bid stage. Then perhaps the car's prospective owner would run short of money, which could require drastic last-minute design changes for a 30% cost reduction in order to complete that car.

With the image of the resulting car in mind, it becomes easier to understand why mechanical systems in buildings are so likely to have problems which waste energy after installation, no matter how elegant their original design concept.

Mechanical systems are designed and built by many groups of professionals—an HVAC and electrical designer, hundreds of different component and controls manufacturers, plus separate contractors for sheet metal, air conditioning, building automation programming, hardware, sensors and wiring, and separate plumbing and electrical crews. It's not surprising that there will be real-world problems integrating all of these functions.

Energy-related integration problems occur even in rather simple systems. Consider a field study of light commercial buildings performed for the State of California. The investigators looked at packaged rooftop air conditioning and heating systems.[14] Their field observations found that of 71 installations, 80% were equipped with automatic economizer dampers that were either not connected to electrical power or were wired backwards. In other words, the economizer dampers would (automatically) pull in large amounts of outdoor air during weather conditions when such "free cooling" *adds to* the internal loads rather than removing them.

One could argue that this example is simply a matter of correct installation on the part of the air conditioning contractor. But in these different buildings, built for different owners by different contractors, it was not always clear whether installing correct components and connecting wiring inside the AC units themselves was the responsibility of the electrical contractor, the controls contractor, the sensor manufacturers, the unit manufacturers or the general contractors.

Keeping in mind that such energy-wasting problems often occur in *simple* systems. As the number of components increases, the potential for energy-wasting integration problems also increases.

For example, figure 4.14 appears in ASHRAE's reference and textbook for new HVAC designers: *HVAC Simplified*.[15] The graph shows one short summary of the massive amounts of data collected during the nation-wide *Commercial Building Energy Consumption Survey*, performed every 10 years by the U.S. Department of Energy and the U.S. Energy Information Agency.[16]

The unsettling results show that in this sample of tens of thousands of buildings, the simplest systems used the least energy. They cost less to run than the larger systems which are usually assumed to be more energy-efficient.

Specifically, buildings equipped with simple unitary AC systems, such as the through-wall air conditioners and packaged terminal air conditioners often used in budget hotels and nursing homes, apparently cost about 30% less to operate per square foot than did central chilled water systems, and about 25% less than buildings with variable air volume systems. Both of these more complex and "more engineered" systems are expected to cost less to operate than unitary equipment. But if they are not operating as designed, as was probably the case with the buildings surveyed, they can actually be less efficient than simpler systems, as seen from these survey results. To be fair, older unitary equipment has been notoriously ineffective at removing humidity, so the results may reflect the fact that the surveyed unitary equipment was doing less work.

But the main point remains valid. HVAC systems can often consume more energy than either the equipment manufacturers or the system designers expected. So one of the keys to actually achieving the *hoped-for* energy reductions lies in commissioning.

What commissioning is, how it works, who does it and how you know it's complete

The goal of commissioning is to have a set of building systems which work together smoothly and reliably, as their owner and designers intended. With effective commissioning, this will be true not only on the day they have been tested, but throughout the life of the building—because they can be easily maintained.

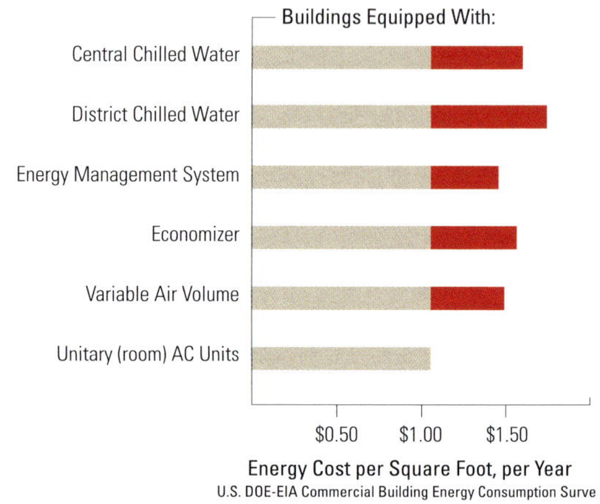

Fig. 4.14
Energy-efficient designs do not automatically save energy
According to the Commercial Building Energy Consumption Survey performed by the U.S. Department of Energy, buildings equipped with energy-efficient measures sometimes use more energy than buildings with very simple equipment.[16]

Commissioning helps those elegant designs actually achieve their expected savings.

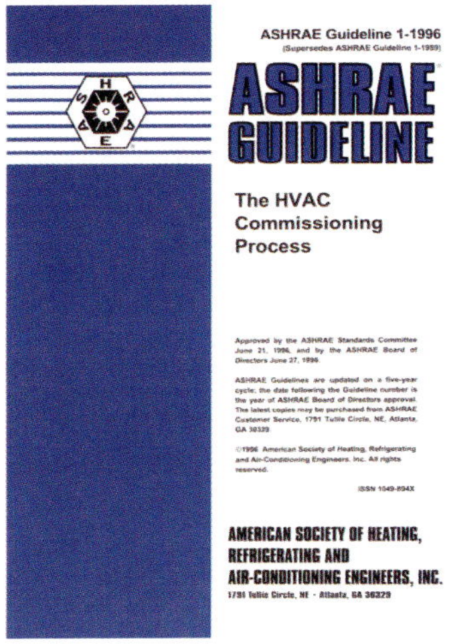

Fig. 4.15 HVAC Commissioning
ASHRAE Guideline 1 provides the information needed by owners, architects, HVAC designers and system operators who choose to invest time and money in the commissioning process. [17]

Commissioning provides the forum for the key decisions, the matrix for quality control and the repository of system understanding that are needed to really achieve the benefits of energy-efficient designs.

As explained in detail in ASHRAE Guideline 1 - The HVAC Commissioning Process, commissioning is a process rather than an event.[17] Like most of the other energy-saving measures discussed earlier in this chapter, the commissioning process provides the forum and the structure for all construction and operation team members to jointly discuss mechanical systems at an early stage. Commissioning also requires additional discussions during key moments at later stages, examining the many mechanical system decisions from the perspectives of all of those team members.

To begin the process, an owner usually hires a third-party "commissioning authority" to develop and oversee the commissioning plan. Ideally, this happens very early in the planning process, because one of the benefits of commissioning is to help the owner clarify and define the overall requirements for the systems and for the building enclosure. Further, the commissioning plan lays out who will be responsible for which functions, and how all functions will be integrated, throughout the construction and start up process.

For example, such a plan can make it clear to the architect and to the interior designer that daylighting is an owner's requirement, and that therefore the window design and the interior finish must facilitate daylighting so that loads will be reduced, allowing the mechanical system and the lighting systems to be smaller and use less energy.

Later in the process, the commissioning plan serves as a repository, recording the design intent for each system. Also, during construction the commissioning plan usually requires that the general contractor provide the owner and the rest of the team with an integrated operating and maintenance manual, and a "certificate of readiness" for the systems. This is a document stating that all equipment and systems have been correctly installed, operated as specified; tested, adjusted and balanced, and have been verified as being ready for functional performance testing and other acceptance procedures. The commissioning plan will go on to spell out what those functional tests must be, who will perform them, when and how they will be performed and how the results will be documented.

After all the systems have been functionally tested, the plan will typically define what documentation and training will be provided to the owner's operating and maintenance staff so they have a clear understanding of the design intent and of the maintenance needed to keep the systems operating in accordance with that intent. The plan will also define who provides that training, along with when and how it will be provided and how the training process itself will be documented. In short, the commissioning process helps ensure that "all the pieces come together and stay together."

The cost of commissioning is usually between 1 and 3% of the cost of the mechanical systems, depending on what services are provided.[18] The benefits include reduced energy consumption, better comfort and better indoor air quality. These benefits occur because the systems have been designed and field-tested as an integrated whole, and because the operating staff understands the design intent, so they can continue the initial energy-efficient operation into the future.

Commissioning, not just "factory startup" and TAB

Commissioning is the overall process of integration during both design and construction, as well during the operational testing, under loads, of that integration.

Commissioning is sometimes assumed to be included in a project when the designer calls for factory startup of equipment, plus testing, adjusting and balancing of the air and water systems (TAB). But commissioning begins much earlier and has a much broader scope. The commissioning process adds much more value than TAB by itself, and it achieves greater energy savings.

Factory startup only addresses individual components, which might be tested before or after other components are tested. The startup of any *single* component does not reflect what inputs that component needs from other components, nor the output that each piece of equipment is required to produce.

Also, to be sure, TAB is a very important part of commissioning the systems. But TAB does not usually require the functional testing of the system, nor does TAB extend to the early stages of the project, when system integration decisions and design trade-offs are made. Nor does TAB extend to training and documenting the system so that operators understand the design intent and understand what will be needed to keep it operating efficiently.

This clarification is not intended to discourage the expert startup of individual components, nor to discourage TAB. *Both expert startup and TAB are absolutely essential*. But by themselves, they will not achieve the smooth integration of overall system operation which is reduces the building's energy consumption to it's minimum—and keeps it there.

6. Seal up all duct connections, air handlers and plenums

Leaking duct connections, leaking air handler cabinets and unsealed air plenums waste more energy than one might expect.

Leaking air accounts for 25 to 30% of the annual cost of operating the HVAC systems in homes and commercial buildings, according to hundreds of field investigations by the Florida Solar Energy Center,[13] the New York State Energy Research & Development Agency[19] and the U. S. Department of Energy's Lawrence Berkeley National Laboratory.[20]

This astonishingly large opportunity for energy reduction is one reason that energy codes in some states, notably California, Washington and Florida, call for sealing up all duct seams, joints and most importantly, their connections to air handlers.

Figure 4.16 indicates the magnitude of the problem. It shows how the whole-building air change rates change when the HVAC systems are turned on. The graphic describes the field measurements of air leakage in 70 light commercial buildings in Florida.[13] When the buildings are at rest, most of them have whole-building air change rates (outdoor air infiltration) of less than 1 air change per hour. But when the systems are turned on and leaky connections begin to influence the infiltration, air change rates skyrocket to 2, 3 and even 10 air changes every hour. All that leakage means energy must be

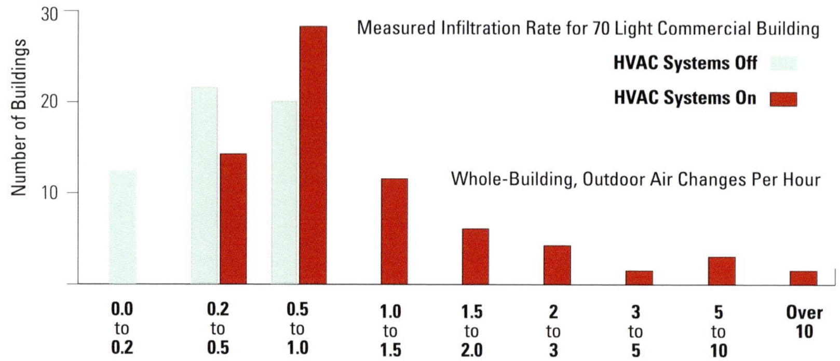

Fig. 4.16 Unsealed ducts, plenums and air handler cabinets increase infiltration loads
Unless ducts and plenums are sealed air tight, HVAC systems actually increase whole-building air leakage, making loads and energy use higher than imagined by the HVAC designer.[13]

invested in two ways: more air must be cooled, using compressor energy, and more air must be circulated to get the cooling effect into the occupied spaces, which uses fan energy.

The air tightness of the doors and panels of the air handlers themselves is equally important. The seams nearest to the fan compartments see the greatest pressure differences in the system, and therefore have the greatest potential for air leaks and energy waste per unit of seam length.

The designer can make a big improvement in energy efficiency by simply requiring that all connections and all plenums be sealed to the standard of "Seal Class A" as described by the Sheet Metal and Air Conditioning Contractors National Association (SMACNA).[21]

Unfortunately, the duct construction and testing standards described by SMACNA only apply to the ducts themselves, not to the air tightness of air plenums, or the joints where ducts connect to other components, or the air tightness of VAV boxes, smoke dampers and air handlers. It is only in recent years that the importance of *overall system air tightness* has been documented. Current SMACNA and ASHRAE standards do not yet reflect this understanding.

But if the designer specifies sealing for all duct connections and all joints to seal class A (high pressure ductwork), the ducts will be sealed with mastic rather than with less effective methods. The sealing specification of seal class A is a good minimum standard for all ducts, all plenums and all connections, regardless of their operating pressures.

In summary, all connections should be air tight, just like all pipes should be water tight, even when the pressure is low. The greater the air leakage, the greater the energy waste.

7. Reduce ventilation air when occupants leave

The graph in figure 4.17 shows how much humidity is carried by the ventilation air into a small-sized office building when that building is fully-occupied on a design dew point day. To avoid mold and comfort problems, the HVAC system must use energy to remove this large humidity load.

However, when the building is not fully-occupied, the volume of ventilation air can be reduced, which in turn can reduce the building's energy consumption.

Three popular methods are often used to achieve this purpose:

- A dedicated ventilation dehumidification system to serve either multiple zones or multiple single-zone units through dedicated duct work (100% outdoor-air system). Equip each zone or unit with either a two-position or modulating damper. Use either time-of-day schedules in the building automation system (BAS) or occupancy sensors such as motion detectors or CO_2 sensors to vary the ventilation provided for each zone or unit. This usually requires a variable-volume fan in the dedicated outdoor-air unit.
- Ventilation control dampers for separate single-zone units. Equip *each unit* with either a two-position or a motorized outdoor-air damper. Use a time-of-day schedule or some type of occupancy sensor to vary the ventilation airflow as the population changes in that single zone.
- Ventilation control dampers for multiple-zone recirculating systems. Equip each system with a motorized outdoor-air damper. Use either time-of-day schedules or occupancy sensors in each space to vary the amount of ventilation air which is blended into the supply air as population changes. The building automation system *gathers and combines zone-level data* to decide how much ventilation air will be needed in the common supply air, such that all spaces have the required outdoor air volume as their individual occupancies vary.

Fig. 4.17 Potential ventilation load reduction when people leave the building

Ventilation is the largest humidity load on the building, by far. Installing a variable-volume ventilation system to reduce air flow when people leave the building can be a very effective energy-reduction measure when there is a big difference in occupancy between day and night or over weekends—as in courthouses and schools. This particular example shows the potential reduction for a small office building in Tampa, FL, only lightly-occupied at night.

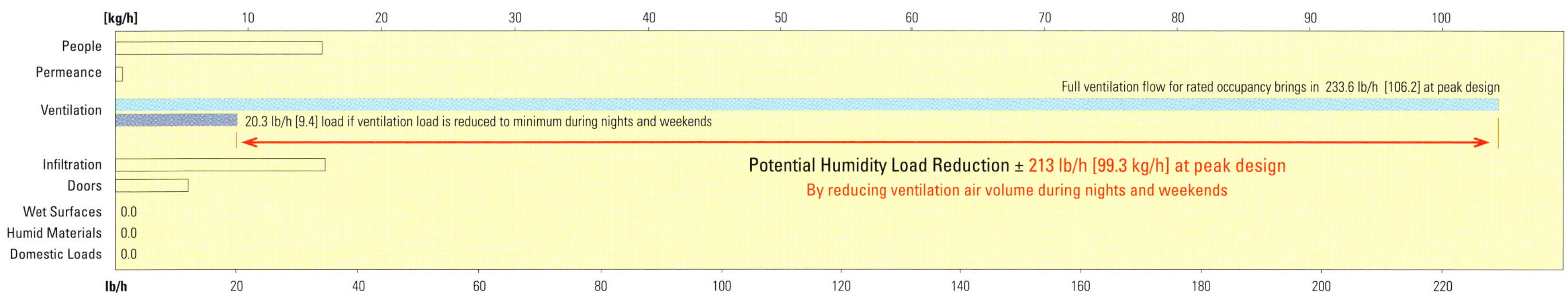

Which of these approaches makes the most economic sense depends on many factors, including the capabilities of the building automation system and the types of cooling systems chosen for reasons other than ventilation. For example, the owner or HVAC designer might have already decided on a dedicated ventilation system for the reasons discussed in other chapters (improved indoor air quality, comfort, mold risk reduction, or because the spaces are cooled by single-room cooling units which may not have a dehumidification mode).

Especially in hot and humid climates, the energy and indoor air quality issues of ventilation control are always linked with the issue of humidity control. Reducing ventilation air flow makes it much easier to keep the indoor dew point low. For example, innovative systems are under development in South Asia to ensure adequate ventilation while also minimizing the energy associated with dehumidification.[32, 33]

The potential energy savings from reducing ventilation air depends on many factors. The most relevant question is the number of hours the cooling systems must operate during periods when the building has only a small percentage of its peak occupancy. If that number of hours is measured in the thousands, a system which automatically reduces ventilation air flow can save a significant amount of energy.

A classic example is a U.S. federal courthouse. The building's ventilation air flow must be sized for maximum simultaneous occupancy of all courtrooms, all judges chambers, all support offices, and all law enforcement and security offices. However, that maximum occupancy only happens once or twice a year, and then only for a few hours. Usually, there are very few courtrooms in use, and those are not fully-occupied. And during nights and weekends, only a few law enforcement investigators, security personnel and overworked law clerks are in the building. Since the building is still occupied during nights and weekends, ventilation air is still required—but probably very little or no ventilation air is needed in the vacant courtrooms, public spaces and most of the offices.

That difference between maximum and minimum required ventilation rates might be as much as 10:1 or 15:1. When the occupancy declines that far, and when a building is occupied by very few people for thousands of hours every year, a significant amount of energy can be saved by reducing the ventilation air flow as people leave the building.

In addition to courthouses, there are other buildings (and zones within larger buildings) which have a very high occupant density at design, but which also have variable occupancy and long periods of very low occupancy. Examples include schools, movie theaters, sports arenas and places of worship, convention centers, conference rooms and function rooms in hotels and other public buildings.

In all of these buildings and spaces, the HVAC designer and the owner would do well to discuss the probable real-world peak and minimum occupancy levels, and to estimate the number of hours when the system will be operating to serve a very low number of occupants. If that estimate is thousands of hours, consider reducing ventilation air in proportion to the reduced occupancy, to save energy.

8. Recover waste energy from exhaust air and condensers

In a hot and humid climate, installing an enthalpy heat exchanger between the exhaust air and the incoming ventilation air can reduce the required peak load capacity of the ventilation cooling equipment by as much as 50%.

This often means the system costs less to install, in addition to saving energy whenever the ventilation system must operate. The energy savings are proportional to the number of hours that the system operates, and proportional to the difference between the indoor and outdoor conditions. Given high outdoor temperature and humidity at nearly all times in hot and humid climates, the key factor for determining energy savings is the number of hours that the ventilation system must operate, and the total air flow of the incoming ventilation air. More hours x larger air volume = greater savings.

Figure 4.18 shows how an enthalpy heat exchanger works. Basically, a desiccant wheel spins rapidly (20-30 rpm) between the exhaust and the ventilation air streams. The heat and humidity from

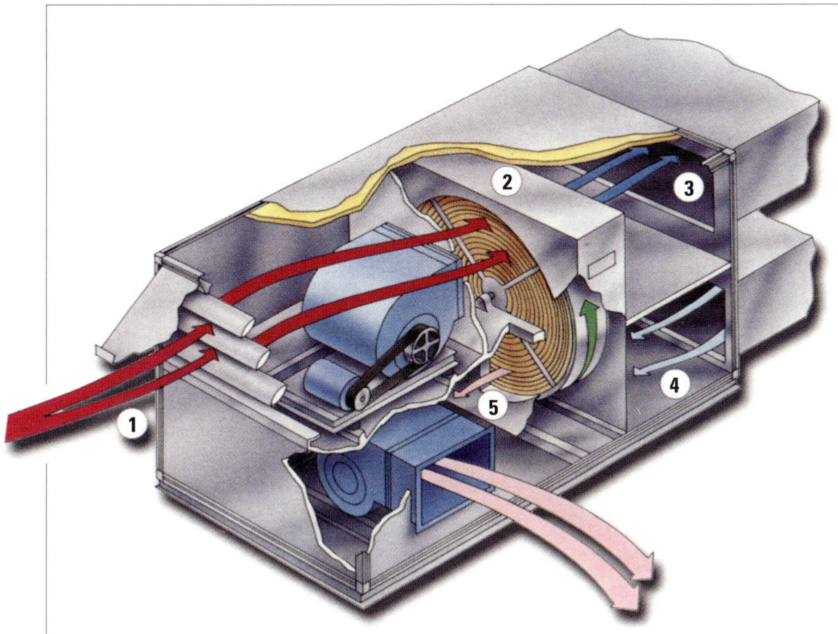

1. Hot, humid outdoor air enters the unit.
2. The air gives up part of its heat and moisture as it flows through the passive desiccant wheel.
3. Leaving the wheel, the ventilation air carries a much smaller heat and humidity load.
4. Relatively cool and dry exhaust air leaving the building strips heat and moisture from the passive desiccant wheel.
5. The exhaust air, now warm and moist, leaves the system, exhausted to the outdoors.

Fig. 4.18
Exhaust air energy recovery

the incoming ventilation air is transferred to the surface of the wheel. Then, as the wheel spins through the exhaust air, it releases that heat and humidity to the outgoing exhaust air stream.

This arrangement keeps 50 to 75% of the ventilation heat and humidity loads out of the building, which means the equipment needed to dehumidify and cool that ventilation air can be much smaller, saving money in the construction budget, as well as energy in the operating budget.

These strong benefits for both energy and cost reduction explain why the 2004 edition of ASHRAE Std 90.1 requires the use of some form of heat recovery when the supply air in a given system is at least 5,000 cfm [2360 l/s] and when the outdoor air portion of that supply air is at least 70%. But those are only the minimum requirements of the 2004 edition. In many systems, the construction cost reduction from using heat recovery is so important that the device is used even on smaller air flows.

In many cases, heat recovery in a hot and humid climate is so beneficial that the HVAC designer will use it as a matter of course. But in some cases, the owner and architect may be called on to reconsider the location of mechanical rooms or vertical duct chases. Using an enthalpy heat exchanger requires both the ventilation and the exhaust air streams to be side-by-side. In some buildings, this can be a challenge.

For example, in an office building, the ventilation air might enter through a side wall, while the exhaust from toilets might exit the building through the roof, or through a different side wall of the building. In those cases, extra duct work and the space in which the ducts must run will have to be accommodated in the architectural design to gain the energy recovery benefit from a rotary heat exchanger.

Reusing condenser heat

Another way to reuse energy is to recover heat from the condensers of the cooling systems. That heat can be used to control humidity in one of two ways: to regenerate a desiccant dehumidification system or to reheat air which has been dried by cooling it. Both of these processes are shown in figure 4.19

A desiccant dehumidification system removes humidity from the air through sorption. The dry desiccant collects humidity as humid air flows across the desiccant's surface. The dry air is sent to the occupied space. After the desiccant becomes saturated with moisture, it must be heated to drive off the moisture it has collected. That's where the condenser heat is used. Heating the desiccant to dry it for reuse is called regeneration or reactivation (the terms are used interchangeably). Both solid and liquid desiccant dehumidifiers can be used for this purpose, and both types can use waste heat from condensers for reactivation.

The diagram in figure 4.19 shows one type of solid desiccant dehumidifier. At first glance, the wheels shown in figures 4.18 and 4.19 would look very similar. Although both wheels contain desiccant, a dehumidifier performs quite differently than a heat exchanger.

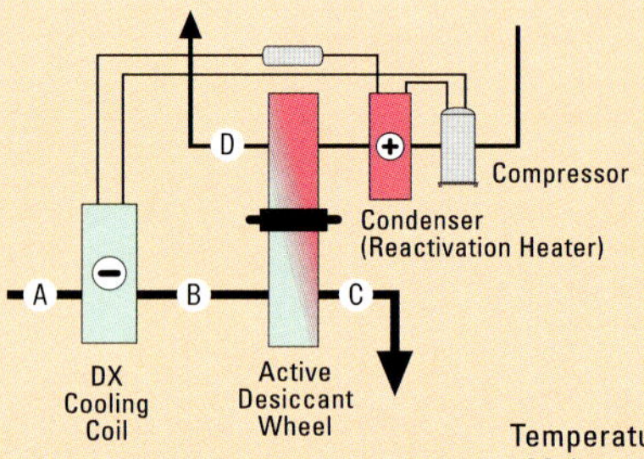

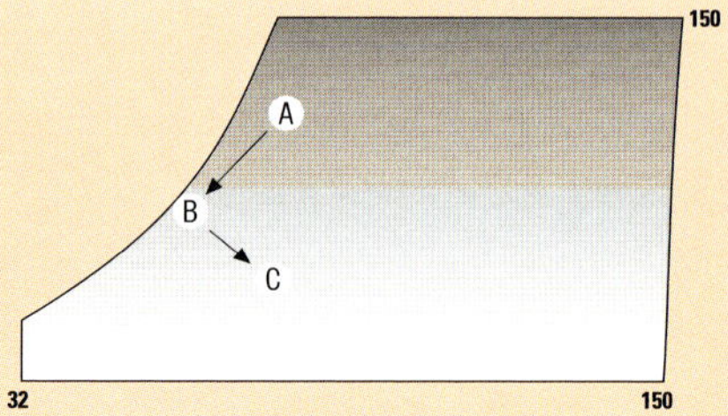

Desiccant Dehumidifier Using Condenser Heat Reactivation

	A	B	C	D
Temperature (°F)	80	60	77	98
Moisture (gr/lb)	124	77	57	144
Air Flow (scfm)	2,000	2,000	2,000	2,000
(°C)	26.7	16.0	25.0	37.0
(g/kg)	17.7	11.0	8.1	20.6
(l/s)	942	942	942	942

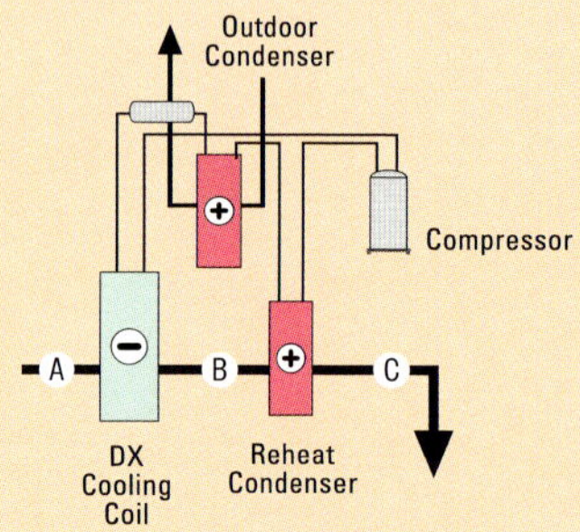

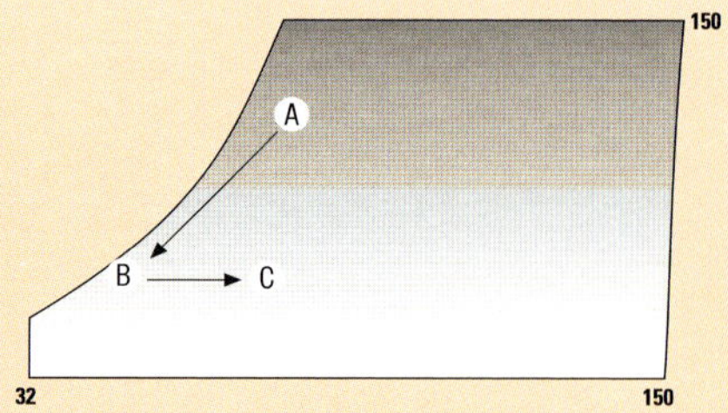

Mechanical Dehumidifier Using Condenser Reheat

	A	B	C
Temperature (°F)	80	55	75
Moisture (gr/lb)	124	65	54
Air Flow (scfm)	2,000	2,000	2,000
(°C)	26.7	12.8	23.9
(g/kg)	17.7	9.2	9.2
(l/s)	942	942	942

Fig. 4.19
Reusing heat from cooling condensers to reduce energy needed for humidity control
In many hot and humid climates, cooling continues all day, all night and all year. The heat constantly ejected from the building by its cooling systems' condensers can be put to use in controlling humidity—the other nearly-constant load.

The rotating heat exchanger in figure 4.18 depends on the dryness of the exhaust air stream to dry the incoming humid air. Without dry exhaust air, the humid ventilation air cannot be dried. On the other hand, no additional energy is needed for whatever drying it does accomplish—the only energy consumed is the power required to turn the wheel plus fan energy used to push both air streams through it.

The desiccant dehumidifier shown in figure 4.19 uses condenser heat to reactivate the desiccant rather than using dry exhaust air. Using heat, the device will dry the incoming air without the need to bring exhaust air back to the unit. And it will dry that air more deeply and quite consistently, no matter what the indoor and outdoor conditions might be. But a desiccant dehumidifier does require that heated reactivation air stream to accomplish its deeper drying. Also, as air is dried its temperature rises, as seen in figure 4.19. The temperature rise depends on the total amount of drying, and on the desiccant unit's configuration. During some operating hours, equipment elsewhere in the system will probably need to remove that sensible heat.

Figure 4.19 also shows another way to use condenser heat. When a cooling coil is used for dehumidification, it cools and dries air deeply. However, in some systems that cold air could overcool the space during part-load operating hours. Condenser heat can then be used for reheating the supply air, avoiding discomfort.

When the condenser is located near the cooling coil, adding a condenser coil (a heating coil) downstream of the cooling coil adds relatively little cost. If the reheating must take place far away from the cooling equipment, as is sometimes the case in larger systems which serve many zones with differing heat loads, a refrigerant-to-water heat exchanger can generate hot water. The water is then pumped out to the reheat locations and back to the condenser.

Reusing condenser heat is an excellent way to reduce overall energy costs, especially those associated with humidity control. In general, the most cost-effective use of condenser heat is achieved when that heat is used close to the cooling equipment which releases it. The further the heat must travel, the greater the amount of energy needed to move that heat, and the greater the cost of additional equipment or piping which transports it to the point of use.

In some buildings, this fact may suggest that the designer should also consider the use of condenser heat for preheating domestic hot water. If it's a long distance from the cooling equipment to either a desiccant dehumidifier or to reheat coils, a water heater might be closer and less expensive to reach. In low-rise buildings which use a great deal of hot water, and where there may be water heaters close to cooling units, such as in hotels and eldercare facilities, this strategy may be more cost-effective than using condenser heat for humidity control. In other buildings, a mixture of all these uses may be appropriate.

But here's the main point: in hot and humid climates there is nearly always some cooling equipment in operation. That cooling equipment is constantly releasing heat to the outdoors. Making productive reuse of this large and constant flow of waste heat can make a useful reduction in the building's annual energy consumption.

Free Dow Jones Sites | As of Tuesday, September 11, 2007

NEWS | TODAY'S NEWSPAPER | MY ONLINE JOURNAL

Home
News
News Main
U.S. Business
Europe
Asia
World News
Economy
Politics & Policy
Campaign 2008
Earnings
Health
Law
Media & Marketing
Entrepreneur
News by Industry
Columnists

Japan Sweats It Out As It Wages War On Air Conditioning

BY SEBASTIAN MOFFETT

Word Count: 1,118 | Companies Featured in This Article: Mitsui Sumitomo Banking

TOKYO – Late last month, the presidents of Japan's three biggest banks gathered to make an important announcement: They were abandoning formal attire for the rest of the summer – and insisting that their 1,630 branches nationwide keep office temperatures at a steamy 82 degrees Fahrenheit in order to conserve energy. In a formal ceremony in Tokyo, young women in cotton kimonos splashed water from wooden buckets on the baking ground -- a traditional way to cool it down without using extra power.

"I want the banking world to get together to promote Cool Biz," said Mitsui Sumitomo Banking Corp. ...

Fig. 4.20 Social factors reduce or increase energy consumption
When occupants decide that warmer indoor temperatures during hot weather are socially admirable, major energy savings become possible.[22]

9. First lower the dew point... *then* raise the thermostat

In September 2007, the *Wall Street Journal* reported that in Japan, there has been a major effort by the government and by building owners to raise thermostat settings during hot weather to 82°F [28°C] in order to save energy.[22] The *Journal* reported that in recent years, corporate culture in Tokyo has shifted from the perception that a cold building was desirable to the view that cold buildings are socially irresponsible, because they use more energy than necessary.

Similar efforts to raise thermostat settings to reduce energy consumption have recently been made in Hong Kong by the Chinese government. Back during the energy crises of the 1970's, the US government used the same strategy to reduce energy consumption in public buildings, because raising the temperature set point can indeed save cooling energy in hot and humid climates.

The cooling load coming through the opaque walls and the roof of the building are proportional to the temperature difference between the indoor and outdoor surfaces of the building enclosure. When that temperature difference is smaller (as when the indoor air temperature is warmer and therefore closer to the outdoor temperature), the amount of heat coming through the enclosure is reduced. With reduced cooling loads, less energy is needed to remove those loads.

The amount of energy saved by raising the indoor temperature depends very much on the specifics of the building and its systems, and its occupancy, operating hours, ventilation rates and a host of other factors. But in nearly all cases, the savings are significant when the thermostat setting is raised when its hot outdoors.

One estimate based on field tests suggested that a school in the Atlanta area (which has a relatively short cooling season compared to hotter and more humid climates) was likely to save more than 20% of annual cooling energy because the thermostat set point was raised from 76 to 78°F [from 24 to 25.6°C].[23] Another field test of a large retail super center in Nebraska (certainly not a hot and humid climate, except for a few months each year) was able to document a 13% reduction of annual cooling costs by raising the set point from 74 to 78°F [24 to 26°C].[24]

On the other hand, there are three very important cautions for owners and HVAC designers when raising thermostat set points:

- When temperature is high, the dew point must be kept low, and air velocity must be maintained to ensure equivalent comfort levels for occupants. If these measures are not part of the plan, occupants are likely to plug in their own coolers and fans, which will increase the building's energy use instead of reducing it.
- To avoid mold risk, keep the dew point low, using some form of dehumidification which operates independently. In other words, it dries air *whenever* the dew point is too high, even when there is no thermostat calling for cooling.
- Some existing buildings may still use old-style constant-volume systems with overcooling and reheat for dehumidification. In those buildings, energy use can *increase* rather than reduce at higher thermostat set points.

The first point is especially important, because it is not always obvious to those who focus on the HVAC design. Occupants strongly influence the building's energy use through what they choose to plug into nearby electrical sockets.

The example reported in the *Wall Street Journal* has not been typical. When worldwide business culture shifts towards the view that thermal discomfort is admired as being more socially responsible, a great deal of energy can be saved. However, until that view becomes typical in all cultures and in all buildings, owners and designers must maintain thermal comfort while saving energy. *Over time, thermal comfort and convenience has tended to override energy concerns.*

For example, a study which measured the actual energy use in highly-efficient single-family homes in three states found a 2:1 difference in energy use in identical homes, with identical occupant demographics.[25] The energy-efficient homes sold at a premium, so presumably the occupants chose the homes partly because they wished to save energy. But energy use varied 2:1 within those same populations. A similar study of high-rise apartment buildings in urban areas found an even greater range of energy use (7:1) in identical multi-story buildings, built in the exact same locations, with the exact same occupant demographics.[26]

The biggest differences are between what occupants choose to plug into their wall sockets, how much power they draw "at rest", and when and how often they choose to operate those devices. In other words, if occupants don't like the indoor temperature, you can be sure that over time, they will find ways to improve the comfort level. And the measures they choose will usually increase energy use.

The way to meet both goals is to keep the indoor dew point low when the temperature rises. The buildings examined during the two field studies referenced earlier, tests which *did* in fact achieve energy savings, used this strategy.[23,24] The indoor dew point in both cases was held to 55°F while raising the thermostat set point to 78 and 79°F respectively. [The dew point was limited to 12.8°C and temperature was allowed to rise to 25.6°C and 26.1°C, respectively]

The second caution is to keep the dew point low to avoid an increase in mold risk when the thermostat setting is raised to save energy. If the indoor dew point is allowed to rise, there are more surfaces inside the building which can either condense water from the air, or will reach a surface relative humidity above 80%. These concerns are discussed in much greater detail in chapter 6 (Avoiding Bugs, Mold & Rot). But to summarize the problem briefly, consider an example from the experience of hospitals in Malaysia as reported by a member of the advisory board for this book.[27]

In those public buildings energy use is always a concern. The practice is to turn up the thermostat in any patient room or in any larger space which is unoccupied. However, the buildings are not always equipped with independent humidity control to keep the dew point low. So humid air from the spaces where the AC system is not removing moisture drifts through the rest of the building, eventually contacting surfaces in rooms which are being kept cool. That humidity is absorbed into those surfaces, and can even condense liquid droplets onto them if they are cold enough. Eventually, enough moisture collects in those materials so that mold growth can be a significant problem.

Schools in the U.S. which raise thermostat settings during summer months to save energy have similar and well-documented problems unless the dew point is kept low, as seen in figure 4.21.[28] Chapter six goes into these problems and their potential solutions in much more detail. In this chapter, it's enough to point out the potential problem, and to caution the owner and designer to keep the dew point low when raising the thermostat settings to save energy.

The final caution about reheat-based humidity control applies to buildings which still might be using old-style constant-volume reheat systems. These overcool a large, constant volume of supply air to dry it, and then reheat all that air with electric resistance coils. Anecdotal experiences during the U.S. energy crises of the late 1970's illustrate the problem.

A Presidential directive ordered all government buildings to save energy. Many responded by raising their thermostat settings to 78°F [25.6°C]. According to the experience of some building managers,

Fig. 4.21
When raising the thermostat set point to save energy—Keep the dew point low
Raising set point temperatures will definitely save cooling energy. But unless the dew point is still kept low, buildings can grow mold, as in this example from a school in Houston. It was closed for vacation. Mold grew when the temperature set point was raised, because the systems had no means of keeping the air's dew point low. [28]

this measure greatly increased energy consumption. The systems cooled air constantly for dehumidification. But then, sensing that the dry bulb temperature was too cold (below the 78°F [25.6°C] set point), the reheat coils would operate at full power, to bring the building back up to the higher thermostat set point.

The Presidential intent was to reduce energy during the cooling season, but with systems which relied on electric reheat, the actual effect of the higher set point was to increase energy use.

To avoid this unfortunate result when raising the thermostat set point to save energy, use some form of independent dehumidification. Keep the dew point low even when the temperature rises.

10. Invest in regular tune-ups (Constant commissioning)

Building owners and operators will have noticed that the previous nine suggestions were focused on advice to designers. Owners know that the real energy losses or gains are made during operation, when the designer's theories and the contractor's best efforts are tested in the real world.

In some ways the life of a building is similar to human life. The design—the building's conception—takes less time and is less work than the months of construction which follow. And construction completion, while very challenging for at least part of the team, requires much less effort than the time, expertise and expense required to get that building through its infancy and adolescence, and into productive maturity.

But unlike our understanding of human development, when construction is complete, people often assume that the newly-born building is fully-mature and ready to make it's own way in the world. As anybody who personally operates buildings knows, this is not a good assumption.

Small and large difficulties happen. Owners find that not all systems are built perfectly, and those which are installed perfectly may not be able to operate according to the designer's assumptions. Sometimes difficulties arise because budget or other compromises had to be made during construction, or because owners, designers and contractors are human, and are therefore are not perfect. Nor can they foresee all of the events which force changes based on the needs of new or different occupants.

These facts seem obvious. But frequently, owners overlook the logical consequence, which is that two identical buildings can and probably will have very different energy consumption levels.

Minimizing real-world energy use will depend on how well the building's operators understand how to adjust and keep the systems operating in response to:

- Initial imperfections, big and small (see figure 4.22).
- Less-than-ideal decisions during planning, design and construction.
- Changes in loads, occupancy and uses.

As with human development, the way to achieve the lowest possible energy use with the least effort is through education and motivation. More specifically, educating the operators about the design intent of the systems and how they can and cannot be adjusted to meet changes. Then prompt and motivate the operators to make changes and adjustments based on a regular flow of measured data which shows the actual energy use of the building as the loads and occupancy change.

This combination of education, information display followed by adjustments is the essence of the very cost-effective process of "constant commissioning" of the building and its systems. Constant adjustments are based on educated judgement, which is in turn based on a clear understanding of how the systems operate and why they operate that way. That understanding allows the operators to decide what changes are needed under what circumstances, and exactly how those changes can be put into action as the loads and uses change each day, each month and each year.

To help owners understand the potential of continuous commissioning, figure 4.23 shows the results of a nationwide study of 224 one-time recommissioning projects.[29] The study was compiled by a team, led by the Commercial Building Systems Group of the U.S. Department of Energy's Lawrence Berkeley National Laboratory. The study was restricted to projects for which measured energy and cost-saving data is available, and in sufficient detail to be credible.

The graph in figure 4.23 shows the cost of each project, compared to the time it took to pay back that investment. All costs were adjusted for inflation, so the values are expressed in constant 2003 dollars. Note that some of the system adjustments and reconfiguration projects paid back their costs in less than one month. Many paid back in less than six months, and the majority paid back in less than a single year.

Apart from loan-sharking backed by costly and it morally questionable collection measures, it is difficult for most organizations to obtain a 100% return on investment (every year, for the life of the building) any other way. Unlike other investment opportunities, with energy savings that rate of return is essentially risk-free. And finally, the annual value of the energy savings is much more likely to rise than to fall over the life of the building.

Fig. 4.22

Not all buildings are built perfectly

Recommissioning existing buildings can save a great deal of energy.

In this example, eight years after construction the ceiling was removed for a renovation, exposing the fact that this major supply duct was never connected. Apparently, a beam interfered with the HVAC designer's plans.[30]

This sort of problem is often responsible for occupant comfort complaints and excessive energy use. Recommissioning the building finds and fixes such problems, saving energy and improving comfort.

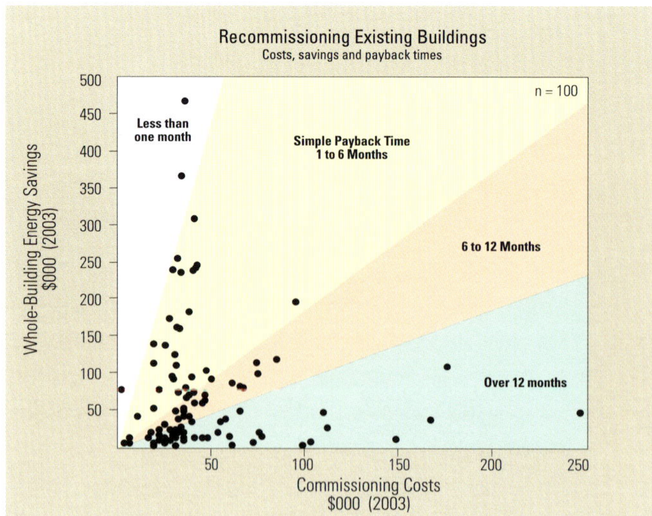

Fig. 4.23

Measured energy cost savings from recommissioning existing buildings

In this study of measured energy savings from 100 buildings during the 1990's, the U.S. Department of Energy's Lawrence Berkeley National Laboratory established the dramatic annual returns accomplished through recommissioning, at essentially zero investment risk.[29]

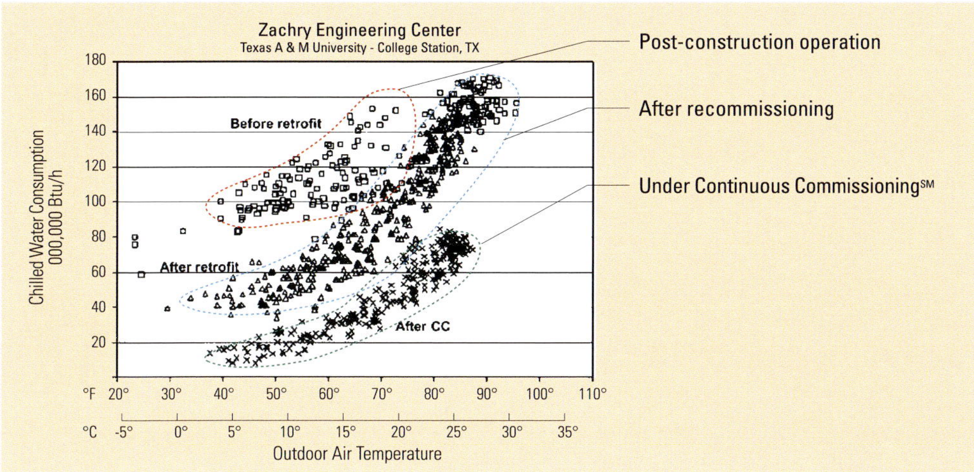

Fig. 4.24 Constant commissioning, for constant improvements

The graphic shows results achieved in two stages.[18] First, the system was "recommissioned," saving chilled water compared to the normal running cost of the fully-functional but un-optimized system.

However, the even more compelling savings come from constant commissioning. Key aspects of consumption are measured continuously, and the systems are changed and adjusted in real time.

Note the difference in chilled water consumption between the recommissioned system and the system under Continuous Commissioning[SM] when outdoor temperatures are between 75 and 95°F [21° and 32°C].

Constant commissioning costs much less than the other measures discussed earlier. But it is often overlooked, perhaps because:

- It is not a one-time purchase of an energy-saving gizmo. For corporations, continuous effort is more difficult to sustain than special, one-time capital expenditures to install miraculous energy-saving hardware. The costs associated with developing and maintaining operator expertise are continuous—but so are the savings.
- Because diagnostic expertise has not always been properly valued and compensated, the necessary skills for building operators are expensive to obtain and difficult to retain.
- The energy and cost-saving benefits are far larger than the costs of obtaining and maintaining the expertise which produces them. But costs are usually more obvious to management than are savings. The flow of measured results from a constant commissioning program helps inform management of the wisdom and financial benefits of the ongoing investment.

Still, the reality is that developing and maintaining large numbers of in-house experts is very difficult for building owners. So a common alternative strategy is to hire and retain a few truly expert operators, and then purchase the needed additional services from outside firms who specialize in recommissioning. The *in-house experts* can provide the system experience needed to shorten the education needed by those *outside* experts in order to accomplish projects at minimal cost in short periods of time.

ASHRAE provides some, but not all of the resources needed by owners who seek to enjoy the energy and cost savings of constant commissioning and recommissioning. ASHRAE documents are largely focused at the early stages of the commissioning process.

Guideline Zero (The commissioning process) describes how an entire building can be commissioned, beginning at the planning stage.

Fig. 4.25 Guidebook for constantly accelerating savings

Constant commissioning is an ongoing process of improvement—not a one-time event. The results can be dramatic, as documented in the Federal Energy Manager's Guide.[18] The guide is based on the energy cost reductions achieved through procedures developed and refined over 10 years in 110 buildings during the LoanStar program for state-owned buildings in Texas.[31]

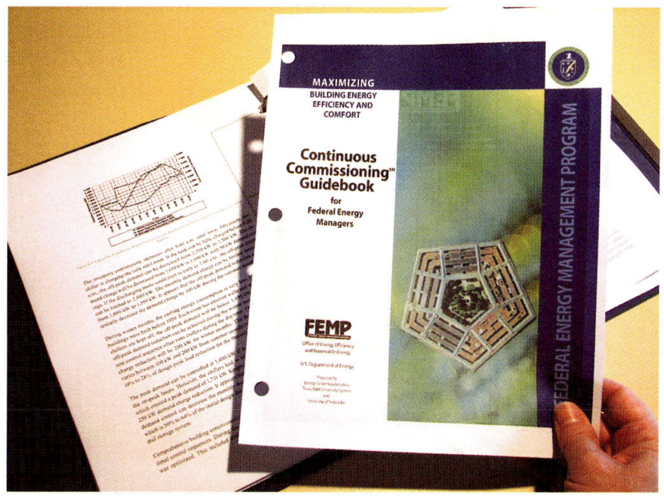

Guideline One (The HVAC commissioning process) provides similar information for the HVAC systems by themselves.

In addition to ASHRAE guidance, the owner will need more detailed procedures for tracking post-construction energy use, and suggestions for the common adjustments which have historically provided the greatest savings for the least effort and cost. For that more detailed guidance, the reader is encouraged to consult the resources provided by the:

- *Continuous Commissioning℠ Guidebook for Federal Energy Managers* - US. Department of Energy - U.S. Federal Energy Management Program (FEMP) - (http://www1.eere.energy.gov/femp/operations_maintenance/om_ccguide.html)
- US. Department of Energy Buildings Technology Program (http://www.eere.energy.gov/buildings/info/operate/buildingcommissioning.html)

Summary

To meet energy goals beyond the basic requirements of building codes and of ASHRAE Standard 90.1-2010, the look-and-feel of buildings will probably be different than in past years. Popular design aesthetics will perhaps change so that energy-wasting buildings such as those with of unshaded, east and west-facing glass are seen as ugly, undesirable and socially unfashionable. And perhaps, as in the anecdote reported by the *Wall Street Journal*[22] the social pressures which influence thermal comfort will also change, so that warmer temperatures in hot climates will be perceived as being more environmentally responsible than chilly buildings. At that point, the suggestions contained in this chapter will need to be updated, to further reduce energy consumption.

In the mean time, by taking these suggestions, along with the guidance defined by ASHRAE Standard 90.1-2004 and by the ASHRAE Advanced Energy Guides, owners can achieve significant reductions in energy consumption compared to buildings built to current codes.

References

1. Huang, Joe & Franconi, Ellen. *Commercial Heating and Cooling Component Loads Analysis* 1999. Report LBL-37208 Building Technologies Department, Lawrence Berkeley National Laboratory. Berkeley, CA 94720
2. ASHRAE Standard 90.1-2004 *Energy Standard for Buildings Except Low-Rise Residential Buildings*. 2004. ASHRAE. 1791 Tullie Circle, NE. Atlanta, GA 30329 ISSN 1041-2336
3. Eley, Charles, Ed. *90.1 User's Manual - ASHRAE/ANSI Standard 90.1-2004.* 2004. ASHRAE. 1791 Tullie Circle, NE. Atlanta, GA 30329 ISBN 1-931862-63-X
4. Jarnigan, Ron, Project Chair. *Advanced Energy Design Guide for Small Office Buildings* 2000. ASHRAE. 1791 Tullie Circle, NE. Atlanta, GA 30329 ISBN 1-931862-55-9
5. McBride, Merle, Project Chair. *Advanced Energy Design Guide for Small Retail Buildings* 2006. ASHRAE. 1791 Tullie Circle, NE. Atlanta, GA 30329 ISBN 1-933742-06-2
6. O'Connor, Jennifer, Lee, E. Rubenstein, F. & Selkowitz, Stephen, *Tips for Daylighting with Windows - The Integrated Approach* Report no. LBNL-39945 1997. Building Technologies Program. E.O. Lawrence Berkeley National Laboratory, Berkeley, CA
7. Gronbeck, Christopher *Window Heat Gain Calculator* 2007. http://www.susdesign.com/windowheatgain/
8. Carmody, John; Selkowitz, Steven; Lee, Eleanor; Arasteh, Dariush and Wilmert, Todd. *Window Systems for High Performance Buildings* 2004. Norton & Company, 500 5th Avenue, New York, NY. 10110 ISBN 0-393-73121-9
9. Straube, John; Burnett, Eric. *Building Science for Building Enclosures.* 2005. Building Science Press, 70 Main St. Westford, MA 01886 www.buildingsciencepress.com ISBN 0-9755127-4-9
10. Cummings, James. "House Airtightness Trends." *IAQ Applications*, Spring 2006. pp.8-9. ASHRAE, Atlanta, GA
11. Persily, Andrew. "Myths About Building Envelopes." 1999. *ASHRAE Journal*, pp. 39-45. March, 1999 ASHRAE 1791 Tullie Circle, NE Atlanta, GA 30329
12. Emmerich, Steven; McDowell, Timothy; Anis, Wagdy. "Simulation of the Impact of Commercial Building Envelope Airtightness on Building Energy Utilization." *ASHRAE Transactions, Volume 113, Part 2. 2007*. ASHRAE 1791 Tullie Circle, NE Atlanta, GA 30329
13. Cummings, James; Withers, Charles; Moyer, Neil; Fairey, Philip; McKendry, Bruce. *Uncontrolled Air Flow in Non-Residential Buildings*. 1996. Final Report, FSCEC-CR-878-96, 1996. Florida Solar Energy Center, 1679 Clearlake Rd., Cocoa, FL. 32922
14. Jacobs, Pete. "Small Packaged System Commissioning—How to Eliminate the Most Common Problems." November 2002 *HPAC Engineering*. pp.60-61 Penton Publishing, Cleveland, OH.
15. Kavanaugh, Steven *HVAC Simplified*. 2006. ASHRAE 1791 Tullie Circle, NE Atlanta, GA 30329
16. Energy Information Agency. *Commercial Building Energy Consumption Survey 2003.* U.S. Department of Energy http://www.eia.doe.gov/emeu/cbecs/
17. *ASHRAE Guideline 1-1996 The HVAC Commissioning Process* ASHRAE 1791 Tullie Circle, NE Atlanta, GA 30329

18. Liu, Minsheng; Claridge, David; Turner, W. Dan. *Continuous Commissioning[SM] Guidebook*. October, 2002. Federal Energy Management Program, U.S. Department of Energy. (http://www1.eere.energy.gov/femp/operations_maintenance/om_ccguide.html)

19. Henderson, Hugh; Cummings, James; Zhang, Jian Sun; Brennan, Terry. *Mitigating The Impacts of Uncontrolled Air Flow on Indoor Environmental Quality and Energy Demand in Non-Residential Buildings*. 2007. Final Report - NYSERDA Project # 6770. New York State Energy Research & Development Authority, 17 Columbia Circle, Albany, NY 12203-6399

20. Delp, William; Woody, Nance; Matson, E.; Tschudy, Eric; Modera, Mark & Diamond, Richard. *Field Investigation of Duct System Performance in California Light Commercial Buildings.* 1998. Report LBNL #40102, Building Technologies Program, Lawrence Berkeley National Laboratory, Berkeley, CA

21. *HVAC Duct Systems Inspection Guide*. 2006. Sheet Metal & Air Conditioning Contractors National Association (SMACNA) 8224 Old Courthouse Road, Tyson's Corner, Vienna, VA 22182 www.smacna.org

22. Moffett, Sebastian. "Japan Sweats it Out as it Wages War on Air Conditioning." *Wall Street Journal*, Sept. 11th, 2007.

23. Fischer, John; Bayer, Charlene. "Failing Grade for Many Schools - Report Card on Humidity Control" *ASHRAE Journal*, May 2003. pp.30-37.

24. Spears, John; Judge, James. "Gas-Fired Desiccant System for Retail Super Center" 1997. ASHRAE Journal, October 1997 pp.65-69.

25. Brown, Rich; Parker, Danny. "Roadblocks to Zero-Energy Homes" 2007. Home Energy Magazine, January-February, 2007. pp.24-28. Home Energy, 2124 Kittredge St., Berkeley, CA 94704 www.homeenergy.org.

26. "The Seven-To-One Dilemma" *Party Walls Newsletter*, July-Aug. 2006. Volume 2, issue 4. Steven Winter Associates, 50 Washington St., Norwalk, CT. 06854. www.swinter.com

27. Personal communication - Suei Keong Chea, Carrier Malaysia

28. McMillan, Hugh; Block, Jim; "A Lesson In Curing Mold Problems" *ASHRAE Journal* May 2005, pp. 32-37

29. Mills, Evan; Bourassa, Norman; Piette, Mary Anne; Friedman, Hannah; Haasl, Trudi; Powell, Tehsia and Claridge, David. "The Cost-Effectiveness of Commissioning" *HPAC Engineering*, October, 2005. Penton Media, Inc. 1100 Superior Avenue, Cleveland, OH 44114-2543 www.penton.com

30. Personal communication - Paul Halyard, Property Condition Assessment, Inc., Orlando, FL

31. Turner, Dan; Claridge, David; O'Neal, Dennis; Haberl, Jeff, Heffington, Warren; Taylor, Dub; Sifuentes, Theresa. "Program Overview - The Texas LoanStar Program; 1989-1999, A 10 Year Experience." Energy Systems Laboratory, Texas A&M University http://txspace.tamu.edu/handle/1969.1/1998

32. Sekhar, S C, Uma Maheswaran, K.W. Tham and K.W. Cheong, "Development of an energy efficient single-coil-twin-fan air conditioning system with zonal ventilation control." ASHRAE Transactions (2004), Volume 110, Part 2, pp 204-217.

33. Sekhar, S C, Yang Bin, K.W. Tham and David Cheong, "IAQ and Energy Performance of the newly developed single-coil-twin-fan air conditioning and air distribution system – Results of a Field Trial." Proceedings of Clima 2007- Well Being Indoors, Helsinki. Finland, 10-14 June 2007.

Image Credits

4.5 Tips for Daylighting. O'Connor, Lee, Rubenstein & Selkowitz. U.S. Department of Energy - Lawrence Berkeley National Laboratory, Berkeley, CA

4.6 Tips for Daylighting

4.13 Mason-Grant Consulting, Portsmouth, NH

4.18 Semco, Inc. Columbia, MO

4.20 WSJOnline.com

4.22 Paul Halyard, P.E., Fellow, ASHRAE. Property Condition Assessment Inc., Orlando, FL.

Chapter 5

Avoiding Bugs, Mold & Rot

By Lew Harriman

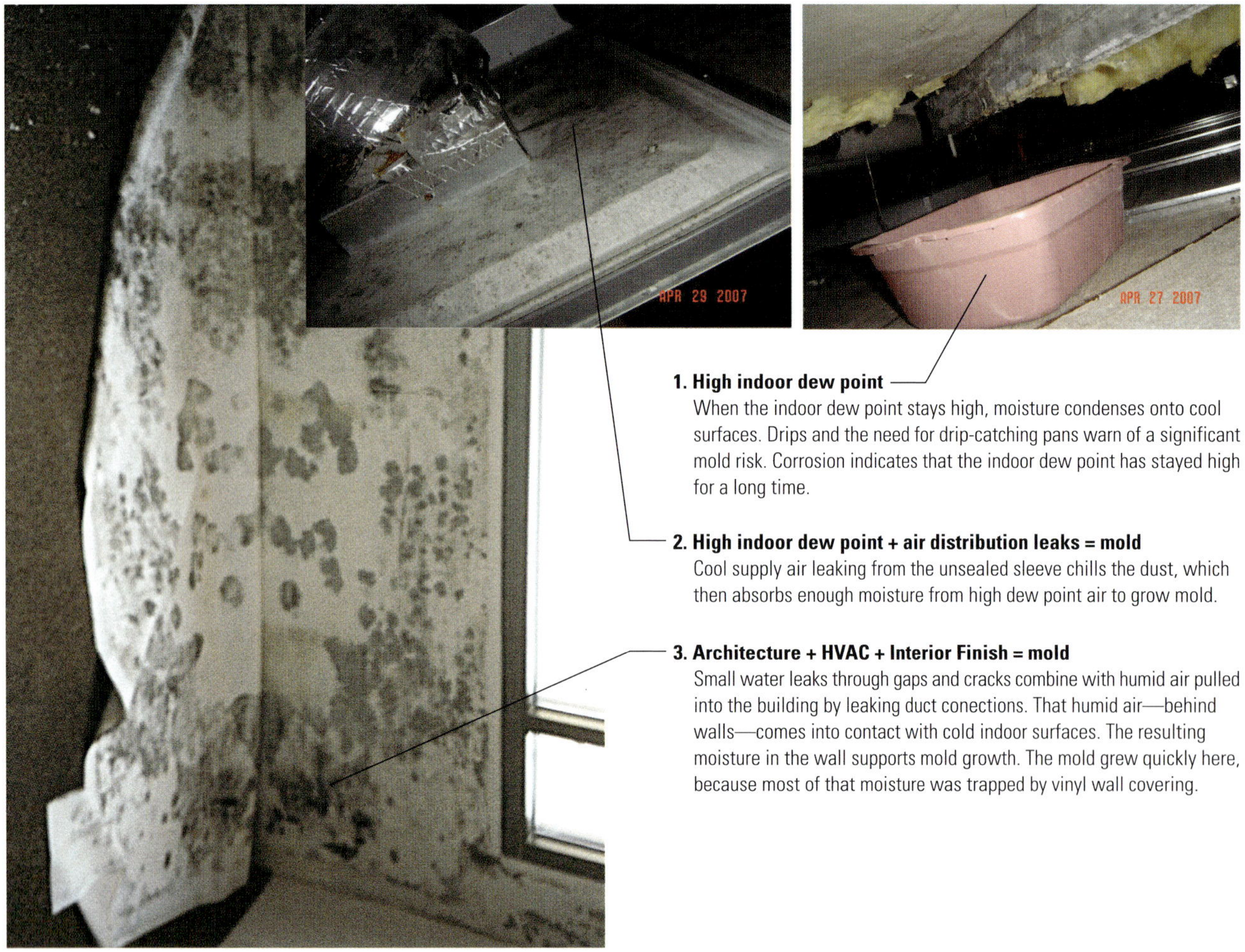

Fig. 5.1 Indoor mold

There is rarely a single cause of mold problems in buildings. To reduce mold risk, make design and operational decisions which keep humid air and rain leaks out of the building. Then, keep the indoor air dry so that any moist material dries out quickly—and stays dry.

Fig. 5.2 Moisture + books = mold
After organic materials absorb enough moisture, some form of mold will be able to digest them. In a school near the Gulf Coast, the building was not kept dry during summer vacation. By early August, this was the result throughout the library.

Key Points

As long as a building's interior stays dry, and as long as condensation does not drip into hidden cavities or become stagnant inside the HVAC systems, the building will not have problems with bugs, mold and rot. But as soon as anything gets moist indoors—and stays that way—the risk of such problems goes up. To reduce that risk, owners, architects, HVAC designers and building operators may wish to adopt these suggestions:

1. Keep rain off the exterior walls and away from the foundation, using roof overhangs and rain gutters.
2. Keep rain from leaking through joints in the exterior cladding with effective flashing under and around windows, doors and all other wall penetrations.
3. Keep rain from wetting the sheathing inside the exterior wall by installing an air gap and water barrier between the exterior cladding and the sheathing.
4. Ensure that exterior walls "dry to the interior" by using vapor-permeable interior wall materials and finishes.
5. Keep indoor air dry, at all times, to avoid humidity absorption and condensation on cool surfaces.
6. Seal all air ducts, air plenums and all connections to air handlers, to limit humid air infiltration and to keep humid air from drifting into contact with cool surfaces.
7. To assess the current risk of microbial growth in a building, measure the moisture content of its materials.

Excess Moisture Leads to Bugs, Mold & Rot

To survive and reproduce, the microorganisms which damage buildings and annoy the occupants all need moisture in their food sources. Some need more moisture, and some survive with less. But they all need more moisture, and for longer periods, than what is normal to have in building materials and furnishings.

Of course, buildings exist outdoors. So rain will frequently soak and flow over their windows and walls. No wall with windows, doors and other penetrations can be hermetically sealed against *all* water leaks. And all buildings have plumbing. So there will always be *some* indoor spills and leaks.

But these small amounts of moisture do not lead to problems in the vast majority of buildings. As one engineer with 40 years of experience investigating building problems notes; "Most buildings will tolerate some amount of excess moisture without too much difficulty. Usually, it's when you have *several* problems at the *same time*, and for a fairly *long time* that you really get into trouble."[1]

The reality is that building materials and furnishings are constantly absorbing and releasing moisture from air, from condensation and from rain leaks. All moisture absorption encourages microbial growth—but subsequent drying inhibits or stops it.

It's when materials stay damp for extended periods that bugs, mold and rot will grow in large enough numbers to become a problem for the building or for its occupants. So to avoid problems, the building and its contents should be kept as dry as possible.

However, no building is perfect, anymore than any architect, engineer or contractor is perfect. So the early design decisions need to reflect that fact. When parts of the interior and parts of the building enclosure inevitably get wet, the architectural design details must drain away that water, and allow any trapped moisture to dry out. And the mechanical system should help that drying process—or at least not add to the problem.

These are shared responsibilities. The owner and architectural designer establish the baseline risks of excess moisture accumulation. The shape of the building determines how much rain lands on its walls. After that initial risk is set, the details of the building's enclosure design—especially the flashing under windows, doors, balconies and other penetrations—determine how much of that rain will be allowed to get into and accumulate in the exterior walls.

Also, the owner and architectural designer control the budget for the mechanical systems. That budget will either empower or limit the HVAC designer in keeping the indoor air dry—the final element in the baseline risk.

Human Health Effects of Bugs, Mold & Rot

The reader will quickly note that this book does not describe or analyze the health-related effects of microbial growth in buildings. ASHRAE is not a medical association. The Society is not equipped with the expertise necessary to provide guidance on health consequences of bugs, mold or rot.

For those who seek information about health, a useful analysis of what was known and unknown about damp indoor spaces and human health as of 2003 is available online from the US National Academies of Medicine.[2] Also, a quantitative assessment of the annual cost of these heath effects (in the USA) was accomplished by the US Environmental Protection Agency and the Lawrence Berkeley National Laboratory in 2007.[3]

Lessons Learned and Forgotten Since 1980

In the USA, the problems associated with "bugs, mold and rot" have been studied intensively since the first conference of that name was established by the National Institute of Building Science in 1991.[4] That was the same year that the Executive Engineers Committee of the American Hotel & Motel Association concluded that mold problems in hotels had become a multi-million dollar problem.[5]

But these investigations and conferences were simply a reflection of the increasing costs and frustration about problems that first surfaced more than ten years earlier. By the early 1990's, water and humidity-related problems in buildings had already become the leading cause of claims against the professional errors and omissions insurance policies of architects and engineers.[6]

Throughout the 1990's and the early years of the 21st century, the problems grew in dollar value, eventually to the point where, in 2003 alone, more than $2 billion in mold claims were filed against property insurance policies.[7] Finally, by 2007 the US Environmental Protection Agency had concluded that in the United States, the financial effect of asthma problems associated with damp buildings has reached an annual cost of approximately $3.5 billion.[3]

Sadly, the agonizingly expensive lessons of preventing mold and moisture problems in buildings have been learned and then ignored or forgotten by two generations of building owners, architects and engineers since 1980.

As of 2007, most professionals who have been personally involved with mold do understand that keeping moisture and humidity out of buildings is the way to prevent problems. However, the actions needed to keep buildings dry are still not part of the standard practices of most owners and design professionals.

Perhaps this is because these measures cost money. Perhaps it's because the required actions cut across the lines of responsibility of several professions. But one suspects that ultimately, there is a reluctance to believe that bad things will happen to "*my* building." Millions of buildings are indeed built every year which do not have mold and moisture problems until later in their useful lives. It's difficult to change standard practices and invest extra money until the decision makers are *certain* that there will be a quick return on more robust building design practices.

Until laws require the practices described here, there will be owners, designers and mortgage holders who will take chances and cut corners, either through ignorance or through intention. Like a new bridge which collapses because it was built at the edge of the tensile strength of its steel, *sometimes* in a leaky building the water and humidity accumulation will be too high, and the building will have multimillion dollar problems immediately. But *perhaps* for that same building, the local rainfall might be less than normal during construction, so that no problem occurs in the early years of occupancy even though the building remains at the edge of catastrophe.

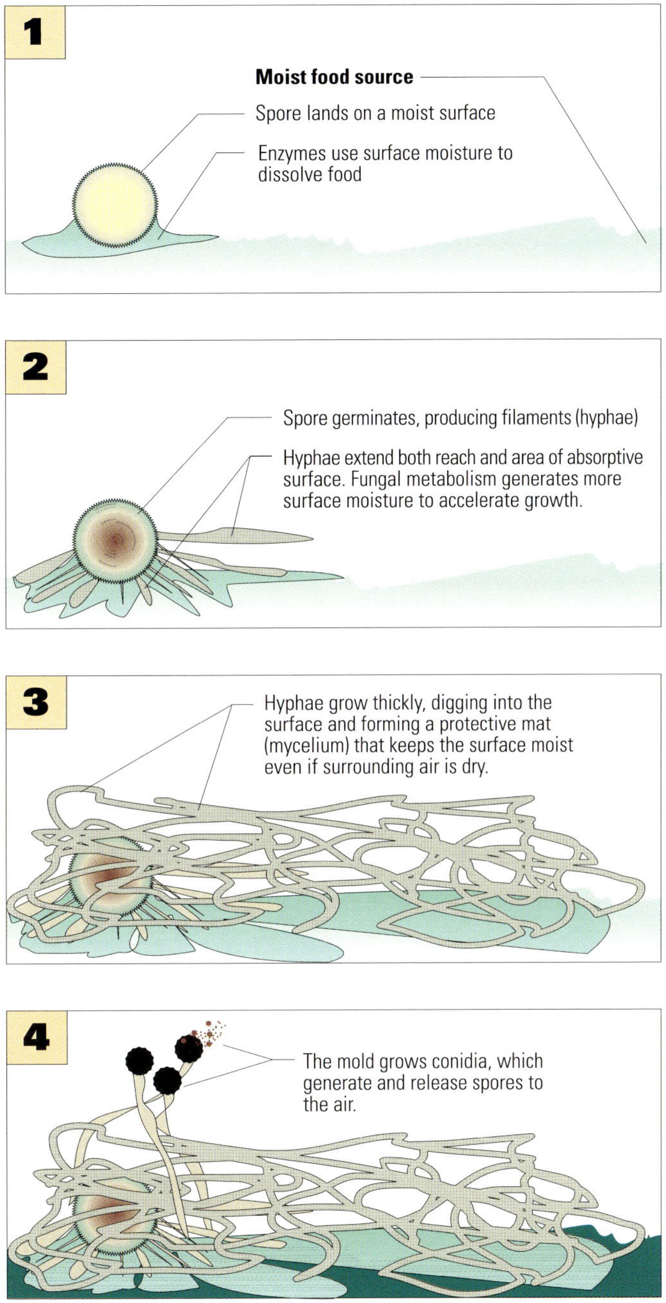

Fig. 5.3 Mold growth cycle
Mold cannot grow until the moisture content of the food source is high enough for the mold's surface enzymes to dissolve its food. Keeping materials dry makes sure that the growth cycle can never begin.

This chapter is for those who want to *reduce the risk* of bugs, mold and rot in their buildings. Others are invited to skip this chapter and move on to more congenial subjects. Because what follows is complicated, it costs money and requires changing what may have been traditional practices for some technical professionals and the investors who fund the construction of their buildings.

Mold growth - water activity in the food vs. rh in the air

The road to less risky practices begins with a clear understanding of what causes mold to grow inside buildings. Mold does not care about the humidity. It only cares about how much moisture is in its food source.

Figure 5.3 illustrates the mold growth cycle. Basically, mold cannot grow until it has enough moisture in its food source to allow the enzymes which cover the outside of its spore casing to dissolve that food, creating a nutrient broth under the spore. Then the spore can absorb the nutrients, as those liquid nutrients diffuse through the rigid wall of the spore, driven by the difference in osmotic pressure between the dry interior of the spore and the dilute nutrient solution surrounding that spore. The liquid diffuses inward, through the walls of the spore, to equalize that difference in osmotic pressures.

Each mold is optimized for digesting a particular set of materials, in a particular temperature range and within a particular range of moisture contents. Some molds have enzymes which digest the chopped cellulose fibers in paper products very well. The notorious *stachybotrys chartarum* is a mold well-suited to digesting paper. Others are better at digesting the sugars on the surfaces of leaves and grasses. These are called phylloplane molds, an example of which includes the *cladosporium* family.

Temperature also plays a role. Some fungi tolerate cooler food, and others need warmer food. But any organic material (including jet fuel, plastic, dust and skin oil) has some sort of fungi which can digest it. And the ideal temperature range for most mold growth is the same range of temperatures which is typical inside buildings.

So the principal variable which limits the fungal growth rate inside buildings is the *amount* of water in each food source. The amount of water also influences which fungus will dominate the food.

For example, *aspergillus repens* can digest some building materials when they have a fairly low moisture content. Other fungi cannot grow as efficiently when moisture content is low. But as the moisture content of the food rises, *aspergillus versicolor* will out-compete *a. repens* and take over the food surface. Later, if the moisture content should rise even further, *alternaria alternata* will overcome both the *a. repens* and *a. versicolor* and dominate that food surface.[8]

But here's another really important point. It's not the absolute amount of water, but rather its biological availability which governs the fungal growth rate. And water availability varies greatly with the internal structure and the chemical composition of the food.

The concept of *water activity* is a useful way for biologists to keep track of how much water is biologically available (accessible to the fungus) in its food source. Water activity (A_w) is quantified by by measuring how much water ends up in the food source *after that food has finally arrived at complete thermodynamic equilibrium with air at a constant relative humidity.*

For example, a water activity of 0.8 refers to the amount of water absorbed into a material when the surrounding air is at 80 %rh—but only after both the relative humidity and the temperatures have stayed constant for *long enough* that moisture is no longer moving either into or out of that material.

This static situation—total equilibrium—can be created in a laboratory, given enough time. But static equilibrium never occurs in a building. That's where the confusion arises between biologists and building professionals. Building professionals have not understood—and biologists have generally failed to clarify—the fact that *water activity of the material* and the *relative humidity reported by the building automation system* are very different.

Nothing is ever at equilibrium in a building. And surface temperatures will vary by several degrees, even on the same wall. So even when the indoor air dew point is constant, those different surface temperatures mean that the surface rh is quite different from point to point, even within a few inches or centimeters.

Therefore, the *surface* rh is quite different from the relative humidity values reported by a sensor located in the air. Figure 5.4 illustrates this point. The rh values reported by the building automation system are not reliable indicators of water activity in the materials. Therefore, the "average rh of the building" is not a reliable indicator of mold risk.

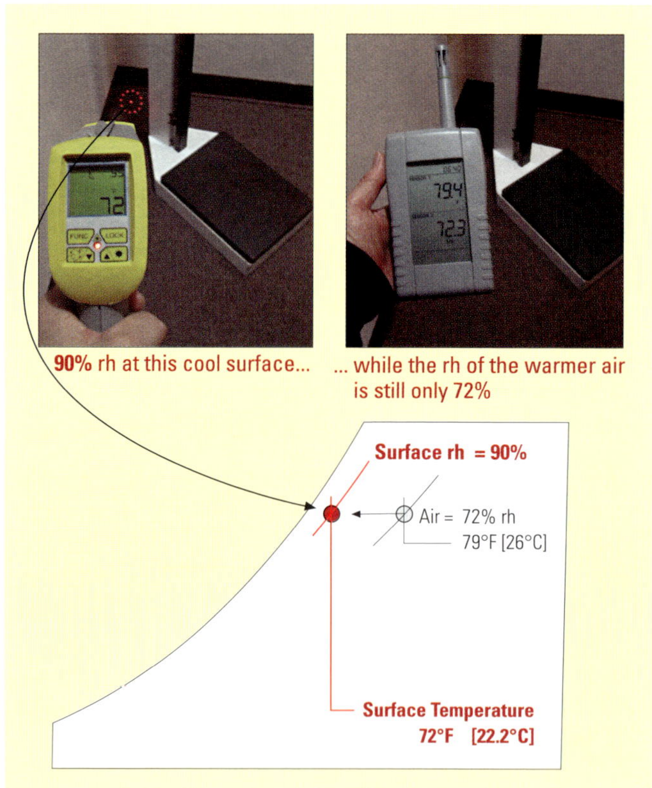

Fig. 5.4

Moisture content is influenced by surface relative humidity

Condensation is the usual problem that leads to mold growth, but materials can also absorb moisture directly from the air when the relative humidity is high at the surface.

In an air conditioned building, the relative humidity is often higher on the surfaces than in the air, because some of the surfaces stay cooler than the air.

This graphic explains how to estimate the relative humidity at a surface. First, measure the temperature and rh in the air, as shown in the photo on the right. Then measure the surface temperature with a non-contact thermometer, as shown in the photo at left.

On a psychrometric chart, plot a line horizontally from the air condition to the dry bulb temperature of the surface. The relative humidity at that point on the chart is (approximately) the surface relative humidity.

When the surface relative humidity stays above 80% for more than 30 days, mold growth is a risk, even without condensation or other wetting to start the growth cycle.[9]

Fig. 5.5
Cold surfaces = high surface rh
The surfaces in an air conditioned building are often much colder than the surrounding air, which raises the local rh, and leads to moisture absorption and then to the risk of mold growth such as that seen in figure 5.6.

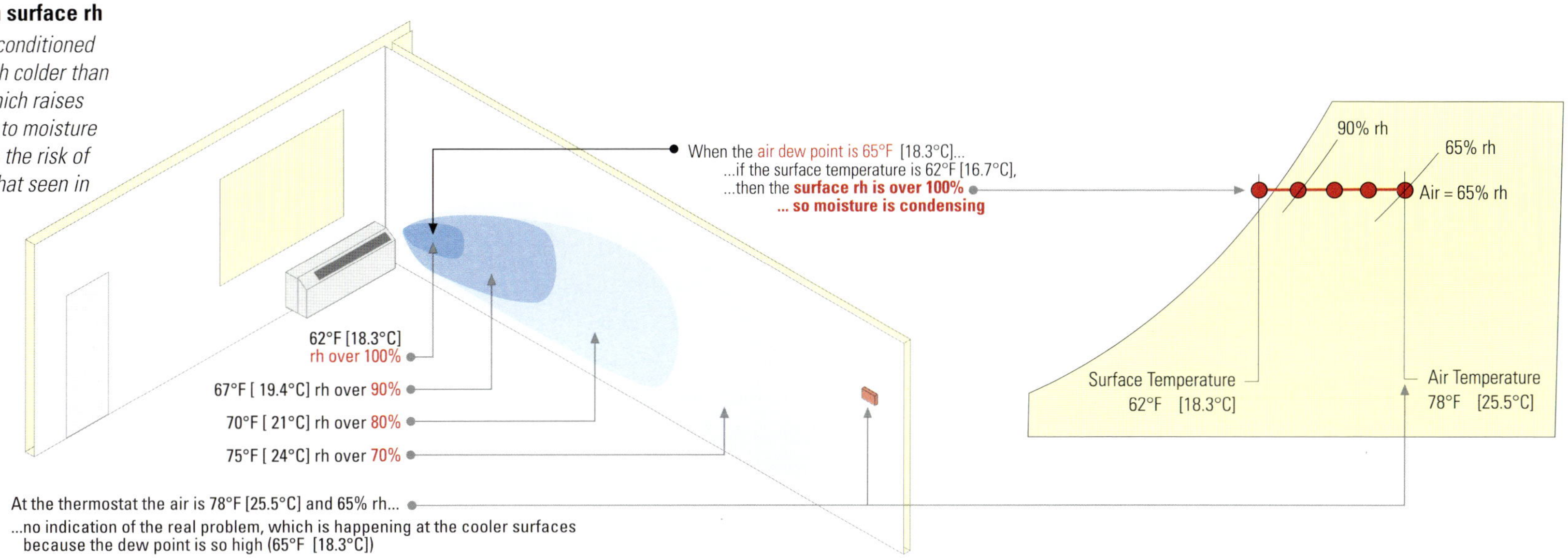

Fig. 5.6 Mold growing on the inside of a cooled wall
When humid air is pulled into the building by leaking air ducts, humidity can fill the cavities behind walls. Then it often supplies moisture to the wall surface when the room-side of that wall is cooled by the AC system. Mold then grows on the moist wall board which faces the inside of the humid wall cavity.

The same shortcoming applies to rh values measured by handheld instruments. RH values from the air are not the same as rh values at the surfaces, which *would perhaps* be approximations of water activity and therefore of mold risk.

To illustrate this problem, consider the diagram in figure 5.5, which describes a wall surface cooled by air from a nearby AC unit. Near the cold air supply outlet the wall temperature is quite cold, so the surface rh at that point is quite high. Further away, the wall surface is much warmer, so the surface rh is much lower.

If *all* those values stayed constant for days (air flows, air temperatures, heat flow through the wall and the air dew point), the moisture content at each point on the wall could be predicted, and therefore the risk of mold could be assessed. But as soon as the cold air supply shuts off, or when humid air infiltration raises the indoor dew point, all the surface rh values change, and therefore all the water activity levels change, which changes the risk of mold growth at each point.

And of course we don't really know what's happening behind that wall. The dew point might be quite a bit higher back there, so the surface rh levels on the back surface of the wall board could be quite different from the surface rh on the room-side surfaces, with a correspondingly different water activity level and therefore a higher or lower risk of mold growth. Figure 5.6 shows a photo of mold which grew behind a wall. High dew point outdoor air was pulled into the building behind the wall, where it added it's moisture to the wall, because the room-side of the wall was cooled by the AC system.

If this all seems very complicated and confusing, that is an accurate perception. But these complexities help explain why buildings which never reported relative humidities over 65% still grew lots of mold, and why similar buildings which might have recorded relative humidity excursions over 80% did not appear to have immediate problems.

Its not the relative humdity reported by the building automation system that determines risk—it's what happens at the surface of the material. The surface rh governs sorption, and therefore it governs moisture content and mold risk. And the surface rh is constantly changing, which is why the indoor air dew point is a better indicator of mold risk.

Knowing the current dew point allows a building operator to estimate the probability of both condensation and high surface rh, as will be shown by examples later in this chapter.

How to Avoid Excess Moisture Accumulation

In a perfect world, one simply plans, designs, constructs and operates the building so that nothing ever gets wet.

No rain leaks in. No humid air leaks in. No plumbing ever leaks or breaks, no groundwater or irrigation spray ever seeps in and nothing indoors is ever cold enough to produce any condensation. The budget is always big enough to accommodate perfection, and the entire building and its mechanical systems are constructed and operated exactly the way the owner and the designers intended.

For those of us who do not live in quite such a perfect world, a prudent risk reduction strategy has two basic principles:

1. Reduce the water and humidity loads on the building to their minimums, through budgeting and design decisions.
2. When parts of the building get wet indoors in spite of everybody's best efforts, make sure the moisture dries out quickly.

The owner—not the law—makes the key decisions

Most importantly, building occupants need to understand that nothing in the law (at least in the USA, at present) prevents owners and designers from constructing a building which grows mold.

This could change in the future. But to date, civil authorities have not yet decided that mold or bacteria present a health or safety risk important enough to change the requirements of building codes. Literally millions of code-compliant buildings are growing mold and bacteria, every day.[10] And because "millions of buildings" suggests these buildings follow design practices which are widespread and therefore "standard" (and therefore acceptable), the building codes and standards of care which govern professional practice are not currently preventing architects and engineers from designing millions *more* buildings which have a high risk of growing mold.

So the decisions described below are hard ones for some owners. At present, both the owner and the designers must decide whether or not to reduce mold risk, on their own, without being compelled to do so by civil authorities. To some owners and designers, the suggestions will be familiar, and may already be standard practice. For others, the suggestions will require some changes from the way things have been done in the past.

Suggestions for owners and Architects

The owner and the architectural designer establish the baseline risk of bugs, mold and rot, because the decisions they make have the greatest and most enduring effects on the amount of water and humidity which get into the building. These suggestions minimize those loads, and therefore they minimize the baseline risk.

Fig. 5.7 Roof overhangs reduce rain contact, and therefore reduce risk

As the photo indicates, it does not take a very wide overhang to reduce the rain contact for the life of the building. The photo also shows how much more rain soaks the wall when there is no overhang.

The volume of rain flowing down the wall surface is the principal risk factor for leaks, and therefore for bugs, mold and rot. Roof overhangs are an economical way to reduce that risk by more than 50%.

Fig. 5.8
Evidence that roof overhangs pay long-term dividends
The short roof overhang sheltered the sculptural detail on this Roman Temple for 2,000 years. That slightly wider roof was a bargain-basement investment, considering how well it achieved long-term resistance to rain water leaks and to water-related building problems.

Roof overhangs, gutters and wall projections avoid 50% to 75% of the rain load on the exterior walls

Since the owner probably knows what the building should look like, he will influence what the architectural designer chooses to present as design alternatives. If the owners insist on roof overhangs, gutters and short projections above windows and doors, the risk of moisture accumulation indoors is greatly reduced.[11]

Look at the differences in rain accumulation shown in figure 5.7. Roof overhangs of about two feet [58cm] can prevent about half of the annual rain load from contacting the building, and therefore from dripping down the wall until it finds an open joint. As one leading building scientist has neatly summarized "If it doesn't get wet—it can't leak."[12]

Reducing the total amount of rain which contacts the exterior walls reduces both the volume and the frequency of any rain leak which will accidentally get into the building through joints around wall penetrations, or through joints where different siding materials meet each other. That's why roof overhangs, gutters and projections above windows make the building more tolerant of minor imperfections in joint design and installation. They keep rain water off the wall.

Conversely, without these details there will be a larger volume of water flowing down over all the joints, for the life of the building. With more water, every joint design and every small error by the installing contractor make the building more fragile with respect to water intrusion, humid air infiltration and indoor moisture accumulation.

For an example of long-term durability enhanced by roof overhangs, consider the Roman temple shown in figure 5.8.

It was built during the first century AD, in Southern France. About 2,000 years later, the sculptural detail under the overhang is still sharp and clear. Two thousand years of durability is a fairly impressive return on investment for the incremental cost of that slightly wider roof.

Install sill pan flashing under windows and doors

One building scientist with over 35 years of experience designing, constructing and investigating buildings has famously said that: "There

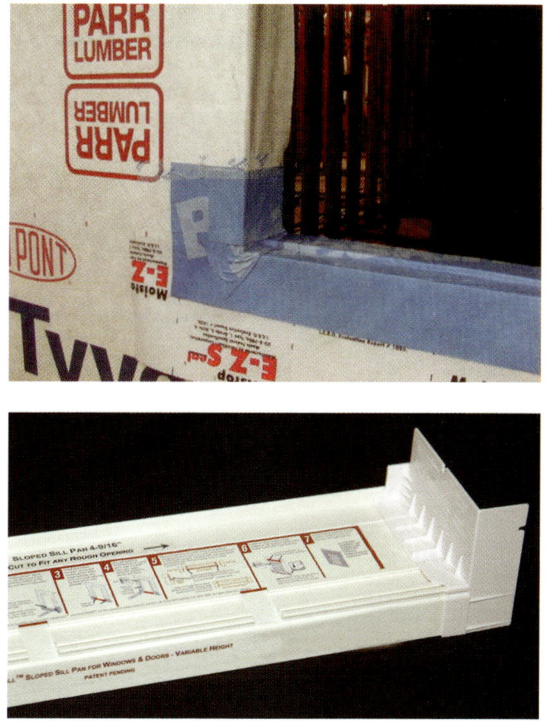

Fig. 5.9
Sill pan flashing under windows
After roof overhangs, the best way to reduce the risk of moisture leaking into the building.

are only two kinds of windows in North America: those which leak... and those which *will* leak, later."[13]

That observation may or may not be overdrawn. But support for that opinion is provided by an engineering collaborative which inspects approximately 25,000 new buildings each year. Within that sample, 35% had water intrusion problems caused either by leakage through the windows themselves, or by leakage around those windows as a result of installation shortcomings.[14]

These facts suggest that—after the roof overhangs and gutters which reduce the water loads—the most important architectural contribution to reducing the remaining risk of moisture problems is to install sill pan flashing under the windows and doors.

Figure 5.9 shows two examples of sill pans. They rest on the blocking which supports the window. They catch any rain water which gets through—or around—the window, before that water can get into the rest of the wall. The outer edge of the pan drains any water leakage *out* of the wall—either outside the cladding, or into a waterproofed drainage cavity between the cladding and the sheathing.

Figure 5.10 is provided courtesy of the building scientist who believes all windows will eventually leak. It shows how sill pan flashing can be built of self-adhered flashing membranes, and how that type of sill flashing can be properly integrated into the layers of the exterior wall. Integration of flashing with the other layers is the next link in the chain of owners' and architects' decisions which keep moisture from collecting in the wall.

Clearly establish responsibility for integrating the flashing around windows, doors and balconies

Another building scientist has suggested that "Water gets into a building through the cracks between the architect and the contractor."[15] Around all exterior wall penetrations for windows, doors and balconies, there are joints. And along every one of those tens of thousands of feet is a crack—through which rainwater can and will leak into the building, unless there is *effective* flashing behind the cracks.

Effective flashing is a sheet of metal or other durable material behind the cracks which stops the water from getting further into the building, and which redirects that water back outside the building, and off the wall, entirely.

The real problem with flashing and with integrating windows and doors into wall layers is that "everybody is responsible" for making the flashing work so that it excludes water. As all readers will recognize from their own experience, when "everybody is equally responsible", nobody is really in charge—so success is unlikely.

Fig. 5.10
Visual explanation of flashing details & window integration reduces risks
When the craftsmen on the job site have diagrams like these, it reduces the probability of leaky corners around and beside windows—the chief cause of moisture intrusion into buildings and therefore the second most common risk factor for bugs, mold and rot after the absence or presence of roof overhangs.

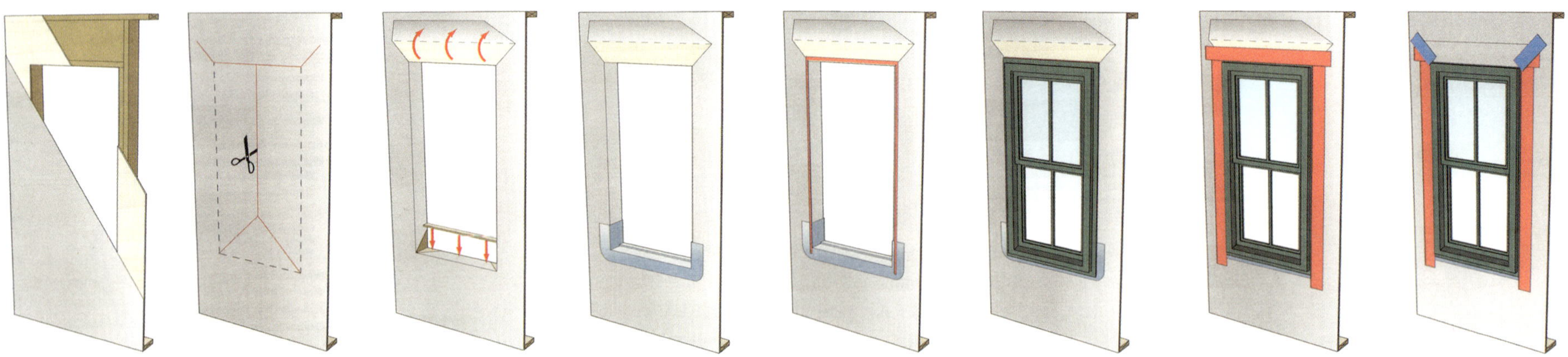

Architects know that all openings and all joints between different materials have cracks, and that all of these must be flashed. The cracks themselves are easy to flash effectively. Single long runs of flashing are very simple to design, and rather easy to install correctly.

But the problems come when the architect and the contractors reach the corners. Excluding water in all of those thousands of corners on a building is a highly complex problem.

Which layer goes over, and which layer goes under? How are the flashing pieces joined and sealed water-tight when they bend, fold and meet at the corners? How does the water get back out of the deep corners under windows when the window leaks? How does water stay out from under sliding doors which open onto balconies? How does water avoid getting into the wall through a mitered outside corner where two pieces of flashing meet, but are not welded? And which trades are responsible for installing the exterior wall framing, the windows, the water barrier and the exterior cladding? Which of these craftspeople is on-site, and when? And who is responsible for making sure that the installation sequence of all these different layers and different components, installed and purchased by different subcontractors, goes together in a way that really excludes water—especially in all those corners? Who makes the drawings which describe that sequence and define which trade is responsible for which layer, at what time? Who inspects the result, and how is water-tightness tested and documented, and when, and by whom?

Diagrams like those shown in figure 5.10 don't happen automatically. Owners who are not experienced with construction realities usually assume that these issues will be all be dealt with by the architect, or the contractor, or somebody, somehow. But millions of buildings with water intrusion around and through windows testify to the fact that flashing in the corners is often left to chance.

Architects are usually not eager to provide the details of how each layer integrates with all the others in the corners. Usually, the architectural drawings show wall sections alone, rather than the more informative isometrics of how each layer meets and integrates with all others in the corners. Such drawings are numerous, complex and costly to produce. So often, the architect just provides wall sections, and leaves the corner integration undefined, assuming that will be the responsibility of the contractor, who is traditionally in charge of the "means and methods" of assembly.

The contractor looks at the wall sections provided by the architect, and then leaves the "details of construction sequencing and trade coordination" of those corners to his site superintendent. The site superintendent tells each crew foreman to "make sure he coordinates with the other trades."

So in too many cases, successful water exclusion is under the control and guidance of the craftspeople who show up on the job site on each day when the pieces have to go together. Without drawings from the architect or from the general contractor, and without physical mock-up wall sections showing what the designers intended in the corners, the craftspeople—whose native languages are often different than the language on the drawings—are now in charge of designing and installing tens of thousands of water-tight corner joints.

Fixing flashing leaks and repairing the resulting damage in a reliable way generally requires removing exterior cladding and perhaps windows, with a cost of millions of dollars and months of disruption. So when such buildings leak water (provided that the leaks are visually apparent), the favorite remedy of the contractor and the architect is the caulking tube, because it is relatively inexpensive and the results are visible, even if seldom effective over time. The owner is left with a fragile building.

Reducing the risk of these problems is difficult, but is most easily accomplished by the owner. The owner can specify in his program requirements which organization—either the architect or the contractor (but not both):

> "...shall be responsible for the design and 3-D drawings which clearly show all layers and their installation sequence for all flashing details *at all corners* in addition

> to all straight joints in the exterior wall. These flashing details and their defined construction sequences shall be effective in excluding water from the building, as measured by the absence of any liquid water or any elevated moisture content of any materials inside the exterior wall. Elevated moisture content shall be defined as any moisture which is sufficient to generate microbial growth, or which reduces the life of the material or assembly in question such that it must be replaced to ensure all of its functions for the useful life of the building, a time period which is defined in the contact documents."

If the architect or contractor is wise, he will ask for additional money in his budget to accomplish this complex and time-consuming task. He will also ask for extra money to construct a mockup wall section on the job site itself, showing the physical reality of these details for the benefit and ready reference of the craftspeople.

If the owner wishes to reduce the risk of the most common cause of bugs, mold and rot in buildings, he will provide those additional funds.

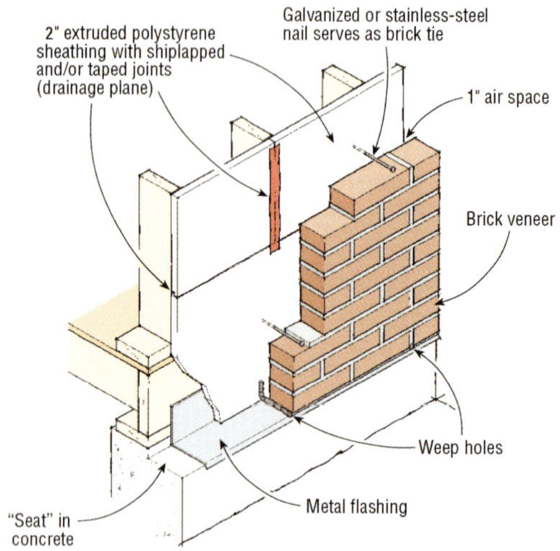

Exterior cladding which drains rain and dries quickly

The selection and design of the exterior cladding is another risk-laden decision which is determined by the owner's look-and-feel decisions, and his budget.

The cladding is the surface which first receives the rain. So first, it should shed most of that rain. Then, when some moisture gets behind it, the back side of the cladding should be an open air space, so that the water will run down the *back* side of the cladding instead of contacting the sheathing, which is usually more moisture-sensitive than the cladding.

Then, the sheathing should be covered by a continuous and completely sealed water barrier, so that when water gets *across* that air gap in some places, the water flows down the face of the barrier rather than soaking and penetrating the sheathing.

The bottom of that air gap between the cladding and the water barrier needs drain holes and flashing. And the top of that air gap needs air vents so a slow current of air can dry out any water that gets past the exterior face of the cladding. One example of brick veneer cladding, air gap and waterproof sheathing is shown in figure 5.11.

That air gap and water barrier is called a drainage plane, and the entire assembly with drains and vents is sometimes called a "rain-screen wall." It costs more money than simply pressing the cladding up against the sheathing. It will also be more complicated to design. But it is far more reliable in excluding water than cladding which does not have that air gap. When all layers are in direct contact, water *will* creep though cracks and along fasteners all the way to and through the sheathing, and then into the building.

Fig. 5.11 Air gap followed by a water barrier keeps water out of the interior wall

The air gap behind the exterior cladding keeps water from contacting the sheathing, reducing the risk of water problems. Behind brick, as shown here, it's also important to place a vapor barrier membrane over any wood-based or gypsum sheathing to protect it from the high vapor loads from the sun-heated, rain-saturated bricks.

Fig. 5.12 Vinyl wall covering problems are notorious

Too often, vinyl wall coverings trap moisture behind walls, leading to mold growth in the face of the wall board and in the adhesives, as seen here. ***Don't put vinyl wall coverings on exterior walls in hot and humid climates.*** *It's just too risky.*

The air gap and drainage are very traditional means of excluding water from buildings. When the exterior cladding system is brick veneer, these features are common practice and are very important to avoiding moisture problems. The owner would do well to look at the drawings for any brick veneer to make sure the air gap and waterproofing are included, and to ask the contractor how he will ensure that no mortar bridges the air gap, and how he will ensure that the water barrier which covers the sheathing is sealed.

With other cladding systems, the air gap and water barrier are not always standard. The more common contemporary practice for stucco, EIFS, clapboards and precast panels is to eliminate the air gap, squeeze and fasten all the layers together and rely on building paper or housewrap to keep the seepage water out of the sheathing. And that practice works for many buildings and saves money in construction. But if there are holes in the housewrap (such as nails), or if the housewrap is not really effective in excluding water[16], or if the housewrap loses its water-exclusion properties over time[17], then any water leaks at the exterior become paths for moisture intrusion into the building.

An air gap and sealed water barrier greatly improve the ability of the exterior wall to exclude moisture. If the request comes after the architect has decided on a design without these features, the costs of adding them may be more expensive. But if the owner asks for an air gap and sealed water barrier at an early stage, there may be very little additional cost.

Interior wall finish which passes water vapor freely

The owner determines what will be used to decorate the interior walls. Common practice in many commercial buildings in the USA has been to use vinyl wall covering to decorate and protect the indoor surface of exterior walls which are constructed using paper-faced gypsum wall board. In air conditioned buildings in a hot and humid climate, this practice has been absolutely disastrous.[5,18,19]

Time and time again, beginning in the early 1980's, forensic investigations have identified impermeable vinyl wall covering as being the principal cause, or the most significant contributor, to mold growth in walls.

The problem is that when humid air leaks into the building behind the wall from outdoors, or when water leaks in around windows or other wall penetrations, the vinyl traps water vapor inside the exterior wall. That wall cavity becomes very humid. Then, because the wall board is cooled by the indoor air conditioning, moisture condenses and supports the growth of mold and bacteria on or inside that wall. Figure 5.12 shows the result.

When planning a new building or when redecorating one in a hot and humid climate, the owner must resist the temptation to use impermeable finishes such as vinyl or vapor-retardant paint on the indoor surface of the exterior walls.

Instead, *use paint or highly-permeable wall coverings attached with adhesives which pass water vapor freely.* Insist on a combined perm value of 10 or higher for that wall. Ideally, the perm value would be greater than 15. Paper-faced gypsum board by itself has a perm value of about 50—it passes water vapor quite freely until it is painted, or covered with adhesive and wall covering.

Indoor surfaces of exterior walls of air conditioned buildings in hot and humid climates should be highly permeable to water vapor. Owners must understand that *im*permeable layers have been proven, over and over again, to be very risky with respect to mold growth.

Installing paper-faced gypsum board—keep it up above the floor

The architect can greatly reduce the risk of mold in paper-faced gypsum board by following a suggestion from the manufacturers: specify a gap at the bottom of the wall, between the gypsum board and the finished floor. A gap of 1/4 inch [6mm] or greater is sufficient.

Figure 5.13 shows an example. The gap acts as a capillary break. In other words, water on the floor or behind the wall will not wick up into the wall, raising its moisture content high enough for mold.

Often, floor mopping and carpet cleaning will wet a floor often enough to be a problem for gypsum board. The edge of the gypsum sucks up moisture, and the base molding keeps the gypsum board from drying out. So mold grows on the wall behind the base molding and inside the wall cavity, on the back side of the wall board.

Providing a gap at the base of the wall eliminates this wicking. It provides protection until there is a water problem big enough to create a pool with a depth greater than 1/4"—a very unusual event.

In most occupancies, this air gap does not significantly affect the fire protection or noise attenuation of the wall. But to keep the wall air-tight for fire protection and noise reduction, the architect can specify that the gap be filled with "smoke seal" foam, or "fire-sealant" foam. A thin bead of these spray-applied expansive sealants will keep the wall air-tight, while still serving as the capillary break which prevents water from wicking up off the floor and into the wall.

Fig. 5.13 Gaps under wall board help prevent mold

By keeping water spills and condensation from wicking up into the wall, a short gap reduces the risk of mold growth behind the base board.

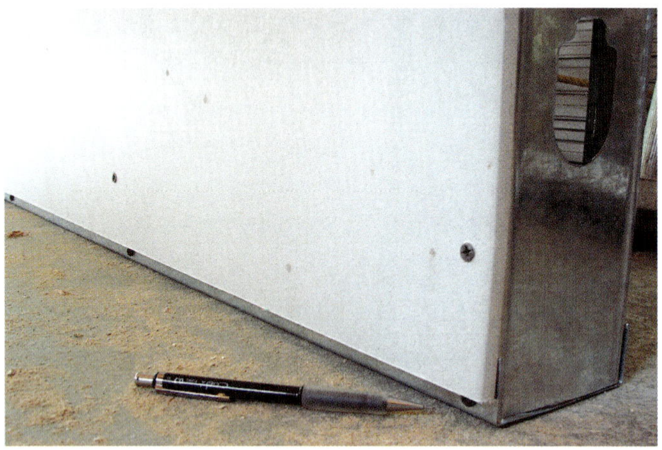

1. Outdoor Air
Hot and humid, it must be cleaned and dried.

2. Ventilation Dehumidifier
Dries the incoming air to a condition below the desired humidity dew point, removing the internally-generated humidity loads

3. Air Inside The Building
Stays at or below the desired humidity dew point

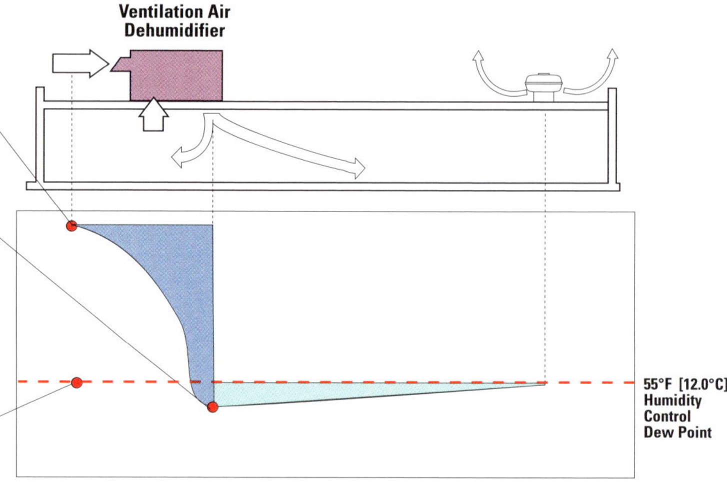

Fig. 5.14 Keeping indoor air dry

By drying the ventilation air, excess humidity never gets into the building. There are also other methods of keeping the indoor air below a 55°F dew point, but all must be able to keep the air dry during unoccupied hours. Generally this requires supplemental equipment and/or controls which can keep the building dry even when cooling loads are low.

A mechanical budget which allows dry indoor air

Keeping the indoor air dry avoids condensation in hidden places. Dry air will also make the building envelope details more forgiving of their typical imperfections and minor moisture leaks. Also, a dry building responds more quickly when the cooling system starts up in the morning, avoiding the usual damp and cold environment before the sun and the occupants heat up the building. And dry air provides comfort at higher temperatures, so thermostat settings can be raised, which saves energy.

However, to keep the indoor air dry enough to achieve these benefits, all of which reduce the risk of bugs, mold and rot, the mechanical system will need equipment which will dry the air to a defined maximum.

Figure 5.14 shows one way this can be achieved—by drying the ventilation air before it puts excess humidity into the building. There are several other options as well, which will be discussed in detail in later chapters.

Mechanical system designers know that equipping a building with dedicated dehumidification capability adds cost beyond the usual "per square foot" estimates of cooling-only systems. And that a cooling-only system was probably the basis of the owner's construction budget.

Fig. 5.15 Sweating supply air diffuser

The air jet leaving the diffuser creates low pressure near the edges of the jet. The low pressure pulls humid indoor air from the room to the cold metal surface, where its moisture will condense. To prevent the problem, keep the indoor air dry enough that its dew point is below the supply air temperature. Usually, an indoor air dew point of 55°F [12.8°C], will be dry enough to prevent this problem.

So they seldom suggest a system which will keep humidity below a defined maximum unless the owner recognizes its value. The mechanical system designer frequently hears that the owner wants many things which are out of the ordinary. Those items that the owner is willing to *actually fund* are the wishes that will be taken seriously and addressed in the mechanical design.

An easy way for the owner to communicate this requirement to the mechanical designer is to specify, in the owner's program requirements, that:

"...the indoor air dew point shall be held below 55°F [12.8°C]. This maximum applies to both occupied and semi-unoccupied modes of operation, and also applies to periods when the temperature is re-set during unoccupied hours."

The easy way for the owner to communicate his seriousness of purpose about this requirement and to allow a meaningful cost-benefit discussion is to ask the mechanical system designer:

1. For the cost estimate for the mechanical systems with and without this restriction.
2. What the maximum dew point will be in the building during part-load and unoccupied periods if the system is *not* equipped with some form of dehumidification that will dry the air independently from the thermostat setting and the size of the sensible heat loads.

Suggestions for the HVAC designer

When the subject of bugs, mold and rot comes up, most HVAC designers immediately think about keeping the mechanical system free from microbial growth in drain pans and duct work. That's understandable. The principal concern of the HVAC designer is the HVAC system.

Fig. 5.16 Sweating Exhaust Duct

As the cool duct passes through the humid attic, moisture condenses on the duct, which then soaks the ceiling. To prevent the problem, seal the attic to prevent humid air infiltration, and seal all duct connections to eliminate the suction which pulls outdoor air into building cavities.

However, over the last 20 years, forensic investigations of mold problems have shown that the HVAC system and its air distribution duct work sometimes have a profoundly positive or negative effect on the building structure and its furnishings.

In the past, HVAC designers have focused on providing the owner with cool temperatures and clean air for occupants, for the lowest possible construction cost and the lowest possible operating cost. These goals will probably always remain the owners' first concerns.

But if buildings are to avoid bugs, mold and rot, the HVAC designer will have to broaden his focus to include measures which avoid the known HVAC-related risk factors for these problems. And that will require the designer to understand how the building and it's materials are affected by the HVAC design and its behavior, throughout its full range of operating conditions.

Since the main concern is how much moisture ends up in the materials, these suggestions are based on the principle that the HVAC system as designed by the mechanical engineer and operated by the building staff *should not add moisture* to the building's materials or furnishings.

Indoor dew point low enough to limit condensation and absorption

When humid air meets a cold surface somewhere deep inside a building, moisture condenses from the air and drips onto nearby material, raising its moisture content.

Also, even when a surface is not cold enough to actually condense liquid, the relative humidity *at that cool surface* could be above 80%, which would allow some materials to absorb enough moisture to grow mold, given enough time (generally more than 30 days above an 80% surface rh).[9]

Sometimes, condensation problems are obvious because they occur in the occupied space and are quite visible. For example, cold water dripping from supply air diffusers onto the heads of diners in a restaurant are a clear indication that the indoor air dew point is too high. See figure 5.15 for an example.

But not all cold surfaces in a building are out in the open. Moisture will condense on cold supply air duct work when it is not insulated, or where insulation jackets are pulled back to make connections to diffusers or branches. Figure 5.16 shows the condensation on an exhaust air duct located above an insulated ceiling, The ceiling space has a high dew point, because the building is not air-tight. The air being exhausted is air conditioned, so the duct surface is cold enough to condense moisture, which then drips onto the ceiling tile.

Beyond duct work, cold water pipes which "sweat" are a visible problem in bathrooms. And there are similarly cold pipes behind walls and above ceilings.

Absorption problems are more subtle and take longer to become visible through mold growth. But absorption problems can be very widespread, and therefore very costly and disruptive to fix. Moisture from high dew point air collecting behind walls and above ceilings is often absorbed into wall board and ceiling tile. This is because the cooling system chills the room-side of those materials low enough that the *surface rh on their reverse sides* is well above 80%. And of course the higher the surface rh, the greater the amount of moisture absorbed by the material, and the sooner mold will grow. Figure 5.6 showed an example of mold growth caused by high dew point air behind walls.

By keeping *all* the air inside the building at a low dew point, the HVAC system will avoid adding moisture to materials. The question then becomes: how low is low enough, and how can the system keep the air below that limit under all operating conditions, including when the building is not occupied?

The consensus of the authors of this book and its project monitoring committee is that a maximum dew point of 55°F [12.8°C] is probably low enough to avoid most problems, in most situations.

A significant minority of this same group holds a more pessimistic view. They believe that a 55°F dew point is so unlikely to be achieved with the usual design and installation practices that a better *target* maximum would be a 52°F dew point [11°C]. The thinking is that if the designer *aims* at that lower target, the 55°F dew point maximum might actually be *achieved*.

With respect to how to achieve that 55°F dew point maximum at all times and under all load conditions, there are many design alternatives. These will be discussed in later chapters where they can be addressed in appropriate detail.

But as a basic principle, any effective solution must be clearly focused on removing the humidity loads brought into the building by the ventilation air, and on minimizing the infiltration of humid outdoor air. Those loads must be removed if humidity is to be held below a 55°F dew point in a hot and humid climate.

Control and monitor humidity based on dew point, not rh

Condensation and moisture absorption by cool surfaces are the problems to avoid. Recognizing and preventing them is easier and more certain when the HVAC system monitors and controls the dew point rather than the relative humidity.

It is not practical to measure the *surface* relative humidity on all cool surfaces throughout the building. And measuring *the air's* relative humidity in the middle of a room with a handheld thermohygrometer, or in a return air duct with an rh transmitter, really provides no useful information about the rh on the cool surfaces, especially behind walls and above ceilings.

The dew point, on the other hand, is an excellent indicator of a potential problem and its probable dimension. One simply has to think about the surface temperatures in an air conditioned building to see how useful the dew point value is in predicting a problem.

Figure 5.5 showed an example in graphic form. If the supply air temperature from the AC unit is 55°F [12.8°C], then the nearby wall surface temperature is probably near 60°F [15.6°C]. One can plot the surface temperature together with the air dew point temperature of 65°F [12.8°C] to realize that the *surface rh* on that wall is well

Fig. 5.17 Leaky ducts = mold

Suction from leaking duct connections pulls humid air into cool walls, where moisture condenses and grows mold. Sealing all duct connections greatly reduces humid air infiltration, in return for a 1 to 3% additional investment at the time of installation. That "insurance cost" is quickly paid back in energy savings. Sealing up duct connections pays for itself quickly, and then continues to pay dividends over the entire life of the building.

over 100%. In other words moisture is condensing on that surface, creating a risk factor for mold, both on the exposed surface and behind it.

Compare this useful information with a rather misleading reading of "65% rh" in the room air. That value—by itself—seems so safe that it creates a false sense of security with respect to mold risk.

After one becomes accustomed to thinking about hidden surface condensation and high *surface* rh, the air dew point temperature is much simpler to interpret as an indicator of current risk than that misleading 65% rh value measured in the open air.

Obtaining a dew point value is quite economical in modern buildings. There are low-cost infrared dew point transmitters available. And any building automation system can easily convert temperature and rh signals to a dew point reading, which can then be used for controlling the system.

By displaying and logging the current dew point, the building operations staff can gain a sense of the current risk. *Higher dew points are more risky and lower dew points are less risky—always.* Tracking the relative humidity is not as reliable.

Air-tight duct connections to avoid humid air infiltration

If all air duct connections and all air handler cabinets are air-tight, the building will have much less risk of mold. To many HVAC designers, this fact will seem rather odd. Why should air-tightness have anything to do with mold risk?

It's because air leaks on the suction side of the system mean that air will be pulled into that system out of building cavities. And those building cavities will in turn pull humid air into the building through construction joints in the exterior wall. The humid air condenses or creates high surface rh, and mold begins to grow.[20]

The same thing happens when exhaust air ducts are not sealed tight. They pull air from the building cavities they pass through on their way out of the building. That slight suction ultimately pulls in humid air from outdoors. And that humid air releases moisture when it contacts the cool surfaces inside the building.

The same problems occur when the cabinets of air handlers are leaky, when the cabinet is mounted inside the building or through the exterior wall. For example, vertical wall-mounted fan-coil units in hotels and apartments are seldom sealed tight. And even more rarely are they sealed tight to their wall-mounted return air grills.

The gap between the cabinet and the return grill means the fan will pull air from the building cavity in which the cabinet is mounted. Since these units are usually mounted near exterior walls, the air which replaces the air pulled from the cavity is likely to come from the humid outdoors. The humid air comes in through the cavity, and drops enough moisture to grow mold, because the room-side surface of the wall is cooled by the air conditioning system.

Further, leaks on the supply air side of duct systems mean that cold air is being blown into building cavities. Surfaces will cool down quickly, because the cavities are small. Moisture is absorbed into these surfaces, or condenses on them or behind them, because they are cool. Further, the cool air wasted in building cavities must be made up by adding more air to the system, resulting in energy wasted in fan power, and in cooling and drying that added air.

In summary, it's been well-established through forensic investigations that leaky duct connections increase mold risk in a hot and humid climate. The appropriate response is to specify that all connections be sealed up tight, using mastic and reinforcing tape, to SMACNA seal class A (the same tightness as is standard for "high pressure" duct work.) Figure 5.17 shows a technician applying mastic to duct joints. Note the size of the gap where the round duct meets the rectangular plenum. Spanning that gap will require glass fiber tape worked into the mastic as reinforcement..

A useful side benefit of this mold risk reduction measure is that it will probably save the owner between 25 and 35% of the annual cost of operating the system.[21,22] In the present discussion the purpose of sealing duct connections is mold risk reduction, but it also saves energy. There is really no good reason not to seal duct connections—tightly and permanently—with mastic.

Fig. 5.18 Use mastic to seal duct connections
It's not pretty—but it saves energy and reduces mold risk. Sealing all connections with mastic, using embedded glass fiber tape to span the wider gaps, is the most effective way to eliminate a well-known risk factor for mold, as demonstrated by the problem shown in figure 5.17.

Economizers which do not flood the building with humid air

When the air outdoors is cool, it makes sense to use it to cool the building... unless that outdoor air is more humid than what you want to maintain indoors.

When the outdoor air dew point is above the target maximum indoor dew point, using outdoor air to cool the building may accomplish some cooling, but it also floods the building with excess moisture, increasing the risk of mold.

Here again is another reason to track dew points rather than relative humidity. Comparing the indoor and outdoor dew points makes it quite obvious when outdoor air can be used for cooling, and when it cannot. In contrast, comparing relative humidities of those two air streams provides no useful information about whether it is safe to use outdoor air for cooling.

No HVAC designer would try to use outdoor air for cooling when it is hot outdoors. Similarly, no designer should allow "free cooling" with outdoor air *until the outdoor air dew point is below the indoor dew point*.

Suggestions for the contractors

Buildings are built outdoors. So they will get wet during construction. That's obvious and cannot be prevented. Also, schedules and budgets are always tight. There are very few projects where cost is no object and the client does not really care when the building will be ready for occupancy. So it's usually not possible for the contractors to let all wet materials dry out as much as one would wish before proceeding to the next task.

The key to preventing construction-related mold is to focus on the moisture-sensitive materials. Keep them from getting wet while they are stored on the job site. And also, don't install them near any damp concrete or wet concrete block.

Store gypsum wall board out of the rain and mud

The paper on the faces of conventional gypsum wall board is very easy for mold to attack. When a stack of it gets wetted by rain as it is stored on the job site in a hot and humid climate, the whole pile is at a high risk for growing mold.

Even when the stacks of wall board are protected from rain on the job site, they will absorb humidity from the air in a hot and humid climate. So in all cases, the contractor should recognize that stored paper-faced wall board in such a climate is a fragile product with respect to mold.

Two appropriate responses are apparent. First, don't let the stacks of wall board get wetted by rain or mud. If they become wet with liquid water, dry them out very quickly—or replace them. Replacing a stack of wall board is relatively cheap and quick. Mold remediation for *installed* wall board is far more expensive and schedule-busting.

Second, keep in mind that when paper-faced wall board is stored on-site, it will absorb moisture from the air. It's already at risk with respect to mold growth. So don't let any *more* moisture get into it by installing it above damp concrete, or over damp concrete block walls. Dry out these huge reservoirs of construction moisture and rain water before the wall board is installed. Otherwise, as soon as the AC system is turned on, the water vapor will move rapidly out of the warm, damp concrete block and into the cool wall board, where it will help grow mold.

Insulated ducts stored out of the rain and mud

Insulated sheet metal duct work is large and clumsy to store, so it often ends up stored out in the open air before it is installed. And because it is made of metal and its insulation is usually inorganic glass fiber, many assume that if it gets wet in storage, there's no real mold hazard. But that's not a good assumption.

A glass fiber insulation jacket or glass fiberboard lining can soak up a great deal of water. After it does, it will not be an effective insulator until it dries out. And the parts of the duct which are located downstream of cooling coils in hot and humid climates are in such a saturated environment that they may not dry out for months, if ever. In the mean time, the wet insulation serves as a moisture source, and the HVAC system conveniently distributes that moisture throughout the building, where it can be absorbed into cool interior surfaces.

Also, sometimes the HVAC system is started without the intended air filtration. When dust and dirt land on that soggy insulation in the ducts, the moisture transfers to the dust. Damp dust supports mold growth on the surface of insulation which might otherwise be mold-resistant.

To avoid these problems, the general contractor should either provide a storage area for duct sections out of the rain, or require that the mechanical contractor do so.

Dry out wet concrete slabs and masonry block walls before interiors are installed

One of the most common reasons for mold which occurs soon after construction completion is water in concrete floor slabs and masonry block walls. In hot and humid climates, regular rain is rather frequent. That rain will saturate masonry block walls.

If the building is closed up before the concrete floors and the block are dry, that water becomes an indoor source of moisture which will help grow mold. It will also generate such a high internal humidity load that cabinetwork and millwork may warp and wooden doors might expand and jamb open or shut.

Fig. 5.19 Moisture meter for concrete

To reduce the risk of moisture transferring from concrete and masonry to wall board and mill work, measure the moisture content, and don't install moisture-sensitive material until the masonry and concrete are dry.

Usually, the flooring manufacturers' specifications call for dry concrete as an underlayment, and they require a moisture content test of a concrete floor slab before they will guarantee the installation. So for floors, the site superintendant is usually attuned to the problem, and will either dry out or seal the concrete before mold-sensitive adhesives and leveling compounds are placed on that floor.

But saturated masonry block walls do not usually receive the same attention. Currently in the USA, neither wall board manufacturers nor coating manufacturers have yet established any *quantitative* definition of what they mean when they require that "the masonry block must be dry before finish is applied."

Without a specific maximum moisture content and without a defined means of measuring that value, the contractor has no guidance other than "his best judgement" about how dry is dry enough.

Figure 5.19 shows one type of meter used to quickly scan masonry walls for excessive moisture. As a very rough guideline, most masonry block holds about 6% of its weight in water before liquid water actually flows out of the block. And when it is completely dry, masonry block has a moisture content of less than 2% of its dry weight. So, until there is specific guidance from the manufacturer of the coating and the wall board, one might reasonably assume that a 2% reading on a concrete moisture meter probably means the block is dry enough to coat or to install wall board. But any reading over 4% means a *great deal of water* will be coming out of that block and into the interior finish during the first few weeks of the building's operation.

Don't start the HVAC system early - use a drying service if necessary

A rather common practice in hot and humid climates is to start the HVAC system early, to provide both comfort cooling for workers and to begin drying out the structure before interior finish is applied. The entire HVAC industry strongly objects to this practice.

From the owners perspective it shortens the useful life of the warranties on equipment and installation. If the system is started before it is fully commissioned, tested and balanced, the equipment can be severely damaged. And the building can be flooded with humid air if the ventilation air quantities are not carefully tested and balanced. And dust from construction clogs cooling coils, reducing both their useful life and their cooling capacity. When condensate drains are not yet piped and connected, water from cooling coils can flood the floors

Fig. 5.20 Construction drying

Rather than damaging the HVAC system through early start-up, use the drying equipment and subcontractors used by the insurance industry after floods, fires and disasters. There are thousands of such firms, and many have the skills needed to keep construction dry, and therefore to keep the schedule on-track when wetting slows the project.

Fig. 5.21 Clogged filters often lead to humid air infiltration

If air can't get into the system through the filters, the building can "go negative," pulling in humid air through construction joints. Keeping the ventilation filters clean reduces the risk of humid air infiltration which would escape the drying effect of the HVAC system.

in mechanical spaces, or cascade down and out of curb-mounted equipment on the roof, saturating materials below.

In short, early HVAC startup is a disaster, or a disaster waiting to happen. Don't do it.

When the building needs to be dried and the schedule does not allow slow open-air drying, the general contractor and owner can invest in drying services. Over the last 30 years in the US and Canada, a large and robust commercial infrastructure has developed to serve the insurance industry, drying buildings after floods, fires and disasters. Figure 5.20 shows a job site in Texas where this technique was used to keep a project on schedule in spite of persistent rain.

These techniques, which use specialized building drying equipment, provide a far more effective alternative than ruining the new HVAC system in an attempt to dry the building.[23]

Suggestions for building operators

To avoid bugs, mold & rot, run the HVAC system and maintain it in a way which keeps the system and the building itself, dry.

Basically, one wants to keep puddles, dirt and dampness out of duct work and drain pans, and keep humid air out of the building. Here are some ways these goals to meet these goals.

Monitor and control the dew point rather than the relative humidity

"You get what you measure" is a familiar saying in engineering and operations. Since the principal risk factors for mold and bacteria are water leakage, indoor condensation and high rh at hidden cool surfaces, the simplest and least expensive way to keep track of those risks is to record and display the indoor air dew point. Then use that value to control the dehumidification functions of the system, keeping the indoor air dew point below 55°F [12.8°C].

The logic of using dew point as the humidity control parameter is fully explained in the previous section, which deals with HVAC design. The suggestion also applies to those who are operating existing buildings.

Replace outdoor air filters to avoid humid air infiltration

If the building sucks in humid air, that incoming humidity is probably going to get into places it shouldn't, such as behind walls and above ceilings.

One of the many reasons a building can "go negative," is because makeup air filters are clogged with dirt, as seen in the photo shown in figure 5.21. A dirty filter restricts air flow. And if the ventilation and make up air can't get in though the filter, the exhaust and return fans will pull air into the building through the cracks, vents and joints in the exterior enclosure.

Humid air behind walls and above ceilings leads to condensation, high surface rh and eventually mold. Keeping the outdoor air filters clean makes it less likely that humid air will be pulled into the building in places where it will do damage over time.

In many locations, a monthly filter change will be adequate for outdoor air filters. But when the outdoor air intakes are near streets or highways where traffic stirs up surface dirt, or near construction and agricultural sites where blowing dust is a factor, the filters will probably need to be changed more often.

Replace filters which keep cooling coils and humid duct work clean

When air from the HVAC system smells "like dirty socks," it means that bacteria and fungus have been able to grow inside the system on dust and dirt which is saturated with moisture.

High relative humidity is a fact of life in duct work and inside AC unit housings. And condensate is a fact of life on cooling coils. So the most effective way to avoid bacteria growing in these locations is to keep dirt out of the system and out of the equipment, by replacing filters frequently.

The filters upstream of cooling coils are especially worthy of attention. They probably need replacement every two to three months as a minimum. But high internal particulate loading, or any lack of filtration on the outdoor inlets would suggest at least monthly replacement.

Run the building "dry" when unoccupied

When the building is unoccupied, make sure it stays dry. Exactly how this can be accomplished will depend on what sort of systems are installed, and how their dehumidification capacity must be controlled.

If the building has a separate ventilation air dehumidification system with a return air connection, then the building can be kept dry by switching that system from "ventilate" to "recirculate."

If the building is equipped with wall-mounted DX cooling equipment with a "dehumidification mode", then simply shutting off the ventilation air and letting the room units operate in response to their internal humidistats may be quite successful in keeping the building dry.

But regardless of what sort of systems are installed, it's very important for the operating staff to understand that the largest humidity load enters the building via the ventilation air stream.

If the ventilation air is not either shut off or dried, moisture will build up in the building during unoccupied periods. This increases the risks of condensation when the cooling system starts back up. It also means the cooling system will take a long time to cool the building, because the cooling system will be struggling to remove all the excess humidity from the air, and from the furnishings and wall board.

So when setting up the system for unoccupied hours, keep in mind that whatever else the system must do during those hours, it's very important that the ventilation air be closed off entirely, or reduced to its code-required minimum and dried. If it is allowed into the building at full volume with its full load of humidity when the building is unoccupied, the risk of mold over time is very great.

Figure 5.22 shows an example of what happened in a school during summer vacation, when the thermostats were re-set to save energy, but the AC units were still set to bring in ventilation air. The AC system must be re-thought and re-set so the building stays dry—not cold—during unoccupied periods.

The authors and the project advisory committee for this chapter are fully aware that most buildings built in hot and humid climates over the last 30 years are not equipped with systems which can easily reduce and dry ventilation air during unoccupied hours. That's an unfortunate risk factor for mold that the operators of these buildings will have to live with, or change as budgets allow.

Clean out condensate drain lines to avoid overflows and bacteria in drain pans

One of the more common causes of moisture damage in buildings is a condensate drain pan which does not actually drain. As the system operates, condensate rises in the pan and then overflows into the building, providing the moisture needed by mold and bacteria.

Rather frequently, condensate drain lines are simply not installed at all, or are installed without traps, or with traps which are too shallow to ensure that the pan will drain freely. These are basically design and installation issues, which, after correction by the operating staff, will stay corrected.

Fig. 5.22 Humid ventilation during summer vacation = mold

The cooling system only ran intermittently, so it did not dry the ventilation air. Mold grew because the air's dew point was high, even though its rh was low.[24]

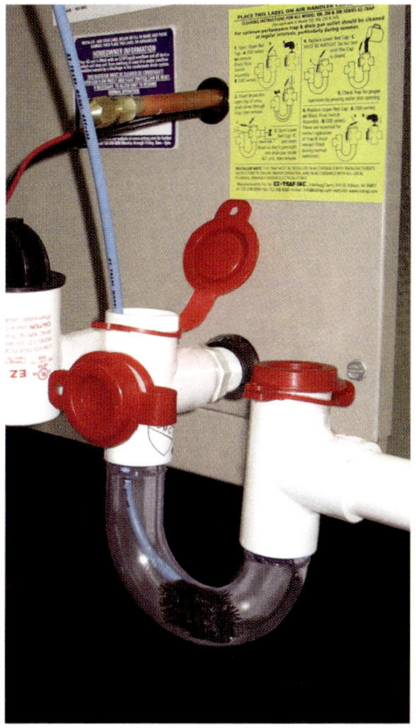

Fig. 5.23
Cleanable condensate drain trap
Every condensate pan needs a trapped condensate drain line. Otherwise, water will stay in the pan and grow bacteria. That trap needs to be engineered. It must be deep enough to resist the air pressures and suction generated by the fans. And like this one, it should be easily cleanable with brushes, from outside the AC unit.

But another common cause of drain pan overflow is simply that either the drain hole or the trap is clogged with lint, feathers, dead insects and rodents, leaves, twigs or some combination of the astonishing variety of crud which ends up in a condensate drain pan.

So one of the more useful and less costly techniques for limiting the risk of mold and bacterial growth is to ream out the condensate drain line on a regular schedule. Flushing out the pan and bottle-brushing the drain line about once a year should be adequate. More frequent cleaning may be needed for cooling coils located at ground level or near trees, where more leaves, twigs, insects and animals can be expected.

Figure 5.23 shows an example of a condensate trap which is designed for easy cleaning; not just of the trap, but of the condensate line and it's opening to the drain pan as well.

Assessing Mold Risk in Existing Buildings

If keeping the building dry is the way to reduce the risk of bugs, mold and rot, how dry does the building have to be, exactly, to prevent these problems? Where and how should material moisture content be measured to assess the extent of the risks, or the dimension of a potential problem area?

The answers to these questions are not yet entirely understood. Also, the tools available for measuring moisture content are not as quick, certain and cost-effective as investigators would like to have. Buildings are simply very complex environments. And no two buildings are exactly identical. But there are some risk factors which are better-understood than others, and progress is constantly being made in this area.

Bacteria: locate any standing water, then drain it or dry it

The most common, well-understood and notorious bacterial problems in buildings are caused by the bacterium *legionella pneumophila*. which causes Legionnaire's Disease. That bacterium will survive in liquid water above 68°F [20°C]. It will multiply rapidly in liquid water which stays between 95° and 115°F [between 35°C and 45°C].

To reduce the risks associated with *legionella* to near zero, make sure there is no standing water in the system, or near the ventilation air inlets. To assess that risk, check the drain pans under cooling coils. If there is stagnant warm water in the pan, there is a risk of *legionella*. If the water drains away completely, there is very little risk.

The other potential source of standing warm water in a building is the sump of a cooling tower.

If cooling tower sumps do not have some form of anti-biological treatment, the risk of *legionella* in a hot and humid climate is high. Then one needs to be concerned about bacteria-laden mist from that tower which can enter the ventilation air of the building, or any nearby building, or the breathing zones of pedestrians passing by the building. It's more prudent to simply make sure that no *legionella* can grow in the cooling tower sump, through flushing and regular treatment with anti-biologicals.

Problems and solutions related to *legionella* are fairly well-understood. And there is extensive guidance from manufacturers of both cooling towers and water heaters to help the owner assess and avoid the risks. Less well-understood are the problems related to bacteria in damp materials, near where fungus is also growing.

There is some suspicion that it is the bacteria as much as the fungus which is responsible for the negative effects of damp buildings on human health.[2] But since neither the causes nor effects nor the damaging exposures are defined at this time, we will not discuss those topics.

Until these issues are better understood by public health authorities, we can simply note that most bacteria need more moisture than most fungi. So perhaps if the building cavities are not growing enough mold to be a problem, they may not be growing enough bacteria to be a problem, either. The reader will recognize that this is not certain. But it *is* certainly true that any measure which reduces moisture accumulation will reduce the risks of both mold and bacteria.

Mold - keep moisture content below 14% WME

When the building operator wants to assess the current risk of mold, he should measure the moisture content of the building's materials. At or below a moisture content reading of 14% WME (Wood Moisture Equivalent), there is little risk of mold on any wood-based or paper-based product.

A 15% moisture content reading corresponds roughly to a water activity (A_W) of 0.8, which represents the edge of concern. An A_w of 0.8 is the amount of moisture that will eventually be absorbed by wood, when the surrounding surface rh is held constant at 80%. That level of water activity, when held for more than 30 days, has been identified by the International Energy Agency as being the probable lower limit for mold growth in most building materials.[9] So, if a wood-calibrated meter reads 14% or less, there is much less risk of mold growth.

Handheld moisture meters are usually calibrated for softwood rather than for all the dozens of other materials in a building. So the meter readings discussed here are called "WME"—the wood moisture equivalent—to make the distinction between what the meter indicates versus the true moisture content for any material other than softwood. For example, when used in wall board, a WME reading of 17% corresponds to an actual gypsum board moisture content of less than 1.1%. While these differences lead to great confusion in reading reports, the wood-based meters are economical, and still quite useful when testing for excessive moisture in typical building materials.

Figure 5.24 shows an example of moisture content readings on a test wall section, and the correlation (in that particular test) between reported moisture content and mold growth.

Between readings of 15 and 18% on a wood-based meter, there is a moderate to high risk of mold growth. Mold will grow, given enough time, and given a substrate—such as paper or cardboard—which is easy for mold to digest. Above 19% WME, framing lumber and manufactured wood products such as oriented strand board and plywood are likely to grow mold, given enough time at a moisture

11% WME, measured on the dry side

15% WME, measured across the visible edge

19% WME, measured on the moldy side

23% WME, measured in the moldy area

Fig. 5.24

Higher moisture content allows more mold growth

This test wall section, intentionally dampened by contact on it's right edge with an earth floor over several months, shows how increasing moisture content leads to mold growth. Note also that moisture content can vary sharply over just a short distance. In this case, a moisture content of 11% WME grows no mold. But less than one inch [24mm] away, the moisture content rises to 19%, which grows mold easily (at least on this particular surface).

content above 19%. And above the fiber saturation moisture content of wood (above 27 to 30% of its dry weight, depending on the species), actual rot fungi will attack framing lumber, plywood and OSB, and most nails, screws or bolts will corrode quickly.[25] Over time, rot creates a structural damage risk from missing wood and corroded fasteners.

All of these thresholds are approximate. Fungal growth can and does occur at higher and lower levels. Also, these numbers do not take into account antifungal treatments, which can postpone the growth of mold. But for building owners and operators who would like to have general benchmark values to use as levels of increasing concern, moisture readings of 14%, 18% and 26% might be appropriate, when measured with a meter calibrated for wood.

From the discussion above, one can also see that when its food is "chopped-up-and-pre-cooked," mold is able to grow with less moisture. Wood is tough. It withstands fungal attack very well. If it were otherwise, trees could not survive. But as wood fibers are crushed and heated, they become easier for more types of fungi to digest. Fewer of the wood's natural defenses are intact. That's why paper (which consists of wood fibers which have been cut, crushed, boiled and re-dried) will usually grow mold at lower moisture contents than will OSB and particle board. And OSB and particle board will usually grow mold at lower moisture contents than will framing lumber.[26]

In assessing the risk of high moisture content, it's important to recognize that moisture moves around quite a bit. The moisture content of building assemblies can vary widely over a few inches or centimeters (see figure 5.24). Also, moisture held in one material can transfer to a different material nearby. Moisture from wet masonry block diffusing into gypsum wall board is a typical example.

So it's useful to understand a bit about the current state of the art in moisture measurement instruments and techniques, to avoid false impressions of safety or risk.

Measuring moisture

Currently, there are two principal types of portable, handheld moisture meters for quick, low-cost measurements in building materials: resistance-based and electrical field-based meters.

Resistance-based meters, also known as "pin meters" or "penetrating meters," measure the electrical resistance between two pins which are set about two or three centimeters apart, and which have been pushed into the surface of the material being measured. The lower the resistance between the pins, the higher the moisture content reading. Most low-cost meters have a single-species wood moisture calibration, usually for Douglas fir or white pine. The instrument shown in figure 5.24 is a resistance-based meter.

Electrical field-based meters, often called "non-penetrating" meters, do not use pins and therefore do not make holes in the material. They create an electrical field, which is modified by any material which comes into contact with the back of the meter. A change in the electrical field characteristics is read out as a change in the material moisture content. Figure 5.19 showed an electrical field-based meter.

Non-penetrating meters are usually calibrated for soft wood moisture characteristics, but they also have multiple scales indicating either actual percent or "relative" moisture content readings for other materials. They are often used for hard-to-measure layered materials such as wood sub-flooring under ceramic tile, or for non-wood materials such as concrete and masonry block. These would be time-consuming to measure with resistance-based meters, because holes must be drilled into the materials to insert conductive nails or pins.

Both types of moisture meters were originally optimized for the wood products industry, and they remain best-suited to measuring moisture content of stacks of lumber and uniform wood products rather than the intricate, complex, difficult-to-access, multi-component assemblies of a building. So most meters have significant limitations when used in building inspection situations.

For example:

- No commercially-available meter can easily reach into deep, dark building cavities, or up into high surfaces such as ceilings higher than raised-hand-height.
- Most meters are not calibrated for more than softwood, so the readings they display when used in gypsum wall board are incorrect by more than an order of magnitude.
- Most non-penetrating meters have multiple scales, some of which may appear to be percent moisture content, but which are in fact only "relative scales", which have no direct or even any defined relationship to actual percent moisture content for any specific material.
- Both types of meters are greatly affected by surface moisture. So a reading may not reflect what is happening inside the material, but instead only indicates a temporary moisture condition at the surface.
- Neither meter technology is accurate above the fiber saturation moisture content of wood. A moisture content reading above 30% on either type of meter just means "too darn wet" in a building inspection situation. Above 30%, differences of 5 or even 15% are neither significant nor accurate for building inspection purposes. Both technologies become quite random above a 30% wood moisture content.

These and many more limitations create a very confusing situation for those reading inspection reports. One hopes that moisture meter technology which is better suited to building inspections will become available in the future. That said, current meters are economical and still quite useful, as long as one understands their limitations.

Building drying experts have found that usually, pin-type meters are more consistent than non-penetrating meters. On the other hand, these experts note that the non-penetrating meters are much more useful for quickly scanning an area for potential problem points. Since non-penetrating meters do not make holes, they are usually used first. Then, a pin-type meter is used when an area seems suspect for elevated moisture content, or when a wet section is being dried and there is a defined target maximum moisture content to be achieved by a drying contractor.

As general guidelines, most investigators would agree there is little concern until meters display readings above 14% wood moisture equivalent. Above that threshold, paper-based gypsum wall board is at higher risk for mold. And above readings of 19% wood moisture equivalent, both paper and wood products are at high risk for mold. When readings go above 25%, surface mold is a near-term certainty unless the material is either protected by antifungals or is dried out quickly, and actual rot and structural damage become long-term concerns above 27%.

In summary, use these meters. But don't take their readings to be more than a general indication of levels of concern. They are simply not very accurate or repeatable in the large, complex composite assemblies typical of buildings.

Locating excess moisture in buildings

Moisture investigations can be very complex, because the paths that water can take inside a building are complex. Water can travel long distances from logical sources, and it can end up in odd locations.

Still, the most productive starting points for an investigation are the rooms which seem to have earthy odors, and the surfaces near the typical sources of rain leaks, humid air infiltration and indoor condensation. These include:

- The inside surface of exterior walls. Especially on the first floor, where irrigation spray might wash the exterior, and at the top floors, where rain will deposit in larger amounts. In particular, look at interior wall surfaces under the penetrations for windows, sliding glass doors, AC units, electrical conduits, piping connections, dryer vents, side-wall exhaust vents and exterior electrical outlets.

Fig. 5.25
In humid buildings, cold ducts condense water and increase mold risk
If the air inside the building is not held to a low dew point, the outside surface of cold supply air ducts can condense quite a bit of water over time, leading to the stains and mold growth seen here.

Fig. 5.26
Cold pipes also condense moisture
In hot and humid climates, even insulated pipes will condense moisture, as seen in this photo of saturated insulation. Keeping the dew point low inside the building reduces the volume of potential condensation on cool pipes, which in turn reduces the risk of bugs, mold and rot.

- Surfaces under indoor-mounted AC units
- Dropped ceiling tiles which conceal chilled water piping, and/or domestic water supply lines, cold condensate drain lines and cold supply air ducts.
- Vented attics above air conditioned spaces. Condensation and high surface rh are often found on the cool surfaces in vented attic spaces, as seen in figure 5.16.

Follow the paths of duct work and piping to find moisture

In addition to looking near typical leak locations, the investigator can follow the path of mechanical and plumbing systems, which usually have cold surfaces and which can leak water or cool air.

For example, when the weather is hot and humid the cooling systems will be operating. So all of the supply air duct work will be cold, along with all of its attendant components like variable air volume boxes, branch ducts and elbows, diffusers, mixing boxes and dampers, filter boxes, access doors and air handler cabinets. If the dew point inside the building is high (above the surface temperature of those cold sheet metal parts), then condensation will form on their surfaces. When the indoor air dew point is high, enough moisture will collect to consolidate into droplets and run down to the nearest low point which can form a drip edge. So if one tracks the path of supply air distribution duct work, it is often possible to locate moisture that does not otherwise seem to be in a logical place.

Often, cold supply air duct work mounted indoors is not insulated. And when it is insulated, it often has great, wide gaps. This is sometimes simply poor installation. But often the lack of indoor insulation is the designer's intention. "It's all inside the thermal boundary" is the common reason for not insulating cool pipes and ducts. The logic is that any loss of cooling capacity to the space above the ceilings and behind the walls will cool the conditioned space, eventually. That's also the common reason given for not sealing the duct connections on the supply air side.

But that logic misses the problem of condensation inside buildings located in hot and humid climates. It also misses the fact that when supply air leaks out to cool the spaces behind walls and above ceilings, those now-cold surfaces can absorb moisture and grow mold. Also, on the energy side of the owner's concerns, lost cooling for the actual occupied spaces will have to be made up by using extra fan energy to circulate more air.

However, the reasons for the problems are not especially important to the moisture investigator. The point is that by following the path of supply air duct work, one can often locate drips coming from that cold duct work, and air leaks which cool nearby areas low enough to absorb humidity and grow mold. Figure 5.25 shows an example of drips from duct work which is indoors, in a building with a high dew point.

Piping systems are another typical source of both condensation and water leaks. Following the path of incoming domestic water supply piping can help track down both condensation points and actual water leaks at joints. The same holds true for cold suction lines connecting the evaporator to the compressor in a split DX cooling system. And of course the chilled water lines which supply fan-coil units throughout the building can leak, and also will have cold surfaces between insulation joints. Figure 5.26 shows an example of a dripping chilled water pipe—under insulation—in a Florida office building.

Also, when cooling units are located deep in the building, the condensate from their cooling coils will be piped to a drain. That drain may be quite a distance from the cooling unit, and the line will be carrying cold water (cold condensate) whenever the coil is condensing moisture. If that drain piping is not insulated, condensation will form on its outside surfaces, and perhaps drip onto wall board or ceiling tile, both of which are easy for mold to colonize when they are damp.

Three examples from forensic Engineer in Florida illustrate these issues.[27] In the first example, a cold condensate drain line, running through a carrier duct under a concrete foundation slab, generated enough condensation over two weeks to entirely fill and overflow the 6" diameter PVC carrier duct [14.5 cm]. The overflow ruined the wood flooring throughout the first floor of the building.

In another example, failure to trap a condensate pan caused water to overflow into the concrete floor of a mechanical room on an upper floor. From there, the flow of water entered an electrical conduit. The investigator was made quickly aware of the problem when inspecting the floor below. He observed that cold water was steadily leaking out of the bottom of a 480 volt electrical panel enclosure connected to that electrical conduit. The water shown on the floor in figure 5.27 is condensate from an AC unit which flowed through the electrical cabinet in the photo.

In the third example, water from condensate pans of through-wall AC units constantly overflowed in every room of a large, multi-story hotel. Condensate held in the indoor side of the units failed to exit through the back of their unit housings. This was because the central roof-mounted toilet exhaust fans generated enough suction that the pans could not drain outward as long as the exhaust fans were operating. The units were not equipped with trapped connections to drain lines, because the manufacturer assumed that the condensate *would* make its way through that casing to be lifted and sprayed against the condenser coil, located on the outboard side of the casing. Because of the suction caused by the exhaust fans on the roof, a steady stream of cold water dribbled out of the *indoor* end of all of the units, constantly.

Figure 5.28 shows the disassembly required to diagnose the problem. Note the condensate which forms a reflecting pool of water in the bottom of the AC unit casing.

As these examples illustrate, the areas around DX cooling equipment, it's refrigerant lines and condensate drain lines are fertile territory for seeking sources of excess moisture in nearby building materials.

Fig. 5.27
Condensate flowing through a 480v electrical enclosure
Failure to specify and install a trapped drain line on a condensate pan in the AC unit on the floor above this location led to water overflowing the pan. In this case, the path of least resistance led down and though a 480 Volt electrical cabinet.

Most building owners believe that engineered condensate drain systems are a better route for liquid water than through high-powered electrical equipment. Hopefully, their HVAC designers will agree, and ensure that all cooling equipment is connected (through traps) to condensate drain lines.

Fig. 5.28
Condensate which cannot drain
Without a trapped condensate drain line, suction from roof-mounted exhaust fans held condensate in the casings of these units throughout a hotel. The condensate overflowed the pans and soaked carpets and walls.

Fig. 5.29 Thermal imaging for moisture detection

The temperature differences seen by a thermal camera can be used to locate moisture problems, as shown here. Odors were present in the hotel, but the moisture source was not obvious until the camera found the leak source—cold water from a leaking condensate pan.

Thermal imaging to locate moist materials

Indoors, damp materials are constantly losing water by evaporation. Evaporation uses energy from the air, and from the material itself. This means the surface of moist material is very slightly cooler than the surrounding dry material.

A thermal camera will show these suspect cool surfaces as soon as they come into the camera's field of view. Identification of potential problem areas is nearly instant—compared to the hours and days needed to use moisture meters to measure moisture content throughout the entire building, in a grid pattern fine enough to locate a problem.

Figure 5.29 shows a thermal image which located the sources of a suspected moisture problem in a large hotel. The investigation was prompted by musty odors. Water was seeping out of several condensate drain pans, because no condensate line was ever connected to those units.

Fig. 5.30 Handheld thermal camera

Water damage restoration contractors and building investigators use these devices to locate suspect areas. But moisture meters are needed to confirm a moisture problem. There are many mechanisms other than evaporating moisture which can make surfaces cool in air conditioned buildings.[28]

Figure 5.30 shows the sort of thermal camera which has become so helpful in locating indoor moisture problems. While relatively costly, these devices have become widely used for building investigations of moisture problems, because they save so much time. Keep in mind, however, that thermal cameras only indicate surface temperatures. There are many other reasons other than moisture for surfaces to be cool inside air conditioned buildings. Meter readings are necessary to confirm the presence or absence of an elevated moisture content, and to quantify the degree of risk.[28]

Surface relative humidity as a risk assessment tool

Figure 5.31 shows an infrared surface temperature thermometer, also known as a spot radiometer or a non-contact thermometer. These are much less costly than thermal cameras. They are not nearly as capable, because they do not display a thermal image. They only display a single value, representing the average surface temperature within a defined circle on that surface. But surface temperature thermometers can still be very helpful in locating suspect areas in moisture investigations, because they can help locate areas where the surface relative humidity is high.

First, obtain the dew point of the indoor air by taking readings from a hand-held thermohygrometer. Then use a surface temperature thermometer to slowly scan the surfaces in the building. Look for the cooler surfaces and compare those surface temperatures to the air's dew point.

The surface rh can be plotted on a psychrometric chart, as shown by figure 5.31. First, locate the indoor air dew point temperature at the saturation curve. Next, locate the dry bulb temperature shown by the surface thermometer. Then draw a horizontal line from the dew point, stopping at the vertical line representing the surface temperature. Read the relative humidity at that point. That value indicates the approximate relative humidity *at the cool surface*.

If the surface rh is above 80%, the material is at risk for mold growth (if the material has been that damp or will stay that damp

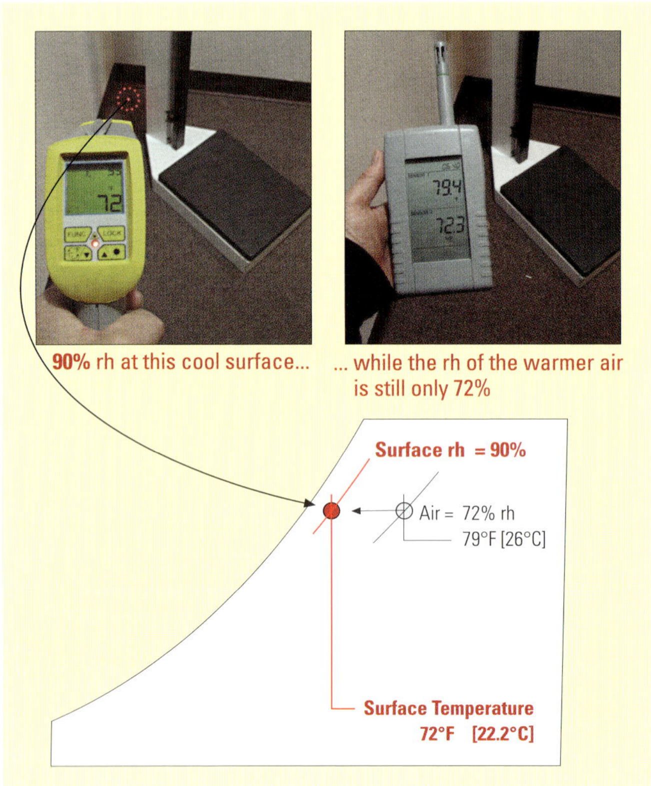

Fig. 5.31 Surface rh

Repeating the information supplied earlier, this graphic shows how to estimate the relative humidity at the surface of building assemblies—a more relevant risk factor than relative humidity in the air.

several weeks). On the other hand, if the surface rh is over 85 or 90%, the mold growth can come much sooner. So it's prudent to actually check the moisture content at that location. Moisture contents above 14% WME (wood moisture equivalent) are a long-term concern, as discussed earlier in this chapter. Moisture contents above 20% WME are a near-term concern, suggesting action to dry out the material right away, before mold can grow.

The next step - hidden surfaces are expensive to investigate

These techniques make the cheerful assumption that the investigator has access to the surfaces in question. In other words, no cabinets block line of sight to the wall, and no furniture is pushed up against it, and no problems are occurring inside the walls, nor above the ceilings. And no heavy equipment such as copiers, plasma TV screens, tiled floors, wall hangings, kitchen appliances, shower stalls or similar obstructions have to be relocated to provide visual access and access for measuring moisture content.

Unfortunately, easily visible and easily accessible problems are a small portion of mold problems in buildings. And when problems are easily visible, it often means that the hidden problems in that building are far worse. So the investigator who really intends to understand the extent of a problem, and to understand what must be done to fix it, will be explaining to the owner that:

- Much furniture will need to be moved or removed.
- Many holes will need to be drilled.
- Access openings for visual inspection will need to be cut into interior walls and ceilings, and sometimes floors and even the exterior cladding.

These are not facts that any owner wants to hear, much less accept. But they are the facts. Without visual access to hidden joints and seams, there is no way to be *certain* of the sources and pathways of the moisture which supports the microbial growth. And without certain understanding and remediation of the causes of moisture accumulation, growth is likely to reoccur.

Along with facing the uncomfortable facts, it's helpful to avoid wasting too much time with the misconceptions and half-truths which can get in the way of an effective solution.

Risky Misconceptions and Half-truths

Several building investigators and biologists have contributed to this list of commonly-heard, but misleading statements about mold. These are correct enough that they cannot be eliminated entirely from discussions—and yet they are incorrect enough that they have often led to ineffective or even counterproductive decisions by designers and owners.

"Mold won't grow until the rh rises above 70% "

This statement is false when it refers to rh measurements of air in the middle of a room, or air inside a duct, and indeed all measurements taken in air which is in any way separated from the immediate surface of the potential food source.

It is only correct when it refers to the air inside the microscopic crevices of the surface of that potential food source. Since measuring rh in micro-crevices on all surfaces is not practical with HVAC sensors, it's best to avoid using 70% rh as the threshold of concern. It creates the misimpression that when the building automation system or hand-held instruments report that the rh is less than 70%, the risk from mold is small. That's not true. There's a great deal of risk at that level. Here's why.

If the surface of any material is colder than the air—such as when cold air from the air conditioning system blows on a wall surface—the rh inside the crevices of the cold wall surface will be far higher than 70%. This fact is explained by figures 5.5 and 5.6.

So in place of the 70% rh threshold for concern, it's more productive for HVAC designers, architects, building owners and building operators to focus on keeping the indoor dew point below 55°F [12.8°C]—at least for building-related mold in hot and humid climates. That's a value which, when reported by the building automation system or by a handheld instrument, is more informative than "70% rh" about the risk of high *surface* rh and therefore about the potential for mold growth in hidden places.

"Paper-faced gypsum wall board is a high mold risk"

This statement is not correct... until one adds more qualifiers. Specifically, this statement is true: "*Wet* paper-faced gypsum board, *which has no anti-microbials in it* and *which does not dry out for days or weeks*, is a high mold risk."

Conventional paper-faced gypsum board provides an economically-attractive combination of fire resistance, sound attenuation, ease of finishing and durability even during prolonged periods of high humidity. And in its unfinished state, paper-faced gypsum board will dry very quickly. So it can retain its structural and other beneficial properties and resist mold growth even with small amounts of periodic wetting, or short periods of actual flooding. That's why nearly all buildings in the US and Canada use it for the interior of exterior walls, and for both sides of internal partitions. Several billions of square feet are installed every year, and the product performs admirably.

On the other hand, it is quite true that when paper without anti-microbials gets wet—and stays that way—it will grow mold rather well. And unfortunately, one type of mold that competes and grows well on saturated paper is the notorious *stachybotrys chartarum*, a fungus which will produce toxic defenses when threatened by bacteria or other fungi which are competing for the same food source.

Also, it is true that high humidity makes paper-faced gypsum board rather fragile from a mold perspective. High humidity puts the paper closer to the edge of a mold problem. Long periods of high surface humidity and/or intermittent condensation allow a nearly invisible layer of fungus to grow on the paper facing and backing. This thin layer "preconditions" the paper facing for a rapid increase in fungal growth as soon as more water becomes available. That's one reason why floods or rain leaks or after-hour spikes in humidity in buildings along the Gulf Coast seem to produce 'explosions" of fungal growth on paper-faced gypsum board within just a few days of the event.

And finally, using vinyl wall covering on the indoor surface of exterior walls made of paper-faced gypsum board is very commonly associated with lawsuits and mold growth. The vinyl traps moisture in both the gypsum and in the paper face, allowing mold to grow in the paper or in the adhesive for the wall covering.

But the fact is that paper-faced gypsum board is and will probably remain the preferred indoor surface for buildings in the US and Canada, and it is becoming much used more widely in the rest of the world as well.

A prudent response to this fact is to avoid using unprotected paper in parts of the building which one would reasonably expect will get damp. For example, bathrooms, kitchens, and laundry rooms will have occasional water spills. Also, when paper-faced gypsum board is used on the interior surface of exterior walls, it would be a very, very risky decision to cover that wall with any vinyl or vapor retardant paint. *Owners and interior designers are specifically warned against this practice, which has led to many agonizingly expensive lawsuits about mold in both commercial and residential buildings in hot and humid climates.*

Further, for all exterior walls, and in institutions where floor mopping and carpet cleaning are frequent, it is wise to install paper-faced gypsum board with a narrow air gap, to serve as a capillary break between the top of the finished floor and the bottom of the gypsum. With an air gap, the gypsum cannot soak up moisture from the carpet or wet tile. Figure 5.13 shows such a gap.

"When you smell musty odors, you need more outdoor air to improve the indoor air quality"

This statement is *not* correct, but it is very widely believed. This statement is often the reason that:

1. Ventilation systems are incorrectly assumed to be at fault, during the early stages of investigating an indoor air quality complaint, and that;
2. A typical response of building operators to musty odor complaints is to increase the volume of ventilation air, beyond the ability of the HVAC system to dry it, *making the problem immediately worse.*

It is certainly wise, when investigating an indoor air quality complaint, to make sure that the space where the complaints originate has an adequate amount of outdoor air. But recognize that adding more ventilation air might simply disguise a problem, and even make that problem much worse—*if the ventilation air is not dried.*

Musty odors indicate that excess moisture is accumulating so that bacteria and fungus can grow, or that humidity absorption is causing materials to react chemically, releasing volatile organic vapors. To solve a musty odor problem, find the moisture and fix the problem that led to the moisture accumulation.

The problem may indeed be inside the cooling coils or an air handler cabinet or the systems' air distribution duct work. Dirt plus water equals bacteria and fungi, which generate musty odors.

But in a hot and humid climate, adding more ventilation air than what is necessary to meet local codes and ASHRAE standards adds an extra and very unwelcome humidity load. And musty odors are usually an indicator that the system is already not removing the current humidity load. Adding a larger humidity load will not improve that situation. Removing *more of the humidity load* may well be part of the answer, but that will cost more money. So it's important to understand where the moisture is accumulating and why, before investing in any additional ventilation.

"To prevent mold, keep the AC system running even when the building is unoccupied"

This suggestion often *creates* mold growth rather than preventing it, and it certainly wastes a tremendous amount of energy. The more accurate and helpful suggestion is: "To prevent mold, *keep the indoor air dry,* even when the building is unoccupied."

Many HVAC designers and building owners assume that running the air cooling system will keep the building dry—but that's usually not the case.

Usually, the operators raise the thermostat set point when the building is unoccupied. But often the outdoor air dampers remain open. Unless the outdoor air dampers are shut, highly humid outdoor air can flood the building. Then, from time to time during unoccupied periods, the cooling system switch on and chill the building briefly—just long enough to create very high surface rh on any wall or ceiling washed by cold supply air. The cool surfaces then condense moisture from the humid air, creating a high risk of mold growth.

Schools, which have long unoccupied periods during nights, weekends and vacations are especially vulnerable to this mechanism of mold growth. This risk is illustrated by the photo in figure 5.22.[24]

Rather than "just running the AC system," keep in mind that the goal is to make sure the indoor air is dry in absolute terms (a low dew point). Often, an AC system can indeed be operated to achieve this goal. But it requires thought, and the specifics depend on the sort of system which is installed in the building.

Keeping the building dry after hours and during shut-downs is rather simple if the system is equipped with a dedicated ventilation air dehumidification system which has a return air connection. The operators simply set the dehumidification system to recirculate rather than to ventilate, and then control that system based on indoor air dew point. When the dew point in the building raises above 55°F (12.8°C], turn on the ventilation drying system to recirculate and dry the indoor air. When the dew point falls below 52°F [11°C], turn the dehumidification system off. Set the thermostat to a higher temperature to save energy, or don't operate the cooling system at all until the occupants return.

For commercial buildings without dedicated ventilation dehumidifiers, and which are not equipped with cooling equipment which has a dehumidification mode, the cooling system might still be able to keep the air dry. It takes more care and thought. The operators and/ or the control system will need to:

1. Close the outdoor air dampers, or reduce the amount of outdoor air ventilation to the absolute minimum required by local codes for unoccupied operation. This is the essential first step. If this is not done, or if it cannot be accomplished given the limitations of the system's air flow controls, then it is best not to operate the cooling system at all. Flooding the building with humid ventilation air will greatly raise the risk of mold.
2. Operate the cooling system in periods of at least an hour, continuously, without regard to temperature, or...
3. Operate the cooling system in response to a dew point signal rather than a thermostat. If the indoor dew point is above 55°F, run the system until the dew point falls below 52°F. [12.8°C and 11°C]. Then turn off the system.

One caution is appropriate for these cooling-based drying strategies. One must be careful not to overcool the space. If the walls and ceilings get too cool, humid air behind walls and above ceilings will condense, feeding mold growth.

For example, if two apartments share a wall and one is cooling down at night while the other stays hot, humid air from the uncooled apartment could condense in the common wall each night. The same problem can happen in a commercial building, where different parts of the building are often served by different systems. When in doubt, monitor surface temperatures vs. the indoor air dew point. If the air dew point is not at least 12°F [5.5°C] *below* the surface temperature, then the operational strategy needs to be adjusted, because the rh at the surface may be above 80%.

There are many types of cooling systems. So these three suggestions are just the beginning of a long list of alternatives. But the principle remains the same: keep the building dry during unoccupied periods. (Below a dew point of 55°F [12.8°C].)

Apartments and condominiums which serve as vacation homes often have long unoccupied periods. Mold can grow during these long owner absences. For vacation homes built with cooling units which do *not* have a dehumidification mode, one way to reduce risk is to place one portable dehumidifier inside the bathtub in the bathroom, and another unit inside the sink in the kitchen. Placing the dehumidifiers inside the bathtub and kitchen sink reduces any risk of condensate overflows. Then, connect a tube to each condensate collection tank, so the condensate flows out of the tanks and into the drains without the need to physically remove and empty the tank. Set the humidistat on the units so they switch on when the relative humidity rises above 50%. Then, the cooling system can be reset to a much higher temperature to save energy costs.

It will take energy to run the dehumidifiers, but not nearly as much energy as cooling the space as if it were occupied, in order to keep the humidity under control. And warmer temperatures reduce the potential for condensation, especially when the dehumidifiers are keeping the humidity low.

"In a hot and humid climate, all that code-required ventilation air causes mold"

This statement is false. It becomes true only when twelve critical words are added. Specifically: "In a hot and humid climate, all that code-required ventilation air causes mold *if it is not dried before it is supplied to the space.*"

Dry ventilation air does not cause mold; it helps prevent it. At the same time, it is quite true that *humid* ventilation air does indeed raise the risk of mold.

The unstated assumption behind this common misconception is that building owners simply will *not pay* to install drying equipment such as that shown in figure 5.14, or will not choose to invest in a cooling system which can also dry ventilation air.

It is quite true that this equipment costs more money to install than equipment not designed to handle ventilation air. But 100 years of ASHRAE experience and millions of dollars of research all over the world strongly suggest that without adequate ventilation, buildings and occupants both have problems. That's why building codes require ventilation air. The appropriate response to the mold risk inherent in ventilation air in a hot and humid climate is to dry that air before it is delivered to the space. If that is not deemed affordable, then the owner and designers must accept the mold risk inherent in humid ventilation air.

To mitigate the risk of humid ventilation air, one can reduce the volume of that air to the code-mandated minimum with a two-stage or variable volume ventilation system. A two-stage system can switch between occupied and unoccupied ventilation air volumes based on occupancy schedules programmed into a building automation system, or based on room occupancy sensors.

For an even closer match between actual occupancy and air volume, CO_2 sensors and dampers can varies the volume of ventilation air. Indoor CO_2 concentration is a useful way to quantify human occupancy. As the indoor CO_2 concentration rises, more ventilation air is allowed into the building. As the concentration falls, the ventilation air volume is reduced.

Such two-position dampers or variable-volume dampers and sensors do add cost and complexity to the system design. But benefits balance those costs. Minimizing the ventilation air flow and drying the air are the best ways to reduce the mold risk while also reducing operating costs.

Again, it's not the ventilation air that causes mold—it's the *amount of humidity* brought into the building which can cause mold. So dry out the ventilation air, and minimize its flow rate.

"All you need is air movement and light to prevent mold"

This statement is accurate, but incomplete, and therefore not useful. It creates the false impression that mold will not grow if the lights in the building are on, and if fans are circulating air. The completely correct statement would be: "If you have enough dry air movement and energy to dry the surface, mold will not grow."

If enough infrared energy (sunlight, usually) is falling directly on a surface, chances are that it will provide enough heat for some of the moisture at the surface to dry out—as long as there is *also* enough dry air flowing across the surface to carry away the moisture.

But it's not true that mold will stop growing when fluorescent lights are switched on while humid ventilation air is brought into the space. In fact, bringing more moisture to a cool surface by flowing humid air across that surface will increase the mold growth rate rather than reducing it.

The key is the moisture content of the food source. More moisture allows more mold. Less moisture allows less mold.

To the extent that adding heat from lights and removing moisture with dry air movement reduces moisture content of the food source, the statement is true. But if the light source adds little or no heat, or if the particular air used to create air movement adds more moisture to the material, the statement is false.

"To prevent mold in basements and crawl spaces, ventilate them with ventilation openings, or with fans."

In hot and humid climates, this advice is only effective for buildings which have no air conditioning. In air conditioned buildings, this advice has been responsible for mold growth problems.[29]

In buildings with mechanical cooling, the surfaces of the structure which form the ceiling of the crawl space or basement are relatively cold. So when humid outdoor air is used to ventilate the crawl space below the ground floor, condensation forms on the cold ceiling of that crawl space, making it vulnerable to attack by mold.

Fans make the problem even worse, because they bring in humid air continuously—far more humidity than would drift in through intermittent wind pressure differences. More humid air means more condensation on any cool surfaces, which means more mold.

To prevent mold in basements and crawl spaces:

1. Keep moisture from getting out of the ground and into the air which fills the basement or crawl space. This requires a durable, well-sealed layer of vapor barrier material to cover any exposed earth or below-grade walls. (Note: Adding gravel without a vapor barrier to cover the earth increases the humidity load rather than reducing it. The gravel provides a greater surface area from which moisture can evaporate.)
2. Keep water away from the foundation. This means the roof must be equipped with gutters with downspouts which drain water more than 3 ft. [one meter] away from the foundation. Also, the basement foundation or area around the exterior of a crawl space must be well drained to ensure that surface water does not flow into the sealed basement or crawlspace.
3. *After* these steps have been taken, mold risk can be further reduced by placing a dehumidifier in the crawl space or basement. Adding a dehumidifier before the moisture contribution of the floor and walls have been minimized wastes both money and energy.

"Unless you have a good vapor barrier, in a hot and humid climate you can expect mold"

The entire subject of vapor barriers in buildings is historically very confusing, and building scientists have found in recent years that much of what building codes say about vapor barriers creates more problems than it solves.

In hot and humid climates, interior vapor barriers and vapor retarders often lead to trapped water, promoting mold rather than preventing it. *Water and air barriers at the exterior* are important for all buildings. But *vapor* barriers are only important on the outside, behind brick or other masonry veneer, to protect wood-based or gypsum-based sheathing.

In a hot and humid climate, brick veneer soaks up rain water. Then when the sun comes out, it heats up the saturated brick. The humidity skyrockets in the air gap between the brick veneer and the sheathing. Because humidity is so very high in that air gap after every rain, it's important to install a continuous vapor barrier over the sheathing, to protect it from vapor permeation in addition to protecting it from water and air penetration. Usually, the vapor barrier can act as the water barrier and air barrier as well, avoiding the need for more than one layer over the sheathing. (Note the use of the term vapor barrier rather than vapor retarder. For purposes of this discussion, we will define a barrier as a continuous layer with a perm rating less than 1.) Just like the need to install a vapor barrier to keep water and vapor out of concrete foundation slabs, vapor barriers are the appropriate choice for protecting sheathing behind brick veneer in a hot and humid climate.

Turning now to other types of exterior walls, vapor barriers and vapor retarders have been a problem, for several reasons.

First, they are often installed in the wrong layer of the exterior wall. Much of the literature on exterior walls and vapor retarders was developed for buildings in cold climates. So designers and contractors who are familiar with cold-climate problems, or those who read literature which *assumes* the reader will know the advice only applies to cold climates, will sometimes install a polyethylene sheet just behind the interior gypsum, over the insulation. This ensures that any water which drips into the wall cavity from outdoors will be trapped and grow mold, because the water leaks and condensation inside the walls (and on the vapor retarder itself) cannot dry out to the interior of the building.

Old habits die hard, new designers always need to be trained, and cool-climate designers and contractors sometimes do work in warmer parts of the world. So the too-general advice to "make sure there's a vapor barrier in the exterior wall" often results in an interior vapor barrier which traps water, rather than an *exterior water barrier,* which is what's needed in a humid climate to avoid mold.

Finally, focusing on a vapor retarder gives an architectural designer a false sense of security. Except for brick veneer, vapor drive is not the real problem. The real problems are infiltration by humid air, and water infiltration through exterior cladding joints which have no flashing, or ineffective flashing.

It's easy—but not helpful—to simply specify that "the contractor shall install a continuous vapor retarder in all exterior walls using good means and methods." It's very difficult—but far more effective—to detail all flashing in layers, in three dimensions, showing the sequence of assembly of each layer, especially in the corners. That's what it really takes to prevent mold problems. Discussing vapor barrier perm ratings and exact locations obscures the real issues and these extra layers often lead to problems no matter where the barrier is located. Focus on the flashing instead.

"The reason we have mold in buildings these days is because we've made them so air-tight in order to save energy"

This statement is false is two important respects, but it has a grain of truth under some circumstances. So the misimpression it creates cannot be corrected by a single simple statement.

First, commercial buildings built recently in the US are—with very few exceptions—not tighter than buildings built 30 years ago. In fact, new buildings tend to leak more air than older buildings. Field measurements performed by the National Institute of Standards and Technology have shown passive air exchange rates of 1 to 2.5 complete air changes per hour even in well-constructed, large-budget governmental and institutional buildings.[30] The popular impression is that buildings have been built tighter in recent years. But the reality—in the US at least—is quite the contrary. Commercial buildings still leak a *great deal* of air.

Next, this statement creates the misimpression that saving energy leads to mold growth. It does not. Excess moisture is what leads to mold growth. And excess moisture is caused by many factors, most commonly rain leaks through construction joints combined with humid air infiltration into exterior walls. If rain and humid outdoor air could be kept out of the exterior walls (if buildings were tighter), owners would both save energy *and* reduce their mold risk.

On the other hand, mold will definitely be a problem if a building allows rain to leak in, and if that water becomes trapped in the walls by extra vapor barriers. It is also true that if a leaky building were positively pressurized with dry air, it might flow more freely outward through the exterior walls, helping to dry out the rain leaks. It is also true that such a strategy would waste a great deal of energy. So in some ways, it's true that reducing energy waste can indeed lead to mold—in poorly-constructed buildings which trap rainwater with extra vapor barriers.

But in a hot and humid climate, buildings which leak air are at a greater risk from condensation on cool indoor surfaces than they are from reduced drying in exterior walls.

In nearly all buildings, humid air infiltration is constant, while rain leaks are intermittent. In air conditioned buildings in hot and humid climates, it's better to avoid air leakage by keeping the building as air-tight as possible. This saves energy while reducing mold risk.

To avoid the mold risk which comes from trapped water, don't rely on sloppy, air-leaky construction which wastes energy. Instead, follow the suggestions outlined earlier in this chapter:

1. Don't let water into the wall. This will require effective flashing around all wall penetrations such as windows, doors, balconies, AC unit sleeves, lighting fixtures and around wall penetrations for plumbing, water and refrigerant piping, electrical conduits and outdoor electrical plugs. There are a lot of penetrations in an exterior wall. They all must be flashed, effectively.
2. When water gets in anyway, drain it out again, quickly. This will require sill pans under windows, and an air gap between cladding and sheathing, with insect-protected drains at the bottom and insect-protected air vents at the top.
3. Place a continuous water barrier (which is also an air barrier, and is often called a 'drainage plane") on the outdoor surface of the sheathing. This prevents both water and humid air from penetrating the sheathing and getting into parts of the exterior wall which are vulnerable to mold growth and corrosion—the sheathing itself, the framing and the interior gypsum board.
4. Don't place polyethylene sheeting (or any other extra vapor retarder layer) on the inside layers of the exterior wall. Instead, rely on the *exterior water barrier* which protects the sheathing to exclude liquid water and humid air.
5. Don't use vinyl wall covering on the inside surface of exterior walls in an air conditioned building in a hot and humid climate. Make sure the *interior* surfaces of exterior walls are *highly permeable* to water vapor. Specifically, ensure a perm rating of higher than 10 for that indoor surface. A perm rating above 15 is even better.

References

1. Lawrence Spielvogel, P.E., L.G. Spielvogel, Inc., Valley Forge, PA. Personal communication, June, 2007.
2. National Institute of Medicine *Damp Indoor Spaces and Health* 2004. Electronic files available at no cost: http://books.nap.edu/catalog.php?record_id=11011

   Printed and bound copies: National Academies Press, Washington, DC ISBN 0-309-09193-4
3. Mudari, David and Fisk, William J.; "Public health and economic impact of dampness and mold." *Indoor Air,* June 2007. Volume 17, Issue 3. pp 226-235. Journal of the International Society of Indoor Air Quality and Climate, Blackwell Publishing, www.blackwellpublishing.com
4. Proceedings of the First International Conference on Avoiding Bugs, Mold & Rot 1991. Building Enclosure Technology and Environment Council of the National Institute of Building Science (BETEC), Washington, DC. www.nibs.org

5. Harriman, Lewis G, III and Thurston, Steven, *Mold in Hotels and Motels—Survey Results.* 1991. American Hotel & Lodging Association. Washington, DC.

6. Odom, J. David, III., DuBose, George and Fairey, Phillip. "Moisture problems: Why HVAC commissioning procedures don't work in humid climates." 1992. Proceedings of the 8th Symposium on improving building design in hot & humid climates. May, 1992. Texas A&M University, College Station, TX.

7. "Mold Risk Management" Presentation by David Dybdahl, American Risk Management Resources Network, Middleton, WI. at the M4 Conference - June, 2003. Building Enclosure Technology and Environment Council of the National Institute of Building Science (BETEC), Washington, DC. www.nibs.org

8. Flannigan, Brian and Miller, J. David. "Microbial Growth in Indoor Environments." *Chapter 21, Microorganisms in the Home and Indoor Work Environment.* 2001. Taylor & Francis, 29 West 35th St. New York, NY ISBN 0-415-26800-1

9. *Report Annex 14 - Condensation and Energy.* "Volume 2 - Guidelines and Practice." March 1991. International Energy Agency. Report coordinated and produced by Prof. Hugo Hens, Director of the Laboratory for Building Physics, Catholic University of Leuven. Leuven, Belgium.

10. Criterium Engineers, Portland, ME. www.CriteriumEngineers.com

11. Straube, John and Burnett, Eric. *Building Science for Building Enclosures* 2005. Building Science Press, Westford, MA www.BuildingSciencepress.com. ISBN 0-9755127-4-9

12. John Straube, Ph.D, P.Eng. Director, Building Envelope Engineering Program, Dept. of Civil Engineering, University of Waterloo, Waterloo, Personal communication, June 2006.

13. Lstiburek, Joseph "Built wrong from the start - 10 blunders that rot your house, waste your money and make you sick" *Fine Homebuilding*, April-May 2004. pp. 52-57 Taunton Publishing, www.FineHomebuilding.com

14. *Construction Quality Survey,* 2003. Criterium Engineers, Portland, ME. www.CriteriumEngineers.com

15. Rose, William B., *Water in Buildings - An Architect's Guide to Moisture and Mold.* 2005. John Wiley & Sons, Hoboken, NJ. www.wiley.com/architectureanddesign ISBN 0-471-46850-9

16. Leslie, Neil "Laboratory evaluation of residential window installation methods in stucco wall assemblies." 2007 ASHRAE Transactions, Vol. 113, pt 1. DA-07-032

17. Lstiburek, Joseph. "Why stucco walls got wet-Designs, methods, codes and workmanship all played a role in Florida's soggy storm experience." 2005. *Journal of Light Construction*. July, 2005 Hanley-Wood Publishing, Williston, VT. www.JLCOnline.com

18. Gatley, Donald P., *Mold and condensation behind vinyl wall covering*. 1990. Gatley & Associates, Atlanta, GA.

19. Shakun, Wallace. "A review of water migration at selected Florida hotel/motel sites." Proceedings of the biennial symposium on improving building practices in hot & humid climates. October 1990. Texas A&M University, College Station, TX.

20. Harriman, Lewis. G. III, G. Brundrett and R. Kittler. *ASHRAE Humidity Control Design Guide for Commercial and Institutional Buildings.* 2001/2006 ISBN 1-883413-98-2 ASHRAE, Atlanta, GA. www.ashrae.org

21. Cummings, James B., Withers, C. R. Withers, N. Moyer et al. 1996. *Uncontrolled air flow in non-residential buildings.* Final report. FSEC-CR-878-96. April 15th, 1996. Florida Solar Energy Center, Cocoa, FL

22. Wray, Craig. *Energy impacts of leakage in thermal distribution systems*. 2006. Report to the California Energy Commission. Lawrence Berkeley National Laboratory. Berkeley, CA. http://epb.lbl.gov/ Report no: PIER II #500-98-026

23. Harriman, Lewis G. III., Schnell, Donald and Fowler, Mark; "Preventing mold by keeping new construction dry." *ASHRAE Journal,* September 2002. pp 28-34

24. McMillan, Hugh and Block, Jim. "Lesson in curing mold problems." *ASHRAE Journal,* May 2005. pp 32-37

25. *Wood Handbook* 1999. U.S. Department of Agriculture, Forest Products Laboratory, Madison, WI Available online without charge at http://www.srs.fs.usda.gov/index.htm

    Printed version published by Algrove Publishing, Almonte, ONT www.algrove.com ISBN 1-894572-54-8

26. Lstiburek, Joseph. "The material view of mold." *ASHRAE Journal,* August 2007 pp 61-64

27. Halyard, Paul, P.E., Fellow, ASHRAE Personal communication, July, 2007.

28. Harriman, Lewis G., III, "A visual moisture detection method. Using infrared imaging to locate moisture in buildings." *HPAC Engineering*, December, 2004. Penton Publishing, Cleveland, OH. www.pentonpublishing.com

29. Dastur, Cyrus; Davis, Bruce and Warren, Bill. *Closed Crawl Spaces - An Introduction to Design, Construction and Performance.* A report to the U. S. Department of Energy, published by Advanced Energy, Raleigh, NC Available online at no cost at www.crawlspaces.org/

30. Persily, Andrew "Myths about building envelopes." *ASHRAE Journal,* March, 1999. pp 39-47

Image Credits

5.1 Rodney Lewis, P.E. Fellow, ASHRAE. Rodney Lewis & Associates, Houston, TX

5.9 E.I Dupont de Nemours, Richmond, VA and SureSill, Ltd., Austin, TX.

5.10 Joseph Lstiburek, P.Eng, Ph.D, Fellow, ASHRAE. *Water Management Guide.* ©2005 Building Science Corporation, Westford, MA Reprinted with permission.

5.11 Joseph Lstiburek "Water-managed wall systems." *Journal of Light Construction*, March, 2003. Reprinted with permission.

5.12 Courtesy of Neil Moyer and Joseph Lstiburek

5.15 Terry Brennan, Camroden Associates, Westmoreland, NY

5.15 Rodney Lewis

5.17 Lew Harriman, Joe Lstiburek and Reinhold Kittler. "Improving humidity control for commercial buildings." *ASHRAE Journal,* November, 2000

5.18 David Hales, WSU Extension Energy Program, Spokane, WA.

5.20 Munters Corporation, Amesbury, MA

5.21 Paul Halyard, P.E., Fellow, ASHRAE. Property Condition Assessment, Inc., Orlando, FL.

5.23 E-Z Trap. Inc. Edison, NJ

5.24 Mason-Grant Consulting, www.masongrant.com

5.25 Rodney Lewis

5.26 Paul Halyard

5.27 Paul Halyard

5.28 Paul Halyard

5.24 Mason-Grant Consulting

Chapter 6

Improving Sustainability

By Lew Harriman

Fig. 6.1 Improving sustainability means that buildings must last a long time

This photo shows what happens when buildings are not sustainable. They get bulldozed and carted off to the landfill. In contrast, sustainable buildings last a long time—centuries rather than decades. And sustainable buildings don't use much energy, which improves the likelihood that the needs of future generations can be met as well as the needs of our current generation.

Key Points

The word "sustainable" suggests a building which has been designed to minimize its impact on the natural environment. And paraphrasing the definition of sustainability adopted by ASHRAE, a sustainable building should "...help its current owners and occupants meet their needs, while not interfering with the ability of future generations to meet their own needs."

These general principles establish the mood and foundation for a great deal of further thought. However, they are not really specific enough to use in everyday decision making. To apply sustainable principles in practice, here are some directly actionable suggestions for owners, occupants and design practitioners:

- Don't build in a flood zone. A sustainable building is one which lasts for a long time, and in a flood zone—it won't. If you must build in a flood zone or in a coastal area subject to tropical storm surges, lift the occupied portions of the building up—high enough off the ground level to minimize the building's impact on the natural environment, and high enough to limit the damage from frequent flood waters.
- Design the exterior enclosure so that the building's shape and its roof overhangs keep rain water off the exterior walls. And design the windows and doors so they keep rain water and solar heat from getting into the building. Otherwise, the HVAC system will be needlessly large, complex and expensive, Also, the building will use far too much energy, and rain water leaks will rot the structure and the building's furnishings.
- Make the building's structure and its interior finishes mostly of inorganic materials like concrete and ceramic tile. These can resist rot and the corrosive effects of sunlight, humidity and rain water for *many centuries*—not just a few years or decades.
- Design the HVAC systems so they are air-tight, and so they smoothly and thoroughly reduce their energy consumption when solar loads are low, and when parts of the building are vacant or lightly occupied.
- Design the HVAC systems so that all their components can be easily accessed for frequent adjustments and maintenance. Otherwise, their components won't be adjusted and maintained, which means the building will waste energy and make the occupants feel uncomfortable.
- Design the building's financial accounting system so that monthly financial reports will highlight and encourage (rather than obscure and prevent) effective maintenance and economical operation of mechanical systems. Also, arrange the compensation of the building manager so that it goes up when the building's energy consumption (not its maintenance budget) goes down—provided that the occupants are thermally content at the same time.

Advancing Beyond Theory To Practice

The word "sustainability" has become so burdened with political implications in its current cultural context that it's difficult to know exactly how to apply its principles to everyday decision making. There are probably many ways to meet the needs of earth's current inhabitants without interfering with the needs of future inhabitants. But one clear path for moving beyond abstract principles to everyday practice is illuminated by two words which have the same meaning as the verb sustain: namely, the verbs "endure" and "maintain."

One senior building scientist, who has long been known for his holistic and economically practical approach to building problems once said: "I'm not sure of *everything* we need to do to make a sustainable world... But I know we'll have to make our buildings last for a lo-o-o-ng time. And then, because our buildings *will* last for a long time, we'll need to make sure they don't use much energy."[1]

This perspective bridges the gap between theory and practice, because it provides a basis for quantification. Longevity and energy consumption can measure at least two important aspects of sustainability. And measurements help lift the concept of sustainability out of the confusion of cultural politics and apply it to real-world decisions, using the power and clarity of science and engineering.

So when a sustainable building is the goal, begin by considering whether this or that alternative will make the building last longer, and which choices will help the building use less energy.

Chapter 6 is an index to sustainability decisions

In part, this book exists because Terry Townsend, P.E, a former ASHRAE President, was concerned about buildings built recently in hot and humid climates which have rotted quickly and which waste a great deal of energy. Consequently, the implicit goal of *every* chapter in this book is to help avoid those problems and improve the sustainability of buildings. So, *this* chapter is principally an index to the suggestions provided in other chapters. There would be no point to duplicating the detailed guidance provided on nearby pages.

More Durable = More Sustainable

Making buildings last longer is largely a matter of reducing the forces which will eventually destroy them. This can be done by avoiding those forces in the first place, through development decisions, and by architectural and HVAC design decisions which minimize the amount of water and humidity which get into the building.

Don't build in flood zones and swamps

The baseline sustainability of any building—its location and its function—is established long before the designers, contractors and operators make any decisions which improve on that baseline.

If a building is built in a swamp on the coastline of a tropical ocean, that building will require a great deal of structure to resist periodic tropical storms. And it may need fresh water pumped to the site from far away; it will need long paved roads which traverse the swamp; and it will be difficult to get rid of human-generated waste without great expense or a great impact on the local environment.

So the location of the building is perhaps the most important decision in determining the cost of making the building last for a long time, and the cost of designing it so that it won't use much energy over that long lifetime. The developer decides the building's location. That decision is guided by economics and guided in some cases by local regulations. But ultimately the location decision is guided by

Fig. 6.2 Unsustainable location

These photos show buildings located on the Bolivar Peninsula on the Gulf Coast of the U.S, just east of Houston, TX. In 2008, Hurricane Ike wiped out the buildings, most of which had been built recently enough to fall under the requirements of the Texas hurricane code. The yellow arrows provide visual reference points between the two photos taken before and after Hurricane Ike.

These photos illustrate the fact that building codes do not guarantee sustainability. The buildings may or may not have been "green"... but their location on the coast, in the path of frequent hurricanes, made them unsustainable.

the developer's preferences and by his or her understanding of the technical issues which govern the building's cost.

Detailed suggestions for site location and functional planning are beyond the scope of this book. But these critical decisions are mentioned here to alert the building owner and its developer to the fact that it will cost much more money to design and build a *sustainable* building if the development decisions impose baseline costs which do not exist at other sites. (See figure 6.2 for an example of what happens when theoretically code-compliant buildings are built in not-very-sustainable locations.)

This is why, when planning a *sustainable* building, the developer and owner would be wise to invest in early advice from architectural and HVAC designers. These professionals are more likely than others to have the technical understanding needed for cost-informed location decisions for buildings which will last for a long time while using very little energy.

Both developers and designers could profit by reading Chapter 10 (Lessons Learned From Tropical Storms). That chapter describes the experiences of building scientists who were asked to assess the adequacy of the Florida building and energy codes in resisting hurricanes. Not surprisingly, it costs more to build durable, low-energy buildings in coastal areas. The structures built near the beaches must be different and quite a bit more robust if they are to endure (if they are to be sustainable). Chapter 10 provides some useful specifics for designing in flood zones when the developer decides that's where the building will be built.

Enclosure design which keeps out water and humidity

After tropical storms and earthquakes, water is the most destructive force acting on buildings in hot and humid climates. Rain and high humidity won't destroy a building immediately. But within just a few years, water leaks and internal condensation can damage a building so badly that repairing the damage costs more than the original construction budget.

So to help the building endure, and therefore to be sustainable over a long time, it's wise to design the roof so it overhangs the walls. Roof overhangs keep most of the rainwater off the walls. And when there's less water flowing down those walls, there's a smaller probability of water leaks, and a smaller amount of water damage when the leaks eventually occur. Figure 6.3 shows the benefit of a roof overhang over time.

Chapter 7 (The Perfect Wall), provides the basic introduction to exterior wall designs which exclude most water, and which can manage any water which happens to leak in despite the designer's careful attention. Chapter 8 (Keeping Water Out Of The Building), provides more details, and more reasons why the general principles outlined in Chapter 7 lead to durable buildings. Chapter 8 will be most useful to the architectural designer, and to the forensic building investigator or building owner who must deal with shortcomings of existing buildings.

Similar logic applies to humid ventilation air. Bring in only what you need for the actual occupancy of the building, so the dehumidification load on the building is reduced to its minimum. And when you can pre-dry the ventilation air before it gets into the rest of the system, you've further reduced the risk of internal condensation and subsequent water damage. With less humidity inside the building, there's a higher probability that the building and its furnishings and finishes will last longer, ie; they will be more sustainable. Chapter 3 (Managing Ventilation Air), provides an overview the ventilation issues for owners, building managers and occupants. Chapter 16 (Designing Ventilation Air Systems), provides the details the designer will need.

Materials & construction which tolerate frequent wetting

No matter how carefully the architectural designer, the HVAC designer and the builder exclude moisture and humidity, there will always be some leakage over the life of the building. Inorganic materials like concrete, ceramics and glass will resist water and humidity longer

Sustained for more than 2,000 years
Roof overhangs have protected the structure and the sculptural detail on this Roman institutional building for 2,000 years—and counting. Overhangs keep rain from hitting and flowing down the walls. They greatly improve the durability—and therefore the sustainability—of any building.

Sustained for more than 200 years
Roof overhangs have protected the cladding and the sculptural detail on this North American Colonial building for 200 years—and counting. The material is wood, not stone. Still, the overhangs have protected the exterior walls for more than 200 years. The photo provides compelling evidence of the durability (sustainability) improvement made possible by roof overhangs.

Rebuilt after fewer than 20 years
No roof overhangs on this nursing home. The architect's fashion preferences did not favor roof overhangs and gutters. Roof overhangs would have reduced the wall's rain contact volume by more than 50%. Less water ***contact*** would have reduced the leakage ***volume***, improving the sustainability of its thermally excellent (but moisture-fragile) exterior insulation and finish system.

Fig. 6.3 The benefit of an overhanging roof
To help a building be sustainable (to help it last a long time), bring the roof out over the walls. That way, there's much less rain hitting the exterior walls, so there's much less water that can leak around windows or through construction joints. As one building scientist has famously observed: "If it doesn't get wet—it can't leak." [2]

than will paper or other processed wood products. But at the same time, concrete and masonry block will collect and retain water, which then migrates to other, more moisture-sensitive materials in the building. So all materials and all concrete and masonry must be dried before the building is closed in, as described in Chapter 17 (Avoiding Mold By Keeping New Construction Dry).

For guidance to designers, Chapter 5 (Avoiding Bugs, Mold & Rot) explains the processes through which mold and rot grow in a building. That chapter will help owners, interior designers and architects make better decisions about materials and construction details which endure over time.

Less Energy = More Sustainable

The less energy the building uses, the less it will cost to operate. Cost-effective buildings are likely to endure longer than buildings which are not cost-effective. The baseline energy consumption of the building is established by design decisions—in particular, how much solar heat will come through the glass, and how much humid outdoor air can infiltrate into the building through construction joints and wall penetrations.

Enclosure design which keeps out heat and humidity

The lower the cooling and dehumidification loads, the less energy the HVAC systems should use. This is largely a matter of designing and constructing the building's enclosure to keep out solar heat, and to make it tight enough to limit the amount of infiltrating humid outdoor air. Those subjects are covered in Chapter 9 (Keeping Heat Out Of The Building). In that chapter, you'll read why and how early glass decisions are critical to overall sustainability.

HVAC design which keeps out heat and humidity

After the owner and architectural designer have settled on an exterior enclosure which limits the cooling and dehumidification loads, the HVAC designer will need to contribute by making sure the HVAC systems don't leak air.

When duct connections and air plenums leak, field studies show that they will *pull in* hot and humid outdoor air through the building enclosure and *push out* some of their cold supply air. The combined leakage raises the annual HVAC-related energy consumption by more than 25%. No amount of "high-efficiency" cooling equipment will compensate for that enormous energy waste, which continues for the life of the building. This subject is covered in Chapter 14 (Designing Cooling Systems) and in Chapter 16 (Air-Tight HVAC Systems).

HVAC design which matches energy to occupancy

Traditionally, the way to design an HVAC system for the lowest possible installed cost is to buy one *big* cooling unit, then circulate a large and constant volume of air through many different spaces, shutting off the cooling when the average returning air temperature from all of the spaces satisfies one thermostat. This approach saves money in the construction budget. But it wastes a tremendous amount of energy, and it does a poor job of controlling temperature, controlling humidity and providing adequate ventilation air as occupancy and cooling loads change in all those different spaces.

The more sustainable approach is to vary the amount of cooling and ventilation air sent to each space, in strict proportion to the constantly-changing cooling loads and occupancies. This approach requires more equipment, and therefore more money in the construction budget. It also demands a more deliberate approach to cooling and ventilation design. But the energy and comfort benefits provided by smoothly modulating both capacity and energy use are substantial.

Modulating ventilation air systems is discussed in both Chapter 3 (Managing Ventilation Air) and in Chapter 16 (Designing Ventilation Air Systems). Modulating dehumidification systems is described in Chapter 13 (Designing Dehumidification Systems), and modulating cooling systems is described in Chapter 15 (Designing Cooling Systems).

More Maintainable = More Sustainable

The basic meaning of sustain is the same as maintain, ie: "to last a long time, to endure." If the systems are not maintained well, and if they are not operated by knowledgeable personnel who are engaged, motivated and adequately funded, then the building will waste a great deal of energy. And obviously, when equipment is not well maintained, it will fail earlier than necessary, which drives up costs and wastes the resources needed to replace it.

Accounting allows—or prevents—sustainability

"You get what you measure" is an enduring truth. When the building management is evaluated based on total operational cost, the quickest way to improve that metric will be to get rid of expensive people, and replace them with less expensive people, or with nobody at all, or only with an emergency-response service contract.

The cost havoc of those alternatives may never be obvious, unless one also *measures and displays the hour-by-hour energy cost of operating the building*, and then adds the amortized cost of periodic replacement of equipment which has failed prematurely, plus the cost of emergency system failure response and the per-event cost of responding to frequent occupant comfort complaints.

Cost accounting is beyond the scope of this book, so no chapter addresses this critical aspect of sustainability. But here, in this chapter, we mention the importance of cost accounting because so often, engineers are told that sustainability in design is critical—then they see the building *operated* by people who are desperately trying their best to keep the systems operating with no resources and little understanding. Many buildings have operations departments staffed by far too few people and a maintenance budget that renders them impotent. Often, these hard-working people have skill levels that are barely adequate for recognizing HVAC components, let alone for optimizing the systems' energy consumption and responsiveness for changing loads in different spaces.

"You get what you measure." If you measure monthly staff count and staff costs, you'll probably end up with very few people, and the ones you have will be cheap. If instead you measure energy costs and comfort, you'll probably end up with a building which uses very little energy and at the same time keeps more people comfortable for more of the time. In other words, you'll have a more sustainable building. The owners' accounting preferences are therefore critical to the sustainability of their buildings.

Budget for constant commissioning—then do it

All HVAC systems need constant attention and readjustment, in order to optimize comfort and minimize energy use (and therefore minimize energy cost).

Complex systems go out of adjustment because they have a large number of components, sensors and controllers, plus overlapping computerized building automation computers—each requiring understanding of different programming protocols. Simple HVAC systems go out of adjustment because they do *not* have the components, sensors and controllers necessary to automatically adapt to widely-changing loads in all the different occupied spaces at the same time. So no matter how simple or complex the HVAC systems may be, they will need constant attention and adjustment if the goals are minimum energy use and maximum comfort.

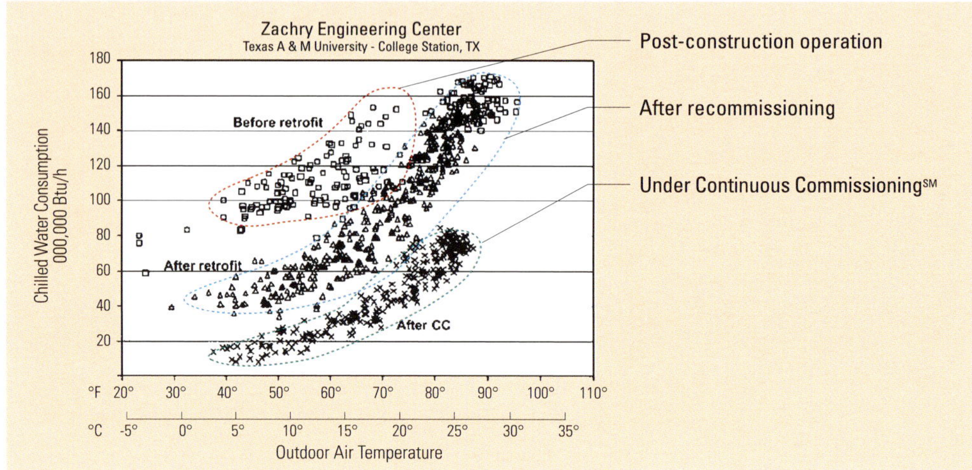

Fig. 6.4 Constant commissioning, for constant improvements[3]

The most compelling savings come when key aspects of consumption are measured continuously, and the systems are changed and adjusted in real time. Note the reduction in chilled water consumption between the recommissioned system and the system when operated under Continuous Comissioning.℠

On the cost side of the ledger, this means that staff must be well-trained and well-motivated. These are expensive people. They must have the skills to make the necessary adjustments. They must also have both diagnostic skills and a clear understanding of the complex interactions between system components. The operations staff also needs instruments and sensors to help them understand what must be adjusted when, and by how much. And finally, they must have authority to act; to employ their skills and make wise judgements as they operate the systems moment-to-moment.

But on the benefit side of the ledger, having these sensors, controls and capable people in place means the systems will cost much less to operate, the occupants will be more comfortable and the indoor air quality will be very high compared to other buildings. In short, the buildings and their systems will last longer and therefore be more sustainable. Figure 6.4 shows an example of the energy reduction which happens when capable operators are allowed to make constant adjustments—the need for which has been made apparent by sensors and controls which were well-placed, well-understood and well-maintained.

Chapter 4 (Reducing Energy Consumption), contains guidance for implementing the principles of constant commissioning, along with examples of the energy reductions which have been achieved by constant attention and system adjustment by capable operating staff.

Access, access, access

Finally, a suggestion which will need attention and action from the developer, owner and architectural designer, in addition to the HVAC designer. Namely: provide enough access to system components to let the maintenance staff service the equipment, adjust sensors and controls; and change the filters.

On the face of it, this suggestion seems needlessly elementary and obvious. But in the real world, lack of access is one of the most complex problems which impede sustainability. Lack of access leads to major problems with energy waste and poor comfort. And lack of access also begins a chain of events which often ends in poor indoor air quality and mold problems. Here are the links in *that* chain:

- Because outdoor air filters cannot be easily accessed and changed, the outdoor air filters clog, which reduces makeup air flow. This makes...
- The building become "negative," because the exhaust from the building exceeds the makeup air volume...
- Which means the systems' fans will pull humid outdoor air into the building *not* through the ventilation unit, but instead through building walls and construction joints so that...
- The humid outdoor air condenses its moisture into cool building cavities behind walls and above ceilings. Then...
- The absorbed moisture provides the solvent that the mold needs to dissolve the nutrients in building materials. Then mold can absorb the liquified food and grow.

All this because the outdoor air filters could not be accessed and changed, as they need to be. Chapter 5 (Avoiding Bugs, Rot & Mold) describes this and other design-related mold problems.

But the larger and often very economically complex issue is *how* to provide that access. The forces restricting access begin at the development stage, and they remain powerful all the way through to final construction documents. It's a matter of the cost of floor space and ceiling height.

In the horizontal dimension, a sustainable building must provide enough floor space for access to components, in addition to the space consumed by piping, duct work and utility connections to the equipment. So adding that last office or that last classroom or last apartment is often the step that shrinks and guarantees problems in the mechanical rooms. The mechanical rooms might still fit the equipment—but they no longer allow non-heroic access for troubleshooting, normal service and adjustment.

Similarly, in the vertical dimension there's great economic pressure to keep the floor-to-floor height as short as possible. Shorter buildings need less structural steel, and shorter floor-to-floor heights mean that more floors can be built in the same overall height limit. One extra floor can make the difference between economic viability and unbuilt dream, especially in dense urban areas where the real estate is very costly and where the maximum height of the building may be limited by law. Again, what's not usually understood by the developer, owner and architectural designer (and sometimes not understood by the HVAC designer), is that if there's not enough space above the ceiling for service access, then the systems' controls cannot be adjusted. Both energy consumption and comfort will suffer accordingly.

Conversely, if the space above the ceiling is *so high* that the components and HVAC controls cannot be accessed without special equipment instead of a simple stepladder, then the components probably will not get the maintenance attention they need to provide comfort at minimum energy cost.

For example, in one high-rise federal courthouse the ceilings of the courtroom floors are so magisterially high that even roll-around platforms are not tall enough for maintenance technicians to reach the variable air volume boxes and their controls.[4] When these need checking and adjustment, the third-party maintenance contractor must locate and rent a scissor-lift which is small enough to fit into the freight elevator and short enough to turn the 90° corners in the narrow service access corridors. Safety regulations require the presence of two people when operating that lift. So to make even one single adjustment safely, the time and the expense are truly astonishing. Consequently, important adjustments don't get made. So the federal judges, attorneys and juries are frequently uncomfortable. They are mystified at why, in a building designed with a sharp focus on sustainability, they must remain so uncomfortably cold. At the same time, the system which provides this year-round discomfort uses far more energy than predicted by its designers.

These problems resulted from a lack of understanding of the nature of service access during those few critical moments when the owner and the architectural designer decided on the floor plan, and on the height of the ceiling in the corridor. Both the aesthetic decision about corridor ceiling height and the budget decisions about floor space combined to prevent mounting the mechanical components low enough so they could be accessed for troubleshooting and adjustment from a stepladder. In theory, the operating staff should be able to overcome such seemingly small obstacles. In reality they can't, because their operating budget does not allow it.

In summary, while it is expensive and difficult to provide enough access for non-heroic access to mechanical system components, it is essential for adequate comfort, adequate indoor air quality and low energy consumption. Therefore maintenance access is essential to the sustainability of the building. This fact is recognized by ASHRAE in Standard 62.1, which requires designers to provide adequate access for service and adjustment. But the HVAC designer cannot comply until adequate space is provided by the owner and architectural designer. Details are explained in Chapter 3 (Managing Ventilation Air), and also in Chapter 16 (Designing Ventilation Air Systems).

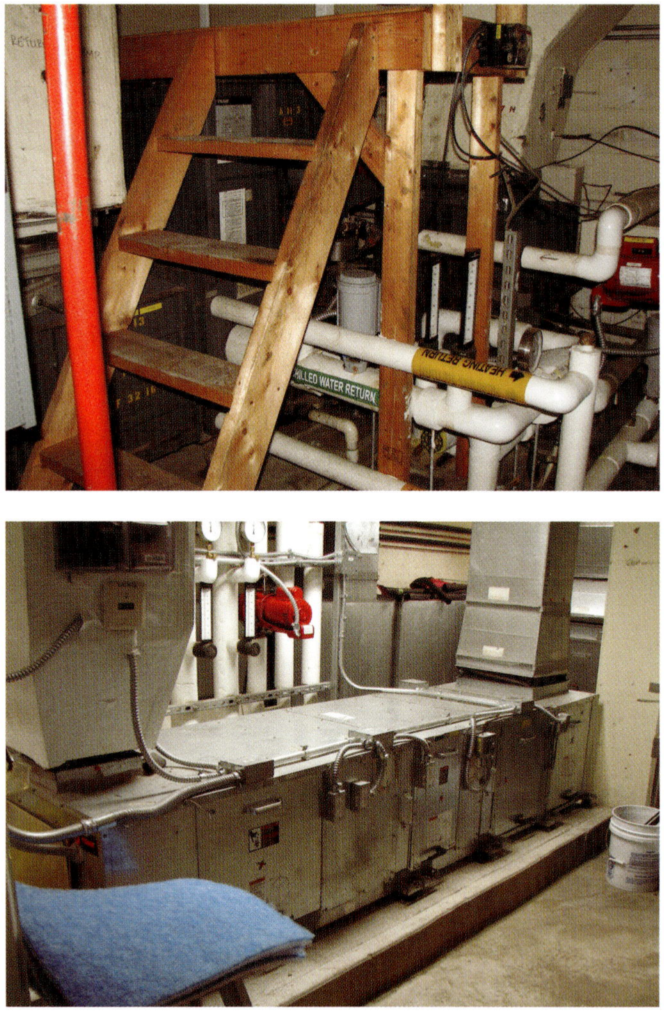

Fig. 6.5 Maintenance access

If maintenance and system adjustments require heroics from the operations staff, they won't happen often. Absurdly obstructed access in grossly-undersized mechanical rooms, like the system in the upper photo, pretty much guarantees the system will go out of adjustment, leading to constant discomfort and needless energy waste.

The lower photo shows a mechanical room large enough to fit the equipment, its duct work and utility connections—and also the service technicians, their tools and supplies. No ladder is required for either inspection or adjustments. Given this adequate space, the HVAC designer was able to locate the gauges and controls at eye level, in plain view. No comfort complaints and no energy waste from this system.

Summary

Improving sustainability is a complex undertaking, especially since many of the decisions which determine environmental impact are beyond the control of architectural and HVAC designers. On the other hand, technical professionals can make a good beginning on the problem. We can advise our clients about the costs inherent in designing a sustainable building in "sustainability-challenged" locations. And we can help them choose architectural and HVAC design features which will improve the durability and reduce the energy use of the building over its long lifetime.

References

1. Terry Brennan, Camroden Associates, Westmoreland, NY www.camroden.com
2. John Straube, Ph.D, PEng. University of Waterloo, ONT, Canada
3. Liu, Minsheng; Claridge, David; Turner, W. Dan. *Continuous Commissioning*^{SM} *Guidebook*. October, 2002. Federal Energy Management Program, U.S. Department of Energy. (http://www1.eere.energy.gov/femp/operations_maintenance/om_ccguide.html)
4. Alfred Arraj Federal Courthouse, Denver, CO - Circumstances demonstrated and explained to the author by maintenance and operations personnel during tours of the building and its mechanical systems - 2004 and 2006.

Image Credits

Fig. 6.2 - Adapted from U.S. Geological Survey images, as displayed by National Geographic Online

Fig. 6.3 - Mason-Grant Consulting, Portsmouth, NH, www.masongrant.com

Fig. 6.4 - Adapted data provided by the Energy Systems Laboratory of Texas A&M University, College Station, TX

Fig. 6.5 - Mason-Grant Consulting, Portsmouth, NH, www.masongrant.com

Chapter 7

Elements of a Perfect Wall

By Joseph Lstiburek

Editor's Note:

Information in this chapter was originally published as the Building Science column of the May 2007 edition of the *ASHRAE Journal*. It represents an edited overview of some of Dr. Lstiburek's extensive advice about walls in hot and humid climates. The chapter describes the issues the architectural designer should be concerned about, and then explains an efficient thought pattern for dealing with those issues during design. We trust the reader will understand that this 5-page chapter cannot describe all aspects of perfect walls, much less how to ensure perfection for all of the possible wall types which are common in hot and humid climates.

Fig. 7.1 Walls must exclude and endure the weather

In hot and humid climates, the architectural designer should expect frequent heavy rain, strong winds and constant sun. Under such loads, some exterior wall designs do much better than others, as shown here. Understanding a few key principles will help the designer create walls which come closer to perfection.

Editor's Introduction

In hot & humid climates, the most energy-efficient and durable buildings would be those which are neither cooled, nor dehumidified, nor filled with filtered air.

But as soon as the owner decides the building will be air conditioned, everything changes. Walls have to become more than simply structurally sound. They must exclude and quickly drain away any rain water they collect. And they must keep the conditioned air in, keep the heat out and avoid condensation on their chilled surfaces. Then, when walls get wet or condense moisture in spite of everybody's best efforts, they must dry out quickly. If exterior walls in air conditioned buildings in hot and humid climates don't do all of these things, the walls might rot, rust, grow mold and collapse.

This chapter is an informally-written overview from a building science perspective, helping owners and architectural designers navigate the complex decisions of implementing the wall system, no matter which type of wall is chosen for the building.

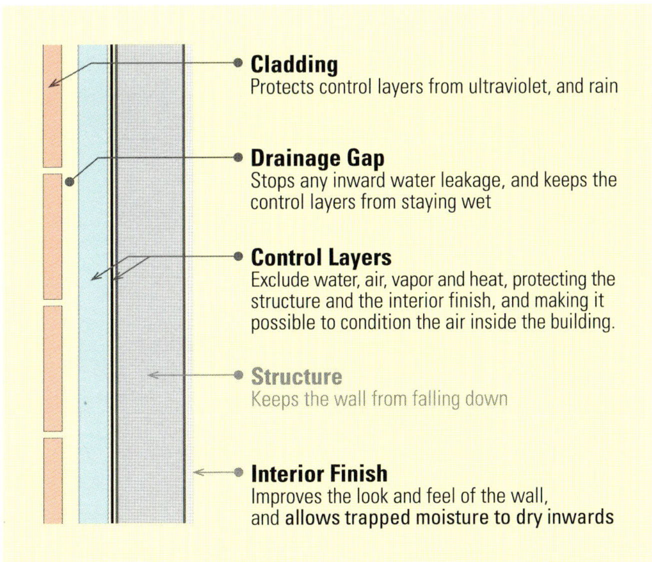

Fig. 7.2 The control layers

In concept, an excellent wall has a rainwater control layer, an air control layer and a vapor control layer—all under the cladding, but all directly on the exterior of the structure. The cladding's functions include shedding rain, but it's principal purpose is to protect the control layers from ultraviolet radiation.

Towards a Perfect Wall

The exterior wall is an environmental separator—it keeps the outside *out* and the inside *in*. To do this, the wall assembly must exclude rain, air, vapor and heat. In the old days, we had one material to do this: rocks. We would pile up a bunch of rocks and have them do it all. But over time, rocks lost their appeal. They were heavy and often fell down. Heavy means expensive, and falling down is annoying. So, construction evolved. Today, walls need four principal control layers, especially when we don't build out of rocks. They are presented in order of importance:

1. Rain control layer
2. Air control layer
3. Vapor control layer
4. Thermal control layer

Note well this order of importance. If you can't keep the rain out, don't waste your time on the air. If you can't keep the air out, don't waste your time on the vapor, etc.

To protect the structure, the best place for these control layers is on the outside of the structure, as shown in figure 7.2. When we built out of rocks, the rocks didn't need much protection. Now we build out of steel and wood, and we need to protect them both. And, since most of the bad stuff comes from outside, the best place to limit that bad stuff is on the *outside* of the structure—before it can get inside.

Also, after generations of building out of rocks, folks got the idea that they wanted to be comfortable and figured out that rocks were not the best insulation. Of course, rocks are not that bad compared to windows. (Note to owners and the designers in the architectural profession: you can't build an energy-efficient green building out of glass. But it does win design awards. That's life: sometimes, you have to choose between design awards and energy efficiency. See chapter 2, figure 2.6 for the consequences.) Back to rocks. They are heavy and you need many layers of rock to make the wall have any decent thermal resistance. So we invented thermal insulation.

But, where to put that insulation? If we put the insulation on the *inside* of the structure, the insulation does not protect the structure from heat and cold. Expansion, contraction, corrosion, decay, ultraviolet radiation, and almost all bad things get worse with higher temperatures. So, all the control layers need to be on the *outside*. Keep the structure from going through temperature extremes, protect it from water in its various forms and protect it from ultraviolet radiation, and life is good.

What about air control? Well, outdoor air carries a lot of water vapor, and water is bad for the structure. So, we must keep humid air out of the structure. Or, if some outdoor air gets into the wall, we must make sure that this air does not get cold enough to drop its water on its way through the wall.[1]

Now, one other item—and this is important: namely, if you intend on living in the building or working in the building or keeping things safe in the building, you'll want to control the interior environment. We especially ought to be concerned about what is *in* the interior air, because the occupants will breathe it.

Here's the basic fact about air conditioning: we can't control the air until we enclose it. So, we need an airtight enclosure if we are to provide low-cost and effective cooling, dehumidification, ventilation and filtration. And, once again, the best place to limit air and vapor infiltration is on the outside of the structure, so that outdoor air—and especially the solar-heated, highly-saturated air in that drainage gap behind the cladding—does not enter and condense inside the cool wall.

Presto! Now we have the elements of a perfect wall: cladding to exclude the rain, a water drainage gap behind that cladding, a thermal control layer next, over an air control layer and vapor control layer, which are all placed directly on the structure. (Figure 7.2).

This was all figured out long ago.[2,3,4] But in the building design business we forget the hard lessons, often in less than a single generation. Experienced people die, or get bored or forgetful, and others loose their jobs during recessions. Also, the publishing world does not let books persist as long as the principles of building physics. The great classics go out of print and become difficult to obtain. So there is always a need for updated presentations of the durable truths.[4,5]

The layers are the same for roofs and foundations

In a beautiful bit of elegance and symmetry, if you lay down the layers of an excellent wall you get a great slab-on-grade, and then when you flip them the other way, you get a great roof (Figure 7.3). The physics of walls, roofs and slabs are pretty much the same.[6]

Notice in the roof assembly, the critical control layer (or mem-

Fig. 7.3 The perfect roof & slab
Flip the wall on its side, and you have either an excellent roof or an excellent slab. The physics of walls, roofs and slabs are very similar.

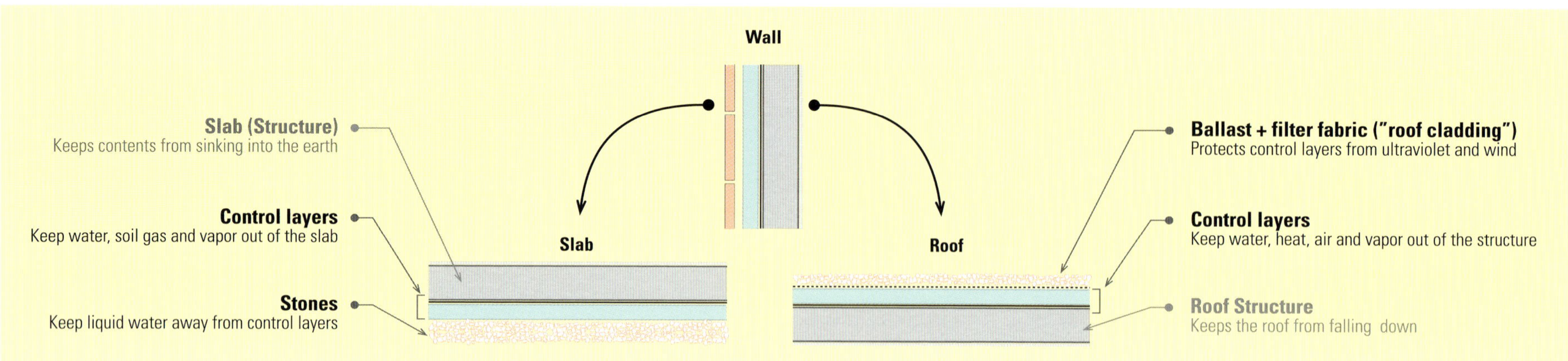

Fig. 7.4 The roof-wall connection

Notice the control layer for rain on the roof is connected to the control layer for rain on the walls. And the control layer for air is connected to the control layer for air on the wall, and so on. Max Baker pointed this out decades ago.[6]

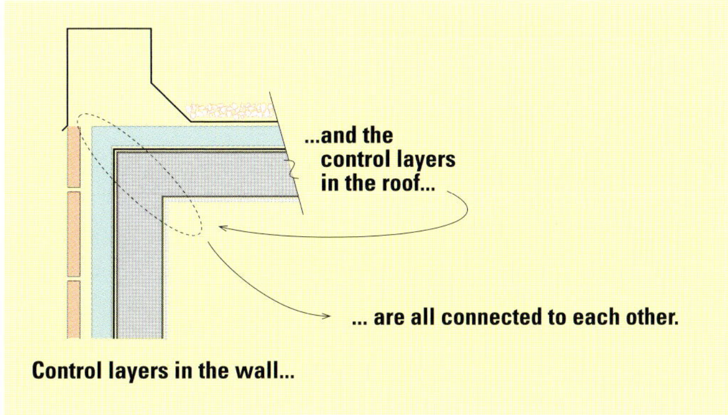

brane) for rainwater control, air control and vapor control is located *under* the thermal insulation layer and under the stone ballast (which could be called the "roof cladding"). Under those ballast and insulation layers, the water, air and vapor control layers are protected from the principal damaging forces of direct rain, solar heat, ultraviolet radiation and the occasional heavy-footed maintenance technician. Why we often expose this most critical control layer—out at the very top of the roof where it can be trashed by these powerful forces—never fails to amaze me. Yes, the control layer is easier to replace when it is located out in the open. But that's an answer which assumes frequent short-term failures, disposable components and unlimited resources.

Wall and roof layers must connect, without cracks & gaps

Most problems in building enclosures occur where roofs meet walls. The classic roof-wall intersection is presented in Figure 7.4 (with both credit and apologies to Max Baker[6]—I took some liberties). Notice that the control layer for rain on the roof is connected to the control layer for rain on the wall. And the control layer for air on the roof is connected to the control layer for air on the wall, and so on. Beautiful. And when it is not, it's ugly.

Translating basic principles into real walls

Time to put some meat on the bones of Figure 7.2. How should an excellent conceptual wall actually be built? The answer depends on your budget, how long you want the building to last, whether your structure is made of steel, concrete or wood, and especially—what can be installed well by local contractors. Here are three excellent choices.

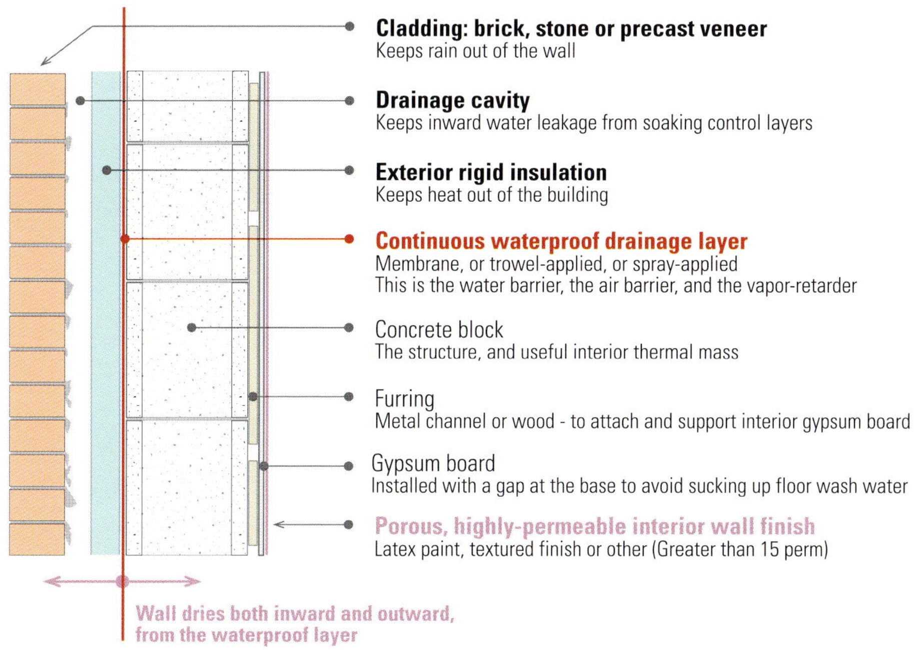

Fig. 7.5 Institutional wall

It works everywhere, in every climate zone. It costs more, but then, it's sustainable. It will pass from generation to generation.

A high-durability wall is shown as Figure 7.5. This is a special wall. It is expensive, but it is often worth it. It is the type of wall that you use for sustainable buildings—those that are passed down from one generation to the next: museums, art galleries, courthouses, libraries and the like. These are institutional buildings, and I call this wall the "institutional wall."

The wall in figure 7.5 is sweet, because it works equally well in any climate zone. The only thing that is changed because of climate is the effectiveness of its thermal insulation. The larger the average annual temperature difference between indoors and outdoors, the more effective the insulation should be. My advice here is simple: whatever you think the right effectiveness of thermal insulation should be, double it. If you love your kids and grandkids, don't argue the point.

The second wall (Figure 7.6) is a "meat and potatoes" wall for commercial buildings. It's a wall any commercial building can use, and has the essential characteristics that our infrastructure should depend on. So, it's no surprise that I call it, yes, you guessed it: the "commercial wall." It does have a thermally conductive structure—metal studs. So, all of the insulation should be located *outboard* of that metal. It is a thermodynamic obscenity to insulate only the cavities inside a conductive structural frame. Again, you can build this wall anywhere, in any climate location. Just consider the insulation levels (see previous advice, particularly the part about loving your kids).

The third wall (figure 7.7) is a residential wall. Notice that the insulation is not entirely on the outside. These structural cavities are also insulated. That's because residential construction often uses wood; a relatively nonconductive structural material. Wood does not conduct very much heat, which is why we do not use wood frying pans, but we do use steel ones.

For this residential wall to work nearly everywhere on earth, we would split the thermal resistance of the insulation at least 50:50 between the *exterior* of the structural frame and the cavities *within* the structural frame. So in an R-20 wall—at least R-10 or more is on the

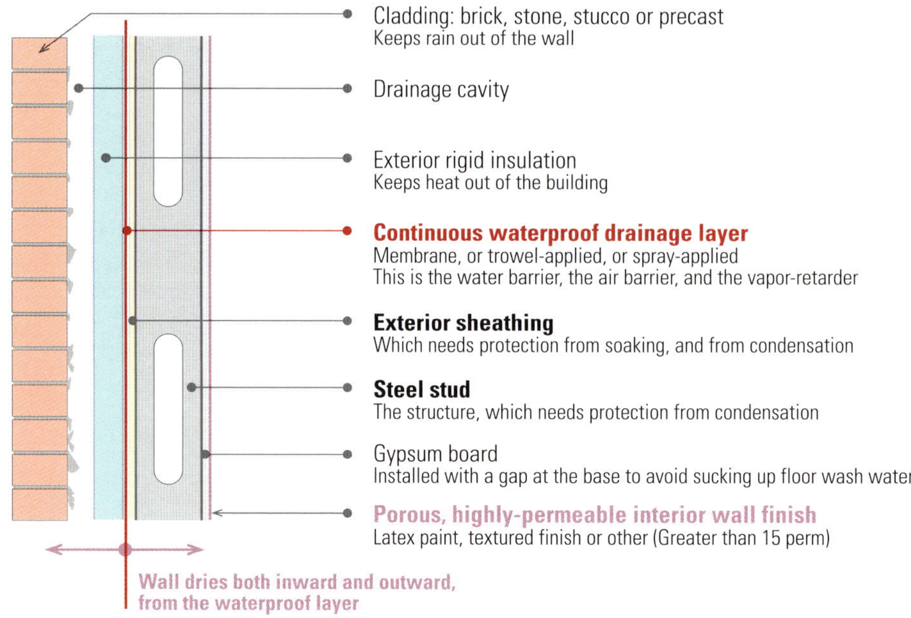

Fig. 7.6 Commercial wall

The almost-best wall we construct today. It's affordable and it works in any climate zone.

outside of that nonconductive structural frame. (Super-cold climates call for more insulation—but that's also a different book.)

Please notice one essential feature of all of these walls. *No vapor barrier is used on the indoor face of the wall.* Repeat after me, *no* vapor barrier is used on the indoor face of the wall. The control layers are all inside the wall. We want the whole assembly to *dry to the indoors* from those control layers—and to *dry to the outdoors* from those control layers. Always. Everywhere.

Avoiding interior-side vapor barriers is especially important in hot and humid climates, because those interior wall surfaces are cooled. High humidity inside the exterior wall will condense, as it moves inward and reaches any vapor barrier which blocks free transmission through that cold indoor surface. For decades, interior decorators have persisted in a nasty habit of ignoring this advice and using vinyl wall coverings on the indoor faces of exterior walls. This habit insults physics—never a good idea—and has been a misery for owners and occupants. Although, to be fair, the addiction to impermeable wall coverings has stimulated the global economy through lots of business for cleaners, doctors, forensic engineers, mold remediators and lawyers.

So what's the bottom line from unpleasant experiences and from building science? We need to have: a water drainage gap outboard of a continuous drainage plane, a structurally sound and continuous air barrier, a thermal layer (insulation) with no major conductive penetrations, and a vapor control layer located to allow drying in both directions when wetting events happen. All of this is outside of the structure. Those are the elements of a perfect wall. We need them now more than ever, because we want sustainable buildings—those which don't use much energy and which last a long time.

Fig. 7.7 Residential wall

One of best residential walls we construct today. It's not cheap, but it works everywhere—even in cold climates, where more insulation inside the structural frame is called for.

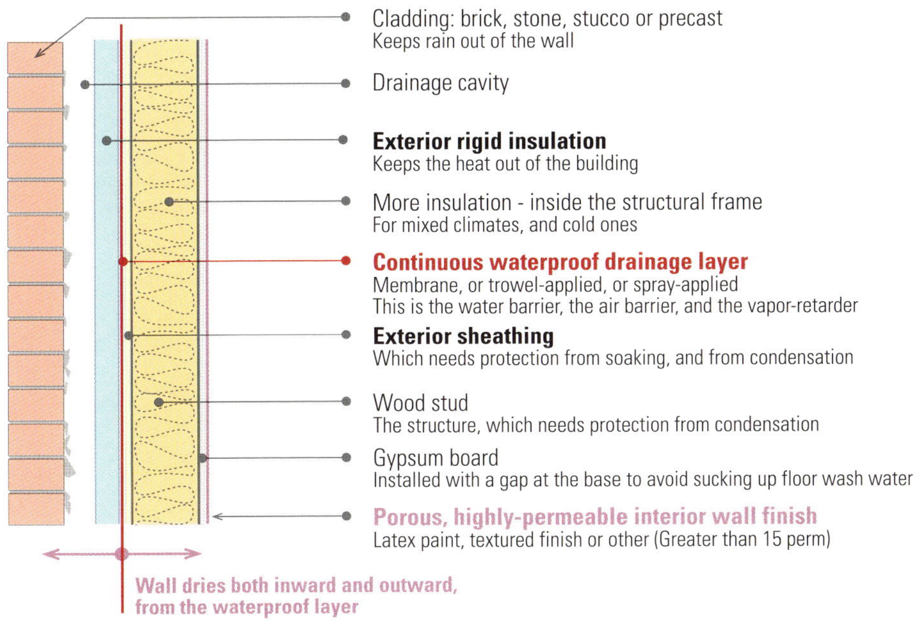

References

1. Harriman, L.G, Lstiburek, J, and Kittler, R. "Improving humidity control for commercial buildings" *ASHRAE Journal*, November, 2000. pp.24-32. www.ashrae.org
2. Hutcheon, N.B. 1964. "CBD-50 principles applied to a masonry wall" *Canadian Building Digest (2)*. National Research Council Canada. Ottawa, Canada
3. Hutcheon, N.B. and G.O. Handegord. 1983. *Building Science for a Cold Climate*. National Research Council of Canada. Ottawa, Canada
4. Straube, John and Burnett, Eric. *Building Science for Building Enclosures* 2005. Building Science Press, Westford, MA www. BuildingSciencepress.com. ISBN 0-9755127-4-9
5. Rose, William B, *Water in buildings; An Architect's guide to moisture and mold*. 2005. John Wiley & Sons, Hoboken, NJ. ISBN 0-471-46850-9
6. Baker, M. 1980. *Roofs*. Montreal: Multi-Science Publications.

Chapter 8

Keeping Water Out Of The Building

By Joseph Lstiburek & Lew Harriman

Members of the Board of Design Consultants appointed to plan the construction of UN permanent headquarters on Manhattan's East River site. New York 18 April 1947

Foreground, left to right: Liang Su-cheng, China; Oscar Niemeyer, Brazil ; Nikolai D. Bassov, USSR ; and Ernest Cormier, Canada. In second row, from left to right: Sven Markelius, Sweden; Charles E. Le Corbusier, France; Vladimir Bodiansky, France, engineer consultant to Director; Wallace K. Harrison, chief architect, USA; G.A. Soilleux, Australia; ***Max Abramovitz,*** *USA, Director of Planning; and consultants Ernest Weismann, Yugoslavia; Anthony C. Antoniades, Greece, and Matthew Nowicki, Poland.*

Fig. 8.1 Keeping water out of buildings... it's not easy, even for experts

"Actually, I am very concerned that the science of building is going to disappear. I wonder if you realize how very few men are left today who are expert in building science. They are very rare and they are passed around among the large offices. You have to dig them out of their holes and revive them. One of them in our office is 80 years old. He passed out the other day and we had to pump stuff into him to get him going again because we couldn't spare him. It sounds like a joke, but we also have one who gets drunk every third day, but we can't fire him."

"One would think we would know whether we can build a marble wall that will not crack and let water in. That sounds simple. After all, they've been doing it for three thousand years. Well, right now we're having a hot argument about it on the United Nations Building. We can't find anyone who will say: "I am sure it can be done this way", or "I am sure it cannot be done." We've asked old builders who have repaired the columns in St. Patrick's Cathedral."

"I am sure many other architects are doing the same thing and that all of us are probably repeating each other's mistakes. If one of us finds the answer, the rest won't know about it. Yet, even if you've created a fine piece of architecture, it's a terrific black mark against your reputation when a simple thing like a leak occurs."

Max Abramovitz, AIA, 1949 (As quoted by William Rose [1])

Key Points

The suggestions in this chapter are based on painful experiences that have led building investigators to a better understanding of moisture problems which are driven by architectural design and construction practices. The suggestions are also based on the cross-disciplinary profession of building science, which views engineering and architecture as a unified whole. Building scientists deal with the fact that buildings are subject to universal physical laws. These are not always conveniently constrained by the technical and commercial boundaries preferred by the legal community and by professional organizations.

So, with great respect for the architectural and construction professions which are the most concerned with preventing bulk water intrusion, this chapter provides some suggestions for consideration during design and budget discussions with the owner:

- Overhang the roof to help keep rain off the building.
- Provide sill pans under all windows and doors.
- Flash all windows, service penetrations and wall joints, especially where different cladding systems come together.
- Provide an integrated waterproof drainage plane to block intrusion and guide water leakage back out of the wall.
- Crawl spaces must be dry, water-tight and not vented.
- Drain the roof and the site in ways which keep water away from the foundation.

Background

The ASHRAE community has a certain level of discomfort when moving outside of the narrow confines of the design, installation and operation of mechanical systems. HVAC professionals hesitate to participate in discussions about the design and construction of the building enclosure. But there has really been little choice in the matter, because of the intimate relationship between water leakage, indoor air quality, energy consumption and humidity control.

Problems of poor indoor air quality, thermal discomfort, excessive HVAC costs and high energy consumption are heavily influenced by moisture accumulation inside buildings. Some of those problems arise because of high indoor dew points, or because of interactions between the HVAC systems and the building enclosure. But often, the really major problems are accelerated by risky practices in architectural design and construction. Excessive amounts of rain water get into the building, soaking insulation, corroding and rotting the structure and supporting the growth of mold and bacteria.

The suggestions described here are not the only ways to keep water out of the building. Each building and each combination of glazing and cladding will have its own critical details. But this basic outline can be helpful during design conferences, when owners, architectural designers and contractors make the key decisions which lead to more water leakage, or less of it.

Roof Overhangs Come First

Barring component failures, the amount of water that ends up inside the building will depend heavily on how much water flows down its walls. Roof overhangs keep a very large percentage of the rain water off those walls, and therefore out of the building. As one building scientist has often observed: "If it doesn't get wet... it can't leak."[2]

The owner and the architectural designer will make this crucial decision very early—at the conceptual design stage of a new building. After the roof overhang decision is made, all the other water-related decisions become either more forgiving and less expensive to implement—or more risky and more expensive to implement.

Figure 8.2 shows the reduction in water-related problems in buildings in a rainy climate, based on the length of their roof overhangs. The wider the overhang, the greater is the reduction in the number of water-related problems.[3]

Interestingly, it does not take a very wide overhang to accomplish this improvement. Even very short overhangs make a big reduction in the amount of water that ends up on the walls during a rainstorm. The actual improvement depends on many factors, including the height of the building, its exposure to wind, the average velocity of wind during rainstorms at that location, its exact geometry on the face which sees the prevailing wind during rainstorms and many other factors. Chapter 12 of Reference 8.2 helps a technical professional make an informed estimate of rain deposition. But a rough approximation is that for many buildings, a roof overhang of 2 ft. or more [600 mm] will probably reduce the net annual rain load on the walls by about 50%.

This can be said another way. When the owner and architectural designer decide *not* to overhang the roof, they should expect about twice the amount of rain water to flow down that building's walls. That doubling in water load will challenge the joints, the flashing and any other cracks and penetrations in the wall—every year, for the life of the building.

Tall buildings especially benefit from roof overhangs

Perhaps unexpectedly, roof overhangs are especially useful for taller buildings. Tall buildings collect more rain compared to short buildings. Also, because they are bigger, they usually have many more joints which can leak water. In particular, more windows equals more joints and therefore more potential leak points.

The taller the building, the greater its annual rain load. That's because the rain usually arrives with wind. Rather than falling straight down from above, most rain whips past the building at an angle, carried by winds. And wind speed, therefore the total rain exposure, increases with distance from the ground.

Near the ground, the turbulence created by trees, bushes and nearby buildings slows down the wind, compared to higher up, where the wind is not obstructed. Since the total amount of rain falling through the air is distributed fairly evenly, the rain load per unit time (gallons or liters hitting the building per hour) depends on the speed of the wind. Higher wind speeds mean that more air will be blowing against the building every hour. Since all the air contains about the same amount of rain, more wind flowing past the building (higher wind speeds) means that more rain water will be flowing by at the top of the building compared to its base.

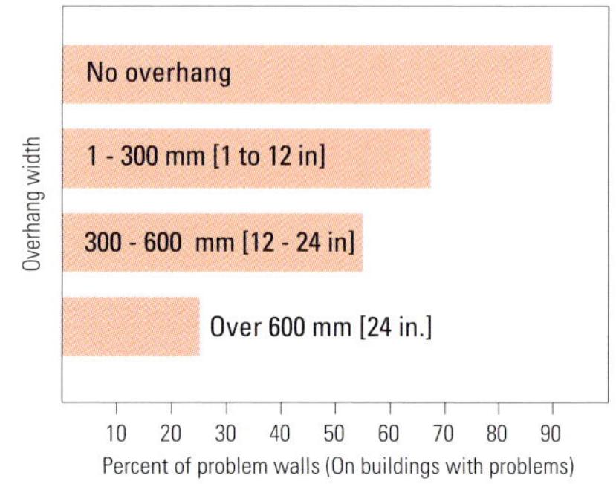

Fig. 8.2 Roof overhang benefits

A field study of water-damaged buildings showed that walls which were not protected by overhanging roofs were the most likely to have been damaged by rain. A wider roof overhang correlated with a far lower percentage of water-damaged walls.[3]

Fig. 8.3 Rain on short building... with and without roof overhang

Note the impressive reduction in water load, from even a very narrow roof overhang.

Fig. 8.4 Rain on tall building.. with no roof overhang

The maximum rain load is deposited at the top and upper edges of the building. That's where the rain-laden wind flow is greater, and where the wind makes sharp turns. The momentum of the rain drops carries them onto the building, while the wind flows around and past it.

This relationship also helps explain why a taller building will collect more rainwater than a shorter building which has the same amount of floor area. The taller building has more of its surface area located high up in the air, exposed to higher wind speeds. Therefore it collects more water during rain storms than the shorter building.

Now consider *where* that rain ends up as it hits the building. As the rain-laden wind hits the top of the building, the wind makes a sharp turn upwards and over the roof line. Most of the rain droplets, however, are too heavy to make that same 90° turn. So the rain drops out of the wind, hitting and sticking to the wall near the roof line. The same thing happens as the rain-laden wind flows around the side of the building. The rain droplets can't make sharp turns. They are carried forward by their momentum, hitting and soaking the corners of the building. Again, the wetting tends to be greatest near the top of the building, where the wind speed is higher and therefore the flow of rain water flux per unit time is greater. The photo in figure 8.4 shows this pattern clearly.

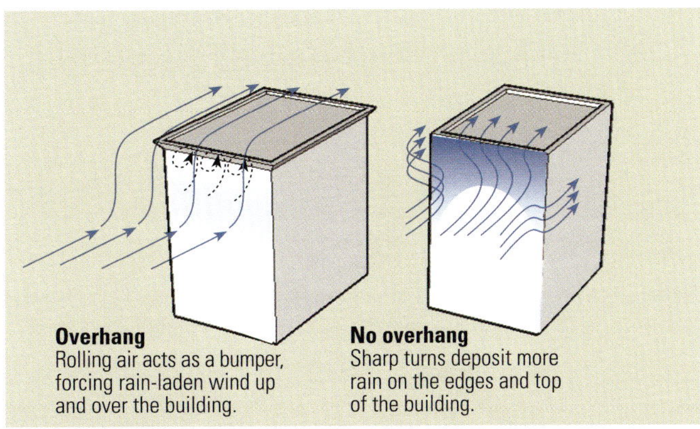

Fig. 8.5 Short overhangs reduce the rain load on tall buildings

Contrary to intuition, a short roof overhang significantly reduces the rain load on tall buildings. Part of the wind is trapped, creating a "rolling bumper," which keeps most of the oncoming rain-laden wind from soaking the upper edges of the building. Reducing the annual rain load reduces the risk of any potential leaks.

The point is that the total annual rain load is not distributed evenly across the face of the walls. Over a year, most of the water load challenges the top of the building and it's corners, as one can see from the image in figure 8.4. Much less of the annual rain load challenges the joints and cracks in the middle of the wall, because the wind speed, and therefore the amount of water per unit time, is much lower in the middle of the wall than at its edges. Structural engineers will notice the fact that the rain deposition pattern on a building is nearly the same as its wind pressure contours.

Now to explain the beneficial effect of the short overhang on a tall building. Figure 8.5 shows how it works. At high wind speed, the overhang traps and forces part of the wind back on itself, forming a rolling mass of air which acts as a smooth bumper for oncoming wind. That way, the oncoming rain-saturated wind *flows smoothly up and over the rolling air*, and therefore carries most of its rain water up and over the building instead of dropping it onto the building's walls. Again, a short overhang—about 2 ft. [600 mm]—is often enough to produce this benefit.

After the owner's and architect's preferences about roof overhangs are set into the design, there's no similarly low-cost way to reduce the water load by half at a later stage. So the decision usually has to be part of the earliest thoughts about the exterior design.

When making this decision, owners and architects might consider one other fact. Note well the rain deposition patterns. These suggest that the people who occupy the more expensive and prestigious floor space (corner offices and apartments which are high on the building) will experience the greatest consequences of the roof overhang decision, for better or worse, for the life of the building.

Sill Pans

Water often leaks through and around windows and doors, so it's prudent to install sill pans below them. Sill pans are also called "pan flashing" and "sill flashing." But the name "sill pan" perhaps best communicates the location of the object and its appearance, which in turn help explain its function.

Figure 8.6 shows two examples of sill pans. One is prefabricated of plastic, and the other is formed on site, using self-adhering membrane. They both have end dams and back dams. Given those dams, water which ends up in the pan cannot get into the wall by dripping out at the sides or back of that pan. Water only drips out at the front, where it can be further controlled.

The purpose of a sill pan is to collect any leakage water which comes through the window, or which comes in above or beside it, and then direct that water to a safe location. Safe locations include either all the way out to the weather, or more commonly just out and onto the waterproof layer which protects the sheathing. In the latter design, the water leaving the front of the pan flows down the face of the waterproof layer, until other flashing catches and directs that leakage back out of the wall to the weather.

Those who believe the impressions created by some in the window industry often do not bother with the cost of sill pans. In theory, manufacturers' designs and their tests assure that windows themselves do not leak. That theory also holds that when "installed in accordance with the manufacturer's recommendations" there won't be any leaks around the sides or above the window, either. There are many impressive ASTM test and installation standards which suggest that, when tested and installed according to those standards, there is no reason to be concerned about water leaks through or around windows.[4,5]

But field experience consistently shows that water leakage through and around windows is very common.[6,7] Indeed, some forensic investigators have observed that: "There seem to be two types of window installations in North America: Those that now leak... and those that will leak later."[8] Evidently, given the combination of window manufacturing and marketing economics, wind-driven rain and construction job site realities, it is very difficult to install windows in a way that ensures that no water will leak.

So the owner and architectural designer have a decision to make. Codes do not currently require sill pans. How lucky do they feel about their building details and about the construction job site superintendant's ability to install them without errors? At the very least, it would be prudent to install sill pans under the windows and doors which face the prevailing winds during rainstorms. But then, to simplify the design and therefore reduce the probability of confusion on the job site, a single detail might be a better choice. Sill pans under all the windows will reduce the risk of water damage.

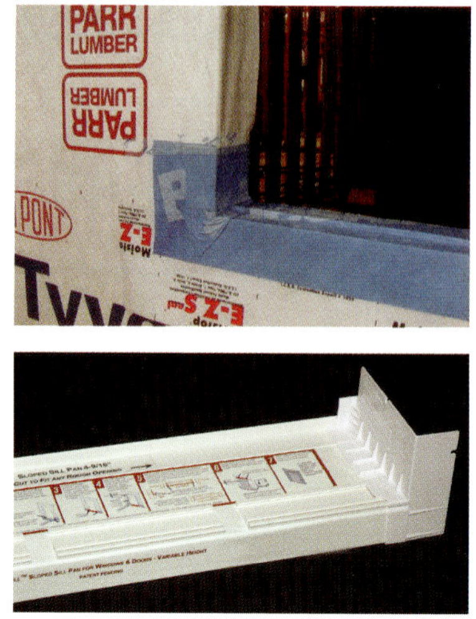

Fig. 8.6 Sill pans protect the walls from leaks through and around windows
Rain leaks near or around windows are very common sources of water damage. Sill pans which are prefabricated or formed-up on site out of self-adhesive waterproof membranes can keep the leakage out of the wall, forcing it back out to the exterior. Note the end-dams and the back-dams which keep the leak from flowing towards the interior of the building.

Flashing

Little if any water gets into a building through solid sheets of building material. The leakage gets in through the cracks, joints and holes. That's why all the joints and penetrations in the exterior walls need effective flashing. Flashing has two distinctly different and equally important functions:

- To keep water from *getting into* the wall through joints and around penetrations through the exterior cladding.
- To guide water *back out* of the wall, when some leakage gets through cracks and joints in spite of everybody's best efforts.

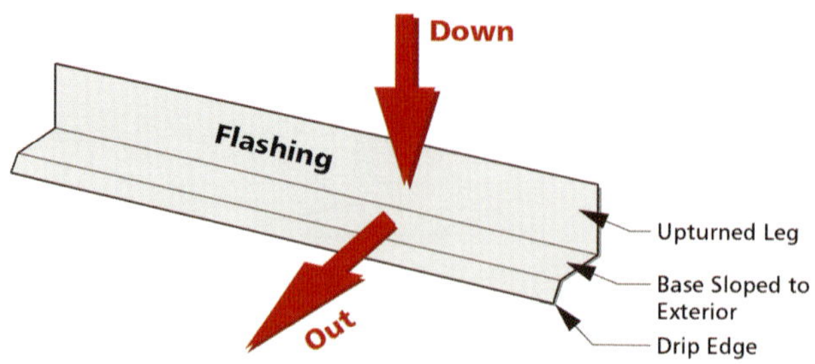

Fig. 8.7 Flashing
Flashing keeps water out, by acting as a dam against water entering at the joint. It also forces any water which got in above, back out to the exterior.

To keep water out of a horizontal joint, the flashing stands behind and above the joint, barring the path the water must take if it is to enter the wall at that joint. Figure 8.7 shows this, in a simplified diagram. Gravity keeps the rain water from leaking up and over the flashing at that joint.

But most joints and penetrations are not as simple as the neat, single horizontal line in figure 8.7. The more common joint is an opening for a window, or a penetration for a pipe or conduit. Those "holes in the cladding" have sides and bottoms as well as tops, as seen in figure 8.8.

Now the flashing has a more complex job. It must still keep the water from getting through the horizontal joint above the opening, as in figure 8.7. But the flashing must also guide the water that comes in *beside* and *below* that opening back out to the weather. So now we need two sets of flashing, one at the sides of the opening, but also another piece of flashing below. The flashing at the side cannot use gravity to push water back out. Gravity only acts down, not out. So the water that runs down the flashing at the side of the window must be:

- drained safely down the face of a waterproof layer, until it is...
- caught by another piece of flashing somewhere below the opening, and forced back out of the wall.

Fig. 8.8 Flashing around penetrations
Flashing is important above and around any penetrations, as well as at any horizontal joints and around windows openings.

Fig. 8.9 EIFS without head flashing
The infrared image inset into the photograph shows the thermal pattern created by rain-soaked insulation above the window. Sealants alone do not prevent this problem... flashing keeps water out of horizontal joints.

These tasks are very complex in a real-world buildings, because all the layers come together in very complicated ways, especially when they meet at corners.

To reliably keep the water out of the wall and guide it back out after it gets in, the architectural designer and the installing contractor will need to think in terms of the entire exterior wall and all of its layers and penetrations. That's quite different from just specifying a few pieces of metal or membrane known as flashing. Further, because different pieces of the assembly cross the boundaries between the construction trades, making a comprehensive and reliable solution to water entry is probably best settled by the architectural designer, rather than by the individual crafts people on the job site, no matter how skilled they might be in their own separate areas.

To help guide a more productive way of thinking about designs which keep water out and get it back out after it gets in, building scientists have settled on a more comprehensive concept, called the drainage plane.

Drainage Planes In Walls

All exterior claddings pass some rainwater. As such, some control of this penetrating rainwater is required. In well-built walls, this penetrating rainwater is controlled by a drainage plane, which keeps water from penetrating deeper, and directs any leakage water back out of the wall.

On face-sealed walls, the bulk of the drainage happens on the exterior face of the cladding. But on other walls, the drainage plane is usually located inside the wall. It consists of three components:

- Waterproof layer
- Drainage gap
- Flashing

The waterproof layer keeps the leakage from penetrating further into the wall, where it could damage moisture-sensitive materials. The drainage gap, located in front of that waterproof layer, allows any leaking water to make its way down the face of the waterproof layer. The flashing catches that downward-flowing leakage, and directs the water back outside the wall assembly.

Drainage planes are interconnected with sill pan flashing at the rough openings for windows and doors, and with flashings at all other penetrations. The goal is an integrated assembly, which drains water to the exterior of the building no matter where that water might have entered.

The materials that form the waterproof layer overlap each other shingle fashion or are sealed so that water drains down and out of the wall. In residential construction in the US and Canada, the most common waterproof layer has been tar paper or building paper (Figure 8.10). More recently, the terms housewrap or building wrap have been introduced to describe building papers that are not asphalt-impregnated felts or coated papers such as polyethylene or polypropylene films (Figure 8.11). Waterproof layers also can be created by sealing or layering water-resistant sheathings such as a rigid insulation or coated structural sheathings. Finally, fully adhered sheet membranes (Figure 8.12), or trowel- and spray-applied coatings (Figure 8.13) can act as waterproof layers.

Fig. 8.10 Drainage plane
A drainage plane includes a waterproof layer, a drainage gap to permit downward water flow, plus flashing to force any downflowing water back out of the wall.

In commercial construction, waterproof layers are sometimes referred to as air barriers. This is somewhat misleading, since their principal function is rainwater control. A more informative description for these types of commercial membranes would be waterproof layers that also act as air barriers and vapor barriers. On drawings they should be referred to by their multiple functions—but in order of importance. For example: "*Waterproof* air and vapor barrier."

In fact, waterproof layers can be either permeable or impermeable to *vapor*, depending on climate, location within the building enclosure or its several required control functions. Building papers and housewraps are typically vapor permeable (more than 10 perms), whereas fully adhered sheet membranes and trowel-applied coatings are typically impermeable (less than 0.1 perms). A few recently developed spray and trowel applied coatings that are vapor semi-permeable (1 to 10 perms) are likely to see wider application in the near future.

Functional sequence: Draining water down, then out

The fundamental principle of rainwater control is to eject water by layering materials in such a way that water is directed downwards and outwards out of the building.

Gravity is the driving force behind drainage. "Down" harnesses the force of gravity and "out" gets the water away from the moisture-

Fig 8.11 Waterproof housewrap

Fig 8.12 Waterproof sheet membrane

Fig 8.13 Waterproof coating

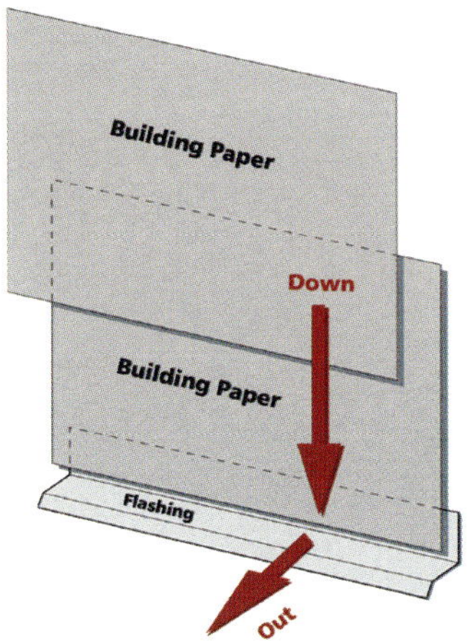

Fig 8.14 Down and out

Gravity pulls any leakage down over the waterproof layer. Flashing forces that water back out to the exterior.

sensitive building enclosure assemblies, openings, components and materials. In general, the sooner water is directed out, the better. On the other hand, immediate ejection is not always practical. For example, at window openings, draining a leak from the window itself into a drainage gap behind the cladding is often more practical than draining that window leak all the way out to the exterior face of the cladding.

After downward drainage, flashing ejects the water. (Figure 8.14). Flashings may be the most underrated building enclosure component, but arguably the most important. They keep most of the potential leakage water out, when it would otherwise get in through construction joints. Also, flashings eject the water which flows down the waterproof layer, when water gets in at other locations.

Flashings are integrated with the waterproof layers and the drainage gaps, creating a drainage plane for the entire assembly. A screen or cladding is installed over that drainage plane to provide a pleasing appearance and protection from both ultraviolet radiation and mechanical damage.

Drainage and drying

For drainage to occur, there must be a gap between the cladding and the waterproof layer. The width of this gap varies according to cladding type and function.

Effective drainage of rainwater can occur in drainage gaps as small as 1/16 to 1/8 in. [2 or 3 mm] For example, drainage is even effective between two layers of building paper in a stucco cladding, as seen in Figure 8.15. The water absorbed by the felt papers causes the papers to swell and expand. When the assembly dries, the papers will shrink, wrinkle and de-bond from the stucco rendering. This creates a tortuous, but still reasonably effective drainage gap. At the base of the stucco wall, and at its control joints between floors, a piece of metal called a weep screed supports the edge of the stucco. The stucco also pulls away from the metal slightly as the stucco cures and ages, so water can drain out between the stucco and the screed, as seen in figure 8.16.

With brick veneers, the width of the drainage gap has been based more on tradition than physics. A 1 in. [25 mm] airspace is more or less the width of a mason's fingers, hence, the typical requirement for a 1 in. [25 mm] airspace. However, historical experience with stucco and other cladding systems show that spaces as small as 1/16 in. [2 mm] will drain water. Note: no matter how wide or narrow that gap, it must be backed by a waterproof vapor barrier.

Going further, a wider, vented gap is an improvement because it allows drying in addition to drainage. Effective *drying* may require gaps as wide as 1 in. [25 mm] (Figure 8.17). The appropriate width depends on the amount and evaporation rate of moisture stored in the cladding.

Also in favor of even wider drying gaps, the wider it is, the less frequently it will be bridged by excess brick mortar. (Figure 8.19)

Fig 8.15 Stucco drainage

Two layers of paper provide a narrow but effective gap which allows any leakage to drain down the wall. The weep screed protects the edge of the stucco, and also acts as flashing, forcing leakage water back out of the wall.

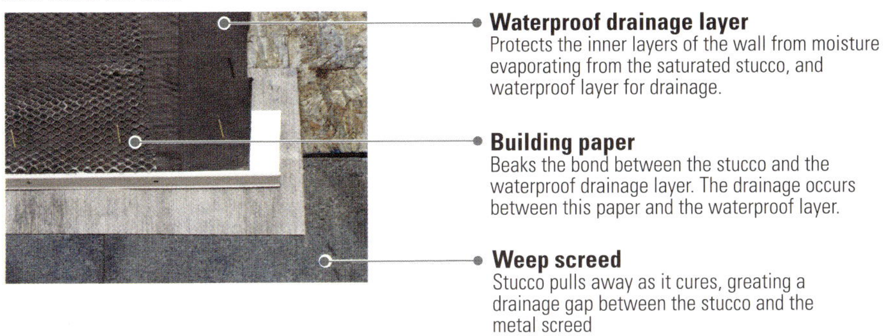

Fig 8.16 Drainage at the weep screed

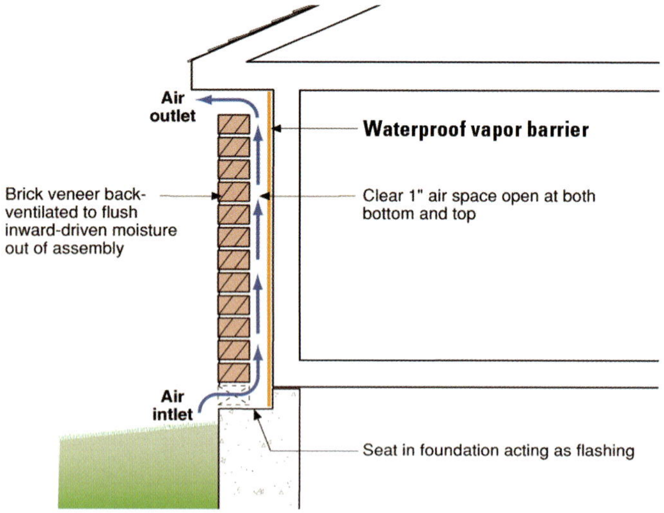

Fig 8.17 Vent cavity behind brick veneer

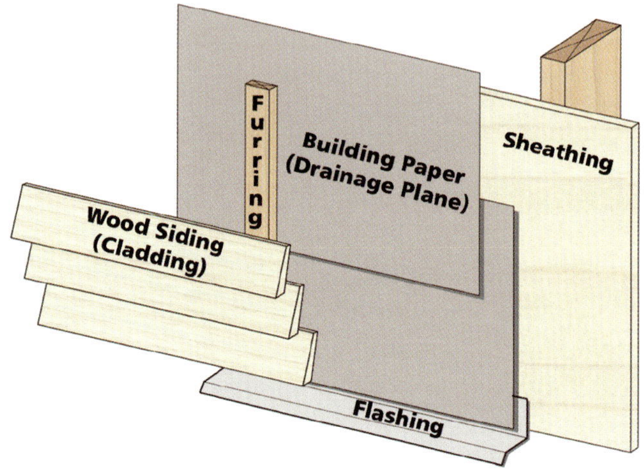

Fig 8.18 Vent cavity behind wood siding

In a brick veneer wall, most of the water leakage will drip harmlessly down the inside face of the brick. However, any mortar bridges will carry leakage water across the gap into contact with the waterproof layer, increasing risks.

Finally, any *drying* gap needs to be vented at the top and the bottom, as shown in Figure 8.17. That way, outdoor air can flow upwards through the drainage plane to remove evaporating moisture.

With wood siding, the drainage gap between the wood and the waterproof layer is probably not entirely open. The horizontal obstructions depend largely on the profile of the siding. Ideally, wood siding should be installed over furring, to create a continuous drained and vented gap between the wood and the waterproof layer (Figure 8.18).

With vinyl and aluminum siding, the cladding material does not absorb water. Also, the surface area at its contact points is quite narrow compared to typical wood siding. Consequently, the drainage gap behind vinyl and aluminum siding is less obstructed, as well as being semi-vented at each horizontal joint, so furring is not necessary.

Why encourage drying air to flow up behind a cladding in addition to providing drainage? It's because many claddings which are popular in hot and humid climates act as large water reservoirs. After such claddings become soaked, the stored water often migrates inwards, causing problems in the wall or inside the building itself.

Common reservoir claddings include brick and stone veneer, masonry block and precast panels. These claddings, because they are often uncoated and have a large storage capacity, can generate serious problems. Their stored moisture migrates inwards, driven by the solar radiation of hot and humid climates. The problems can come quickly and can be very severe when there is no vapor barrier behind the veneer to stop inward migration. And the problems are accelerated when there is also an *interior* vapor barrier installed, such as vinyl wall covering (Figure 8.20). Interior vapor barriers are never a good idea in air conditioned buildings in a hot and humid climate. Indeed, they have been responsible for untold thousands of expensive building problems. On the inboard side of the waterproof layer, the exterior wall needs to be able to dry freely to the interior.

Fig 8.19 No drainage gap, an incomplete waterproof layer and no flashing

Without a drainage gap, water seeping through brick joints will cross to the framing, in this case not fully protected by a waterproof layer. Brick joints always leak. The dimension of the future water damage problem now depends primarily on time, and on the amount of rain on the wall.

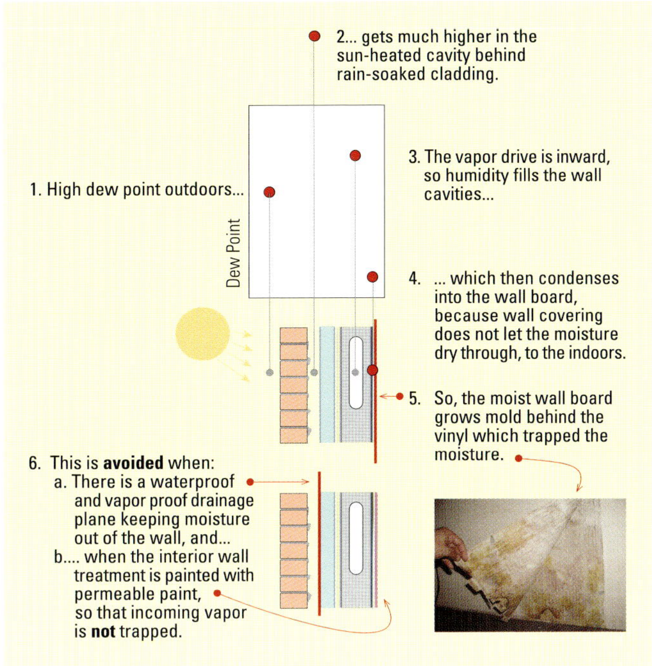

Fig 8.20 High humidity behind reservoir cladding

When damp brick or stucco or precast panels are heated by the sun, the humidity in the drainage gap skyrockets—far above the dew point in the outdoor air.

That's why it's important to protect the wall behind any reservoir cladding with a waterproof and vaporproof layer—but don't mis-locate that layer on the interior face of the wall!

Back-venting a reservoir cladding uncouples (isolates) the reservoir cladding from the rest of the wall assembly, preventing such problems. The greater the reservoir, the greater the moisture load, and the larger the drying air flow must be. In some cases, it is difficult to provide a vented gap wide enough for effective drying. In those situations a reservoir cladding must be isolated. This can be done in a thin wall by installing a durable condensing surface, such as impermeable insulating sheathing, behind the cladding. (Figure 8.21) Another solution is a fully adhered, impermeable sheet membrane (a vapor barrier) which covers the sheathing behind the cladding. These are attractive approaches where it is not possible or practical to provide a vented gap which will be free from mortar bridges.

Drain the window and door openings

Drainage planes should be integrated with window and door openings. This is because the seal between a window component and the drainage plane is rarely perfect. Even if it is perfect at the time of installation, it certainly will not be perfect forever.

Furthermore, the window or door component installed within the opening is rarely perfect. It can and often does leak. Window and door openings should be drained to the exterior using the same principles used in the design and construction of wall assemblies in general.

Sill pans, self-adhered membranes lining the openings, precast sills with seats extending under window and door units, formable flashings, building papers and housewrap-lined openings are all methods of providing drained openings.

These measures acknowledge and honor the real world. With sill pans or other measures in place, sealants can age or be imperfect, without catastrophic failure of the assembly. A leak is not truly a leak if it exits to the exterior without wetting a water-sensitive material.

Fig. 8.21 Narrow drainage gap

When a waterproof layer covers the moisture-sensitive parts of the wall, there's no need for a wide air gap for drying the cavity. Just make certain the air gap is wide enough to allow water to drain down and out.

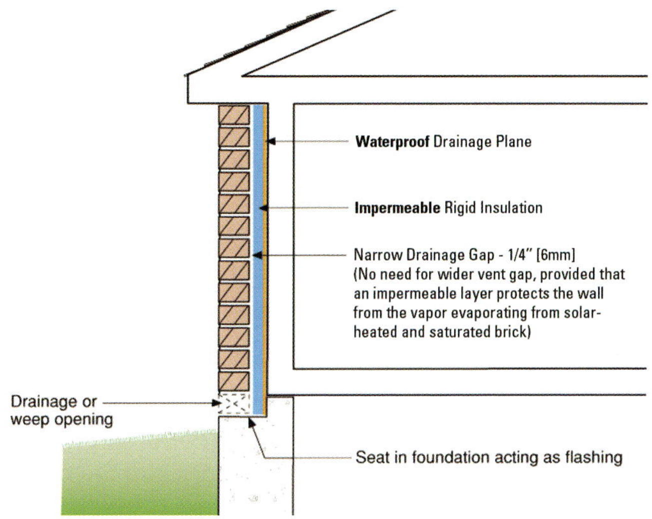

Drain the windows and doors

Windows and doors are complex assemblies. Where their component parts are joined, they have the potential to leak water. Ideally, windows and doors themselves should be designed and assembled in such a way that all of their components shed water to the exterior, and each joint between those components also drains water to the exterior.

Sealants and gaskets at window and door joints should not be relied on to provide the only defense against water entry. The principle of drainage applies to the design and construction of window and door components just as it applies to walls and wall openings and penetrations.

A conversation with the window supplier can be useful. If the supplier claims the window will not leak (a highly probable response), ask what happens if, by some mischance, it leaks anyway. What happens to the water?... how does it get back out to the weather without accumulating inside the window assembly, and without dripping into the wall itself? Does the supplier recommend sill pans? If so, what details are recommended and who supplies the components other than the window itself? The same concerns apply to doors.

Use coatings to keep water out of materials

After water gets absorbed into a material, the simple, cheap and effective force of gravity will not be enough to remove that water from the wall. Absorbed water will have to be evaporated, and the vapor will have to be vented. Evaporation and venting are much slower, less certain and more complex than simple gravity drainage of liquid water. So, as much as possible, keep the water from being absorbed into materials, as long as those efforts don't prevent the material from drying out if it does absorb moisture. A delicate balance, to be sure.

Materials that are outboard of drainage planes should not absorb water or should be treated to shed water. For example, wood trim and wood siding should be coated on all surfaces to repel water. Think of this as a capillary break for the *materials* that are located outboard of the drainage plane.

The repelled water should be also drained away. Ideally, these materials should also be back-vented so that any absorbed water can evaporate and be vented to the exterior.

Claddings that are water absorptive such as stucco and brick should be separated from the rest of the assembly by a capillary break. This capillary break can be an airspace and/or a material that sheds water rather than absorbing it or allowing it to pass though.

Stuccos can be coated on their exterior faces to reduce water absorption, or mixed in proportions or with additives to reduce water absorption. ASHRAE research has shown that stucco coatings can be very effective in reducing the humidity deep in the wall, at its more moisture-sensitive layers.[9]

Window and door elements should also be treated to repel water or coated to repel water. Or, they should be made with materials that do not absorb or transmit water. For wood windows, this means that all wood pieces should be coated and treated on all six surfaces, with the most critical surfaces being the ends. End grain surfaces are the most prone to water absorption.

Drain everything, actually

The logic of drainage should be applied to every joint and every seam in the entire building enclosure. Deck, balcony and railing connections should be designed and constructed to shed or drain water to the exterior. Roof-wall connections and roof-dormer connections should be designed and constructed to shed or drain water to the exterior. Garages, decks, and terraces should be sloped to the exterior and drained.

Drainage summary

Assemblies should be designed and constructed to shed or drain water to the exterior. The unifying concept is a drainage plane. Window and door openings should be designed and constructed to shed or drain water to the exterior and they must be integrated with the drainage plane. Windows and doors themselves should be designed and manufactured to shed or drain water to the exterior. Building materials should drain or be treated so that they drain rather than store water. Building connections should be designed and constructed to drain to the exterior. And, of course, the sites themselves should be graded to shed or drain water away from building perimeters.

Whenever moisture or liquid water accumulates, the risk of damage increases. So, avoid accumulation through drainage, as much as possible.

Crawl Spaces

Crawl spaces need to be built like modern basements. In other words, they need to be well-drained, and they also need to be sealed against water leakage, humid air infiltration and vapor permeation from the earth, and insulated to comply with energy codes.

This is a change from traditional practice in the U.S. Until recently, typical designs for crawl spaces filled the floor area with crushed stone, and called for air vents with insect screens around the perimeter wall. Indeed, some building codes still require venting a crawl space.

Actually, however, venting was never a good idea in hot and humid climates, and there was little or no science to support that practice.[10] But until air conditioning became nearly universal, venting did not result in spectacular failures. Now, problems have arisen which argue strongly in favor of dry, sealed crawl spaces. When air conditioning cools wall surfaces and floors, small problems become big ones.

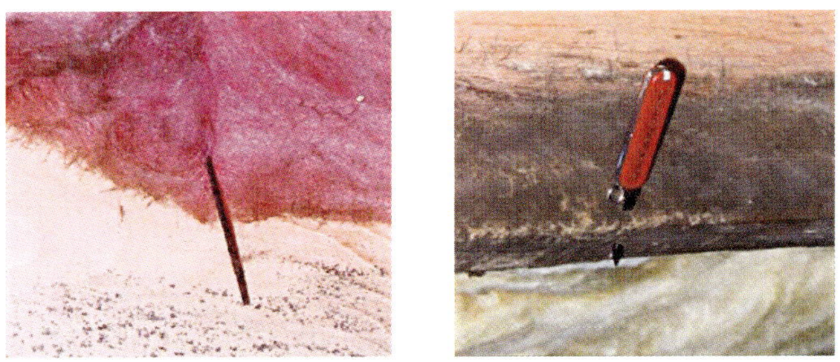

Fig. 8.22 Venting rotted the structure in a crawl space

An owner noted the minor mold growth shown at left. A contractor enlarged the foundation vents in an effort to dry the structure. After another year, the increased ventilation had rotted the structure, as shown in the photo at right. Venting with humid outdoor air allowed even more moisture to condense on the cool indoor structure. The better design for crawl spaces is to seal them up tight and keep them dry.[12]

The first problem is humid air infiltration. Warm, humid air from the crawl space makes its way up through the building through joints and penetrations, rising through the cooler air conditioned air in the building, and loading the structure with moisture. Field measurements have established that in traditional construction, about 30 to 45% of the air contained in a low-rise residential building actually entered the building through the crawl space.[11] As outdoor air enters through the dirty crawl space, it picks up unhealthy particulate, which eventually reaches the occupants' breathing zones.

The second problem is moisture absorption from incoming humid air. Humid air inside the crawl space condenses moisture into the cool, air conditioned surfaces, leading to mold, insect infestation and eventually to rot and structural failures. Figure 8.22 shows an example of what happened when an *unvented* crawl space *became vented* in an air conditioned building in a hot and humid climate. A small mold problem became a structural hazard. Basically, the structure absorbed moisture from the humid air, and then it rotted. The better solution would have been to seal up the space and keep it dry.

For detailed guidance on closed crawl spaces, the reader is encouraged to consult a useful report of the comprehensive research performed for the U.S. Department of Energy.[12] That report, which is available on line without cost, also contains useful guidance for any governmental agencies seeking to improve building codes in hot and humid climates. The suggestions below briefly summarize a few key points from that report.

Drain the crawl space

Liquid water is the first concern. Sooner or later, some amount of ground water is likely come up from below the building and stagnate in a crawl space, unless the earth is well-drained. To prevent this eventuality, install a sump pump at the low point of the space (or at each of the low points, in the case of a crawl space with varying levels).

A further improvement is a layer of large-diameter crushed stone under the vapor retarder. The stone will act as a capillary break, keeping water from wicking upwards to contact the all-important vapor-tight liner, and also allowing water to run freely down to the low point of the foundation, where the sump pump is located. A stone layer also makes it easier to collect, capture and vent any unhealthy or undesirable soil gas that may be present on site, from under the vapor-tight liner.

Crawl spaces need vapor-tight liners

Above the drainage layer, the space must have a robust, sealed, vapor-tight layer which covers the ground, and is tightly sealed to the vapor retarder which covers the foundation walls. (Figure 8.23).

The architectural designer's chief concern should not be with the exact vapor permeability rating of the material. Instead, focus on the more important goal of a *sealed liner*. Clearly detail the seams between the sheets, the joints around the inevitable penetrations, and the joints between vapor-tight floor covering and the covering for the foundation walls.

The DOE report suggests that as a minimum, ground sheets should be overlapped by 10 inches [240 mm] and then sealed with the tape specified by the manufacturer. At the foundation walls, the floor vapor retarder should overlap and run up over the vapor retarder on the walls by at least 4" [102 mm]. Then, these two layers must be sealed together, creating a continuous vapor-tight liner for the whole space.

A typical sealing detail for the floor-wall connection uses glass fiber tape embedded between two layers of duct sealant mastic to seal the overlapping layers, pinned in place by a termination bar which is mechanically fastened to the foundation walls. Duct mastic and glass fiber tape can also be used to seal the liner to all the inevitable penetrations. (Figure 8.24) The ground layer will probably be penetrated by utility connections, such as power and sanitary lines. And more penetrations will be needed through the foundation walls, for items such as dryer vents, refrigerant lines, irrigation plumbing or power conduits. These must all be sealed.

In addition, if the crawl space is likely to be used as storage space, or if it contains mechanical equipment (which needs periodic service attention), the sealed liner will be at risk for punctures and tears. When people must enter and leave the crawl space as part of the normal operation of the building, the designer can specify a thin layer of unreinforced concrete over the floor, to cover and protect the liner after it has been laid down and sealed up.

In flood zones, install air-tight flood vents

In areas where building codes permit construction in flood zones and other flood-prone locations, building codes often call for flood vents in crawl space foundation walls. Flood vents allow surface water to pass through the crawl space, relieving pressure at the foundation and perhaps saving the building in case of a catastrophic flood.

In older design practice, this requirement was met by installing *air* vents. But to avoid the problems discussed in this section, flood vents should be air-tight, yet still able to open under pressure from flowing water.

Fig. 8.23 Sealed and insulated crawl space

The vapor retarder on the floor also covers the walls behind the foil-faced insulation. That vapor retarder layer is continuous, and sealed to the foundation walls.

Fig. 8.24
Penetrations & termite control

Duct sealant mastic does a good job of sealing the many awkward penetrations through the floor and walls of a sealed crawl space. The paste-like material is highly adhesive, and structurally strong over time.

The wide gap between the insulation and wooden structure helps slow any future termite progress upwards into the building, and the white paint allows quick visual inspections for termite or other insect infestation.[12]

Such devices are commonly available. The designer should simply be aware of the need to ensure that the specified flood vents are air-tight, matching the air tightness of the rest of the foundation wall of the crawl space.

Site and Foundation Drainage

To any architectural designer, it is quite obvious that a building should not sit in a pool of water. It is also obvious that foundations, no matter whether they are simple slabs, crawl spaces or full basements, need to have waterproofing, along with a vapor barrier and a drainage system beneath the building and all around the foundation walls.

Given those measures, ground water is prevented from moving into the building through capillary suction, and even in constantly saturated soil, water vapor from the earth cannot diffuse into the building through the foundation's concrete. There is no need to belabor these obvious design imperatives. In large, highly engineered and buildings, most of the obvious problems are avoided, because they receive so much professional attention, and because so many building codes apply.

But in smaller commercial and multi-unit residential buildings, some problems might be overlooked, including:

- Roof drains discharged too close to the foundation.
- Landscaping which collects irrigation water at the foundation.
- Parking lots and driveways which drain rainwater near the building.

Roof drainage - Beware and take care

The roof collects and drains a tremendous amount of water, concentrating that flow at relatively few locations. What happens to the rainwater after it leaves the roof requires some care.

Often a pitched roof is allowed to shed water evenly over all of its edges, so that the rain collects at the foundation wall. As long as the soil is well-drained and the foundation is well-waterproofed, this might not cause problems. But if either of these features is missing, or if rain is very frequent during some seasons of the year, or if the landscaping holds water near the building, draining roof water all around the foundation can lead to serious water intrusion. Gutters and downspouts are a step in the right direction, because they allow the concentrated flow to be managed. On the other hand, because the flow is greater at each downspout than it would be around the entire edge of the roof, water leaving the downspout requires attention.

The gutter downspouts should guide the water *well away* from the foundation walls. The appropriate distance will depend on the drainage rate from each downspout, and on the drainage capacity of the soil. But this does not usually need to be a complex calculation. Releasing the drained water on a splash block located more than three feet [1 meter] away from the foundation will avoid most problems, provided that the foundation is waterproofed and drained.

For a more useful solution, the rainwater collected from roof drains can be captured in storage tanks, and then used to irrigate the landscaping through gravity and slow-release soaker hoses or drip irrigation nozzles.

Landscaping must drain water away from the building

Planter boxes attached to the building, and decorative borders around ornamental plantings at grade sometimes cause water intrusion problems through walls and foundations. Also, the decorative earth berms used to help visually separate buildings which are sited close together are sometimes so close to the buildings that rain and irrigation water cannot escape.

Basically, any landscaping feature which holds rainwater or irrigation water in contact with the foundation adds significantly to the risk of water intrusion. In many real estate developments, land is so expensive that the buildings will necessarily be built close together.

So in all of these situations, the architect can help reduce the risk of water intrusion by making sure that the rainwater and any irrigation water is not trapped against the foundation of the building. Instead, either drain that water safely way from the foundation, or collect it in purpose-built waterproof reservoirs and tanks, and use it to irrigate the landscaping during dry weather.

Parking lots and driveways concentrate water

Pavement collects and concentrates rainwater, just like the roof of the building. In many parts of the world, building codes dictate the slope, drainage and rainwater catchment basin requirements for parking lots and driveways. But in other locations, good design practices may be less well-established, or less well-enforced.

Newly-urbanized areas are not always well-regulated in this respect, because until there's a great deal of pavement, the ground can absorb the rain. Both in developed and developing countries, newly-urbanized areas are prone to problems of excessive rainwater accumulation near buildings caused by nearby parking lots or streets.

Another classic environment for pavement-related water accumulation is on hillsides in densely-built residential or commercial zones. Rain water accumulates and accelerates as it runs down the pavement on a hill, perhaps augmented by more rainwater coming off of roofs and draining to the hillside street from between closely-spaced buildings. The buildings closest to the bottom of the hill are at greater risk in this situation.

At a very basic level, the architectural designer should remain aware of the issue, and make every effort to make sure the water that collects on the pavement is not forced against the foundation of the building. Further detailed design guidance is provided by building codes, landscaping design guides,[13] and rain rate data for U.S. locations is available from industry sources.[14] Alternatively, the designer might choose to specify *porous* pavement, an alternative that can avoid both risky accumulation, and perhaps the expense of a catchment basin.

Porous pavement avoids risky concentrated flows

There are several benefits to the owner when the pavement of the parking lot and driveway is porous enough to allow rain water to pass through it freely.[15] First, there is less risk of accumulated water flowing against the foundation. Porous pavement spreads the water load more evenly across the site.

Also, because water is not drained off the site on pavement, the water table in the ground is recharged each time it rains. That means there is less need for purchasing water for irrigation. And with the right plant selection, perhaps there may be no need for the expense of irrigation at all—saving operational funds as well as further reducing the risk of irrigation water accumulating near the foundation.

Fig. 8.25 Porous pavement

Allowing rain to permeate concrete and asphalt pavement reduces storm water runoff, reducing the risk of rain water flowing towards, or pooling against the foundations of buildings.[15]

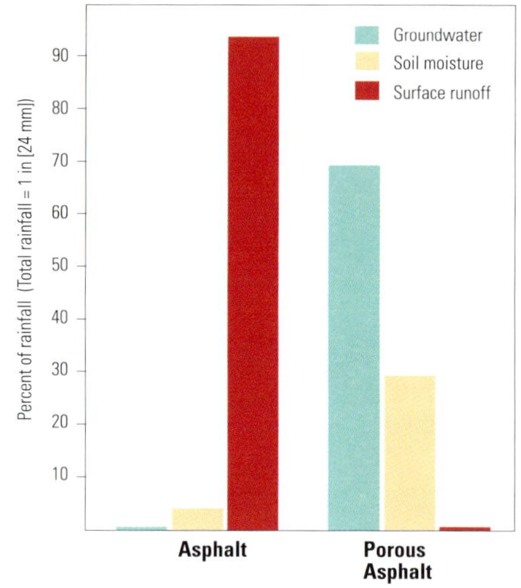

Fig. 8.26

Porous pavement reduces surface water risks to buildings[15]

Porous pavement also allows rain water to flow into the ground, reducing the need for landscape irrigation, which is a common source of water accumulation near foundations.

Summary

This chapter is intended to provide interested professionals with a few basic points with which they can begin a productive discussion with architectural designers and building contractors, about ways the design will ensure that excess moisture will stay out of the building.

The limitations of this brief treatment of the subject will be painfully obvious to many forensic investigators, and to most architectural designers and general contractors. Indeed, some of the reviewers for the drafts of this chapter have said this discussion is so oversimplified that it may do harm by implying that the task is simpler than it really is. That result would be unfortunate. This chapter is not comprehensive. Nor do we wish to imply that these suggestions are "ASHRAE requirements." They are not.

LAWSUITS

MIT Sues Over Alleged Building Flaws

The Massachusetts Institute of Technology filed a four-count lawsuit on Oct. 31 in Suffolk (County) Superior Court for unspecified damages and charging Partners LLP and with negligence in design and construction of the $300-million Center. The suit claims flaws caused leaks, cracks, mold and drainage problems, which need remediation.

▲ Shambles. building program, l

According to the complaint, a flawed drainage design caused efflorescence in the amphitheater's brick masonry by mid-2004. There was no drainage mat under the seating, thereby subjecting materials to excessive cyclic saturation, MIT says. Masonry cracking also appeared, caused by improper spacing and the number of control joints, it claims.

MIT hired a designer and contractor in 2006 to repair the damage at a cost of $1.5 million. Work included removing existing masonry and installing a drainage mat and cavity wall.

Fig. 8.27

Complex walls increase rain risks

Complex architectural designs usually increase the risk of rain leaks. This is especially true where different cladding systems come together, and where there is no effective drainage behind the cladding.[16]

This chapter cannot address all such cases. But it provides a useful agenda for conversations between owner and architectural designer about water leakage risks, and how these can be reduced rather than increased, through architectural design.

On the other hand, these few pieces of advice will be new information to many owners, and also new to many HVAC designers. Both of these groups are confronted, too frequently, with moisture problems which are not of their own making, and over which they have little control. Perhaps with more moisture-aware discussions between owners, architectural designers and contractors, the results described by Figure 8.27 will be less common, or at least less expensive to fix.

References

1. Rose, William B, *Water in buildings; An Architect's guide to moisture and mold.* 2005. John Wiley & Sons, Hoboken, NJ. ISBN 0-471-46850-9
2. Straube, John and Burnett, Eric. 2005. *Building science for building enclosures* Building Science Press, Westford, MA. ISBN 0-9755127-4-9. www.buildingsciencepress.com
3. Canadian Mortgage and Housing Corporation. 1996. *Survey of Building Envelope Failures in the Coastal Climate of British Columbia.* Report by Morrison-Hershfeld to CMHC, Ottawa, Canada www.cmhc.ca
4. *ASTM Standard E2319-04 Standard Test Method for Determining Air Flow Through the Face and Sides of Exterior Windows, Curtain Walls, and Doors Under Specified Pressure Differences Across the Specimen.* ASTM International, West Conshohocken, PA. www.astm.org
5. *ASTM Standard E2112-07 Standard Practice for Installation of Exterior Windows, Doors and Skylights.* ASTM International, West Conshohocken, PA. www.astm.org
6. Criterium Engineers 2003. *Construction Quality Survey, September 2003.* Criterium Engineers, Portland, ME. www.criterium-engineers.com
7. CMHC 2003. *Water penetration resistance of windows: study of manufacturing, building design, installation and maintenance factors.* Research Highlight 03-124, Canadian Mortgage and Housing Corporation, Ottawa, Canada www.cmhc.ca

8. Lstiburek, Joseph. 2004 "Built wrong to start". *Fine Homebuilding Magazine*. April/May 2004. Taunton Press, Newtown, CT www.taunton.com/finehomebuilding

9. Final report - *ASHRAE Research Project 864-RP "Controlling moisture in walls exposed to hot and humid climates"* ASHRAE, Atlanta, GA www.ashrae.org.

10. Rose, W. B. and A. TenWolde. 1994. *Issues in crawl space design and construction – a symposium summary. Recommended Practices for Controlling Moisture in Crawl Spaces*, ASHRAE Technical Data Bulletin volume 10, number 3. American Society of Heating, Refrigerating and Air-Conditioning Engineers, Atlanta, Ga.

11. Salonvaara, M., Zhang, J. S., and Karagiozis, A., *Combined Air, Heat, Moisture and VOC Transport in Whole Buildings*, Proceedings of the 7th Healthy Buildings Conference (CD), Singapore, National University of Singapore, Singapore, December 7 - 12, 2003 ASHRAE, Atalnta, GA www.ashrae.org

12. Dastur, Davis & Warren. 2005. *Closed crawl spaces; An introduction to design, construction and performance*. A report prepared for the U.S. Department of Energy under contract DE-FC26-00NT40995. Advanced Energy, Raleigh, NC. (PDF available at no cost at http://advancedenergy.org/buildings/knowledge_library/crawl_spaces/)

13. Ferguson, Bruce K, 1998. *Introduction to stormwater: concept, purpose, design*. John Wiley & Sons, Hoboken, NJ. SBN 978-0471165286

14. Woodson, R. Dodge. 1999. *Chapter 10 - National Rainfall Statistics*. Plumber's and Pipefitter's Calculations Manual. McGraw-Hill ISBN 0-07-071857-1 www.books.mcgraw-hill.com

15. Ferguson, Bruce K, 2005. *Porous pavements* CRC Press, Taylor & Francis Group, London, New York. www.crcpress.com ISBN 0-8493-2670-2

16. "MIT sues Gehry and Skanska over alleged building flaws" *Engineering News-Record*, November 19, 2007. enr.com, McGraw-Hill, New York, New York.

Image Credits

Fig. 8.1 UN Photo/DPI - NICA 76197 - photolibr@un.org

Fig. 8.3 Mason-Grant Consulting - www.masongrant.com

Fig. 8.4 John Straube - www.uwaterloo.ca

Fig. 8.6 Joseph Lstiburek - www.buildingscience.com

Fig. 8.7 Joseph Lstiburek - www.buildingscience.com

Fig. 8.8 Neil Leslie - www.gastechnology.org

Fig. 8.9 Mason-Grant Consulting - www.masongrant.com

Fig. 8.10 Joseph Lstiburek - www.buildingscience.com

Fig. 8.11 Joseph Lstiburek - www.buildingscience.com

Fig. 8.12 Joseph Lstiburek - www.buildingscience.com

Fig. 8.13 Joseph Lstiburek - www.buildingscience.com

Fig. 8.14 Joseph Lstiburek - www.buildingscience.com

Fig. 8.15 Joseph Lstiburek - www.buildingscience.com

Fig. 8.16 Neil Leslie - www.gastechnology.org

Fig. 8.17 Joseph Lstiburek - www.buildingscience.com

Fig. 8.18 Joseph Lstiburek - www.buildingscience.com

Fig. 8.19 Mason-Grant Consulting - www.masongrant.com

Fig. 8.21 Joseph Lstiburek - www.buildingscience.com

Fig. 8.22 Advanced Energy - www.advancedenergy.org

Fig. 8.23 Advanced Energy - www.advancedenergy.org

Fig. 8.24 Advanced Energy - www.advancedenergy.org

Fig. 8.25 Bruce Ferguson - www.uga.edu

Fig. 8.27 Engineering News-Record - www.enr.com

Chapter 9

Keeping Heat Out Of The Building

By Lew Harriman & Joseph Lstiburek

Fig. 9.1 Looking for better comfort?
Keeping heat out of the building is the key. The architectural design controls the cooling loads, and the most influential decisions will be all about the building's glass.

Key Points

In hot and humid climates, when less heat gets into a building, its occupants are more comfortable, and its cooling systems are smaller. They also cost less to operate and are easier to maintain. So when you are interested in ways to keep heat out of your building through design, here are some suggestions.

The suggestions are arranged in order, beginning with decisions which come first during the planning and design stages. The earliest decisions have the greatest effects—for better or for worse—on the amount of heat that gets in. Implementing the suggestions in the order presented here can also reduce construction costs for the building enclosure and for its mechanical systems—provided that one of the designer's goals is to build a low-energy building.

First, some suggestions for the owner and architectural designer, who control the most influential heat-excluding decisions:

1. Reduce the glazing and then shade the remainder, especially on the west side of the building.
2. Make the remaining glazing effective for daylighting.
3. Control lighting power in proportion to measured room daylighting and occupancy.
4. Install continuous insulation outboard of the building's structure, along with an airtight waterproof membrane.
5. Allow enough money in the mechanical budget for demand-controlled ventilation.
6. Allow enough ceiling height and enough money in the mechanical budget for ducted supply and return air.

Next, some suggestions for the HVAC designer:

7. Seal up all air-side connections and joints.
8. Don't use building cavities for supply or return air. Instead, use hard-connected and well-sealed duct work.
9. Install demand-controlled ventilation.
10. Don't let outdoor air economizers fill the building with humid air.
11. Use exhaust air to precool and predry ventilation air.
12. Keep the indoor dew point low, allowing warmer indoor temperatures.

Owner & Architectural Designer Decisions

For owners and architectural designers, by far the most important decisions concern glass. More glass means more heat getting into the building and less glass means less heat getting in. If you want to save money in construction and in operation, use far less glass than in the past, and use it cleverly, for daylighting. A maximum of 30% of the exterior wall as glass is a useful rule of thumb, but less is better, from the perspective of excluding heat.

This point is worth stressing, because it is not well understood by all architectural designers. The drama of large glass sheets, and the many exciting advancements in glass technology seem to have embedded the unhelpful misimpression that huge glass walls save energy and increase the building's sustainability. Such is not the case in an air conditioned building in a hot and humid climate. To keep heat out of the building, the glass decisions will need to be guided by the useful principles that "Less is more" and that "God is in the details."[1]

Reduce the glazing and shade the remainder, especially on the west side of the building

Glass transmits far more heat from the hot outdoors than does an insulated wall. Their different heat transmission rates (their respective U-values) quantify this key point. The U-value of single pane, untreated glass is about U=1.0, compared to walls at about U=0.05. Said another way, 20 times more heat comes through a sheet of conventional glass than through an insulated wall.

Even the heat transmission of expensive triple-glazed, argon-filled glazing with low-emissivity coatings is still a huge U=0.25 compared to

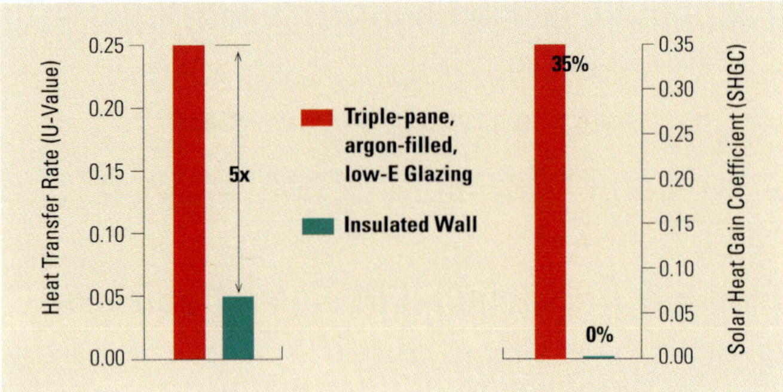

Fig. 9.2
Insulated walls keep nearly all the heat out... glass does not
The U-values and solar heat gain coefficients show the far greater heat exclusion effectiveness of insulated walls compared to high-tech glass. Minimizing glass minimizes the building's heat gain.

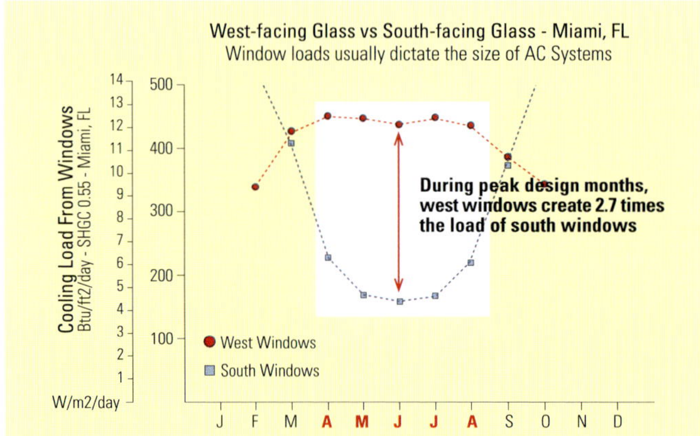

		Jan	Feb	Mar	Apr	May	Jun	Jul	Aug	Sep	Oct	Nov	Dec
South	Btu/ft2/day	609	537	404	236	173	159	174	221	380	543	616	604
	W/m2/day	17	15	11	6	5	4	5	6	10	15	17	16
West	Btu/ft2/day	271	339	424	465	458	433	453	432	392	345	276	237
	W/m2/day	7	9	12	13	12	12	12	12	11	9	8	6

Fig. 9.3
Keep the windows off the west face
During the peak load months, the west-facing glass lets in more than 2.7 times more heat than the same glass on the south and north sides of the building. Keeping the windows off the west side of the building allows the cooling systems to be smaller, simpler and less complicated to operate.

the wall at U=0.05. In other words, even expensive glass passes five times more heat than the insulated wall. But it actually gets worse.

Even more important in hot and humid climates, the solar heat gain through glass is very high (about 35% for low-E glass). Compare this to the solar heat gain of insulated walls, which is basically zero. With walls, solar gain is largely excluded. But with glass, between 30 and 70% of solar radiation will enter the building, along with the heat moving through that glass by convection. (Fig. 9.2)

That's why it is important to *reduce the amount of glass to the absolute minimum* needed for the building to be successful. Obviously, some buildings need more glass than others. A design without *any* windows would be more successful for a prison than for a hotel or eldercare facility. But there are better or worse ways to design the glass to satisfy the building's functions.

In particular, west-facing glass increases the size and complexity of the cooling system more than does glass on the other three faces. This is because the sun streams its heat through the west-facing glass at the end of the day, after the entire building has been heated up. Also, as shown in figure 9.3, the west-facing glass passes about 2.7 times more heat during the peak summer months than does the same glass on the south or north faces of the building.[2] (So too does the east-facing glass, but that only happens in the early morning, when there is plenty of cooling capacity. So east-facing glass does not usually govern the size of the cooling system to the same extent as west-facing glass.) If there is less west-facing glass, the cooling systems can be smaller and more economical than if the west face has lots of glass.

Next is the issue of shading those virtuously small windows which are located mostly on the north and south sides of the building. Even when the glazing is expensive and has an excellent, low solar heat gain coefficient, the windows will still need shading to achieve the reduction in solar gain required for low-energy buildings (those which comply with ASHRAE Standard 90.1).

For example, an excellent, modern, high-tech window may have a solar heat gain coefficient (SHGC) of 0.35. In other words, only 35% of the sun's radiant heat streams through the window, unlike the 70% solar heat gain coefficient of single pane, untreated glass. However, Std 90.1-2004 calls for a solar heat gain coefficient of 0.25 for all vertical glazing in hot and humid climates. Usually, the most practical way to achieve that last reduction in solar heat gain is to shade those already excellent windows with a horizontal projection over the whole window of about 3 ft. [about 1 meter].

All of these values will need to be calculated exactly during design development, to size the cooling systems and to ensure compliance with energy codes. But to help the owner and architectural designer during the conceptual design phase, it's enough to keep three simple rules of thumb in mind:

- Less glass is better (30% of total wall surface or less).
- Avoid the west side (entirely, if possible).
- Buy really good windows (solar heat gain coefficient below 0.4), and then shade them all.

These rules of thumb are simple ways to limit costs for low-energy buildings. But in most code jurisdictions they are not laws. When the need for (or the attraction of) extra glass is *irresistible*, plan to spend more money during construction, and each year for the life of the building.

There are many creative ways to use huge sheets of glass for visual drama. But to achieve equivalent energy consumption and equivalent comfort, those creative solutions all involve much more money for glazing, plus more money for larger cooling equipment and heroic engineering (not to mention many highly questionable energy modeling assumptions[3]).

Fig. 9.4 Daylighting keeps heat out
When windows are effective for daylighting, the designer can reduce lighting power, saving energy and reducing internal heat gain, which in turn saves cooling energy.

Next, there's another critical glass design decision which also has an outsized beneficial or negative effect on heat gain and energy use—the decision to design the windows for effective daylighting.

Design high, horizontal glazing for effective daylighting

The shape and height of the glazing with respect to the occupied space has a very strong effect on the amount of heat that ends up in the building, and therefore a strong effect on comfort, cooling equipment size and the building's electrical consumption.

If the glazing is *high* on the walls of the rooms, and if it is *horizontal* rather than vertical, and if it has a *light shelf* under the sill on the outside of the building, the glazing has the potential to make a major reduction in heat gain and electrical consumption. That's because it can be effective for daylighting.

Figure 9.4 shows what such a building exterior might look like. And Figure 9.5 shows how pleasant such designs can be, from the perspective of the occupants.

Horizontal windows mounted high on the wall let sunlight penetrate more deeply into the room, without generating glare at eye level. That way, the lighting power can be greatly reduced for most of the daylight hours.

With daylighting, the power reduction has two components. First and most importantly, the interior lamping consumes

Fig. 9.5 Aesthetic effectiveness of daylighting
Effective daylighting not only reduces heat gain and cooling costs, but it's also very pleasant and free of excess glare at the working level.

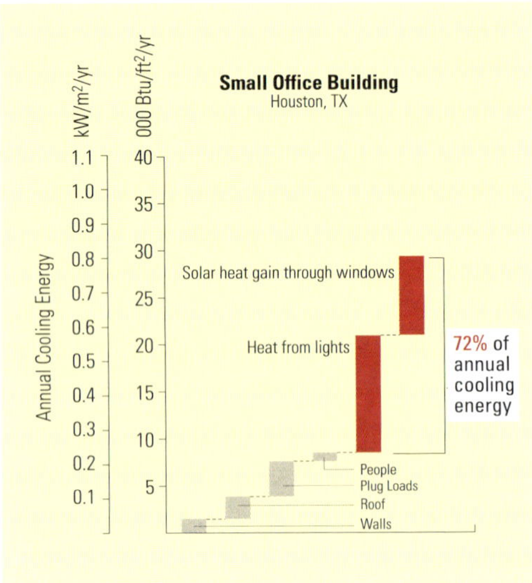

Fig. 9.6 The glass controls the cooling loads

Both the heat gain from lights, and the heat gain through windows are controlled by the glazing decisions. To keep more heat out of the building, use less glass, and make it effective for daylighting so lighting power can be reduced during the afternoon peak load periods.

less electrical power. Second, the cooling systems do not have to remove the heat that would otherwise be generated by total-coverage electrical lamping.

Figure 9.6 shows the approximate cooling loads for a small office building in Houston, TX. Note that 72% of those annual cooling loads are governed by the glazing decisions.

So, to keep the greatest amount of heat out of the building, first minimize the total amount of glass and shade it, as discussed earlier. Then make sure what remains is effective for daylighting, so that the heat generated by electrical lighting is also reduced.

By way of contrast, consider Figure 9.7, which shows a typical building built in the US during the 1990's. The building ignores all of this advice. To some, that sort of glazing may be visually dramatic. But it also leads to higher costs for the building enclosure, a more expensive mechanical system, and a higher annual energy bill. That glazing lets a great deal of unnecessary heat into the building, and also too much light at eye level. So the windows need shades. The shades create the need to turn on more interior lights even during the daytime. The lights in turn generate heat which must be removed by the cooling systems. Voilà—larger cooling systems and more complex ones, plus high electrical bills and higher carbon footprint from both cooling equipment and lighting power.

Effective daylighting design avoids these problems, greatly reducing the building's heat gain. A few words of caution are needed, however, for the interior designer. Make sure the ceiling finish and the wall finishes do not wipe out the daylighting. That is to say, the walls and ceiling must softly reflect and distribute the incoming daylight rather than absorbing it, or reflecting it harshly, creating glare. This means avoiding dark colors and polished surfaces. Instead, specify paints and wall coverings which have well-quantified reflective and diffusive characteristics.

Nobody here = Excess heat from lights
Smaller lighting zones would help avoid this excess heat gain (and wasted power).

Architecture blocking daylight = Excess heat from lights
The valence blocks daylight, so extra lights are needed, increasing heat gain and wasting more power.

Too much glass = Excess heat from lights
The view window is so large that the daylight must be blocked by shades, encouraging excess internal lighting, which generates excess heat.

Fig. 9.7
Glass & lighting decisions which increase heat gain—a lot

The building looks attractive—until you notice the thermal and energy-consumption consequences of the designers' decisions. To keep heat out of the building, don't make these choices.

Control lighting power according to available daylight and actual room occupancy

Next, install controls to turn the lights off when they don't need to be on, and/or controls which dim those lights when they don't need to be providing their full lighting power. As long as the lights are off, they are not generating heat, which saves the cost of removing that heat with cooling power.

This is a key point that is sometimes forgotten during architectural design, because most of the focus is on the windows themselves. Lighting *controls* are a necessary cost to achieve the heat-reduction and power reduction benefits of daylighting windows. And when designing for lighting modulation to take advantage of daylighting, it's also useful to include occupancy-based lighting control. There's no point in spending money to remove heat generated by lights in unoccupied rooms. In this respect, the most important first step is to design for small, rather than large lighting control zones.

In some ways, lighting control design is like HVAC design. That is to say, it's best to design with small zones. Lights can be controlled according to the needs of that one single space. Small zones will save more energy and provide better conditions for occupants. The effectiveness of daylighting changes throughout the day, as the sun moves around the building. And during the evening and night time, very few lights need to be on—but some will. Small, independently-controlled lighting zones can take advantage of these changing needs much better than would a single control for a large zone.

As an example of large lighting zones and their problems, consider the case of a traditional high-rise office building. Figure 9.7 shows the problem. Older lighting control is often floor-by-floor. So at night, when the building is largely unoccupied and only the cleaners are working, entire floors are brightly lit, instead of just the small areas being cleaned. Similar waste can happen in schools, when large zones are fully-illuminated even during the daytime over weekends and vacations, when most of the school is not occupied.

Modern lighting controls can take advantage of daylighting windows through ambient light sensors combined with occupancy sensors. If the illumination provided by the windows is sufficient, the controls can cut out a bank of lights, or dim some or all of them. And further, when the space is unoccupied, the controls can turn off all lighting power entirely until the space is reoccupied.

During design, the key point is that for the daylighting strategy to work in the real world, the lighting control zones must be *small enough* that the:

- ambient light sensors accurately report the light throughout that zone, at the relevant working height.
- occupancy sensors do not miss the fact that part of the zone is occupied, when most of the zone is unoccupied.

Occupant annoyance with lighting controls happens most frequently when the zones are so large that the decisions of the control system are illogical, and get in the way of the occupants needs. The potential power reduction and heat reduction benefits of daylighting will not be achieved if the controls annoy the occupants. They will simply bypass the controls, in order to get on with life.

At this point... pause before reading further

After all the key glazing and lighting control decisions have been made, any additional suggestions for the owner and the architectural designer may be largely irrelevant. If the building will look like that shown in Figure 9.7, then don't bother reading the rest of this section. Skip right to the suggestions for the HVAC designer. Better to admit the problems, and give the HVAC designer the larger budget and the larger mechanical space needed to overcome them.

On the other hand, if the building will look more like the building shown in figure 9.5, then the remaining suggestions in this section will continue to provide ways that the owner and architectural designer can keep even more heat out, further reducing overall costs.

Install continuous insulation outboard of the structure, along with an airtight waterproof membrane

Insulation located on the outermost layers of the building, along with an air-tight waterproof membrane, will keep solar heat out of the building, and it will also keep out hot, humid air. Plus, when the insulation is all on the *outside* of the structure, the large mass of the building (its structural frame) is inboard, so it will act as a thermal storage buffer, absorbing some of the excess heat from the indoor air during peak loads. Helped by that buffering capacity, the AC system does not have to work quite as hard during the hottest time of the day, when energy prices are highest and when cooling systems are at their least efficient point of operation.

The benefits of external insulation include less energy in cooling, plus steady indoor temperatures, which keep occupants more comfortable. Steady temperatures in turn help to avoid the thermostat fiddling which drives up energy consumption. The benefits of an airtight waterproof membrane include keeping the humid air out, so that it will not condense and support mold growth in building cavities. Also, with less air infiltration the AC system will use less energy to remove heat and moisture.

In cold climates, the wisdom of these architectural design choices is more constantly and clearly obvious. In areas closer to the earth's poles than to its equator, the annual temperature difference between indoor and outdoors is much greater.

Fig. 9.8
Conductive structure needs external insulation
Too often, insulation is applied between conductive framing members like those shown here. It's far more effective to place the insulation on the ***outside*** *of any such highly-conductive structural frame.*

So in Canada, Scandinavia, Southern Chile and Argentina, the freezing drafts remind designers that it's a good idea to have plenty of insulation, along with a waterproof membrane which is air tight. And in such cold climates, one is also reminded by condensation dripping from indoor surfaces of cold structural framing when the insulation is not outboard of that structural frame.

But in hot and humid climates, the consequences of insulation gaps and air infiltration are not so instantly obvious. Any condensation happens inside the wall, or on the outside of the building, where it is not easily visible. And infiltrating humid air takes a long time to generate an obvious mold problem. But the problems occur just the same. They just take longer to generate the lawsuits.

So consider the simple and compelling logic of keeping all the insulation outboard of the structure: no thermal bridges to waste cooling capacity. Notice the typical, highly-conductive steel framing in the building shown in figure 9.8. That framing is potentially a radiator, moving heat into the building from the hot outdoors.[4]

The architectural designer *could* place the insulation in between the framing members. That would be quite typical, but what a poor design. Imagine two competing approaches to insulating your body. Putting a sweater on over your chest is simple, quick and effective. That insulation is continuous, and it's outboard of your "structural frame." Insulating between structural members in a building is more like tearing that sweater into small pieces, and gluing those strips in between your ribs. You're going to loose much more heat that way, and the installation will take a lot more time.[5]

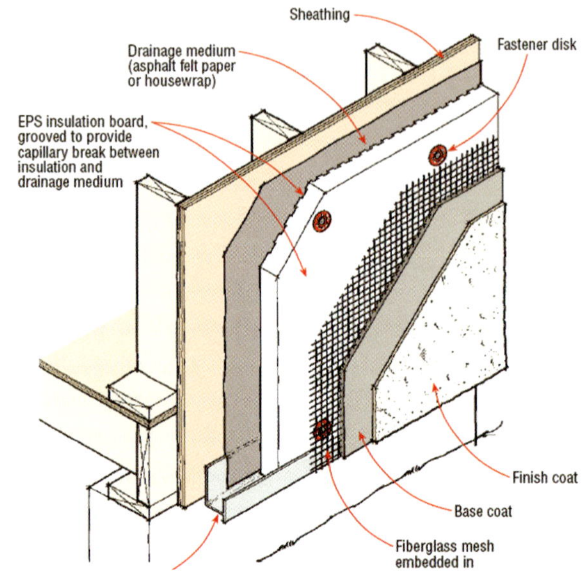

Fig. 9.9 EIFS external insulation
External Insulation & Finish Systems (EIFS) are an excellent choice for insulation in hot and humid climates. Just make sure the system is well-drained, so it does not trap water.
The same approach works for stucco.

Fig. 9.10 Brick veneer with external insulation

Fig. 9.11 Masonry block with external insulation

Better models are shown in figures 9.9, 9.10 and 9.11. In all cases, the insulation is continuous and outboard of the structure. And all of these alternatives have placed an air-tight waterproof membrane near the outer-most layer of the wall, to keep heat, humidity and moisture out of the building. This is a much more effective approach than stuffing foil-faced strips of insulation between steel ribs.

Allow enough money for demand-controlled ventilation

The building shown in Figure 9.7 illustrated many useful points to keep in mind when considering HVAC budgets and ventilation design. Notice that there are very few people left in the building—but there are some. One man is still at his desk, but most of the offices are empty and nobody is in the conference rooms. The image is a reminder of the value of lighting controls, and it also suggests the high value of demand-controlled ventilation.

In typical, low-grade HVAC systems, the HVAC designer does not have the budget to vary the ventilation to each of these spaces. So the system takes in a large volume of hot and humid outdoor air, then cools it, dries it and supplies it to *all* spaces, regardless of whether the spaces are actually occupied.

One could imagine the waste of high-volume ventilation it it were not already notorious. For example, one field study of U.S. Federal courthouses in Florida measured the ventilation rates to be between 400 and 6,000 cfm per person.[6] At that time, ASHRAE standards called for 20 cfm per person. The HVAC designer may have had little choice. Without the money to vary the ventilation air, he probably had to design the ventilation air flow to meet some minimum occupancy assumption. But courthouses are very lightly occupied at nearly all times—even far below the assumed minimum. Imagine the even greater needless ventilation in those courthouses at night. The systems would still be on, because there are usually a few attorneys, security personnel and law-enforcement officers working in those buildings all night long.

With demand-controlled ventilation, the ventilation air can be metered into each space in proportion to its actual occupancy. If a conference room or classroom is not occupied, the ventilation can be reduced to the bare minimum needed to dilute contaminants generated from fabrics and finishes. For example, the current ASHRAE Standard 62.1 calls for ventilating a university classroom with 0.06 cfm/ft^2 plus 7.5 cfm/person. In a classroom measuring 1,000 ft^2, that would mean an occupied ventilation rate of about 540 cfm, compared to the unoccupied ventilation rate of only 60 cfm. In other words, a nine-fold reduction in the heat and humidity load when the ventilation air to that classroom can be modulated down to its minimum.[7] Figure 9.12 shows another example in graphic form.

[The current ASHRAE Standard 62.1 calls for ventilating a university classroom with 0.3 l/s/m^2 plus 3.8 l/s/person. In a classroom measuring 100 m^2, that would mean an occupied ventilation rate of

Fig. 9.12 To reduce latent heat gain, don't ventilate unoccupied spaces

In humid climates the largest load from ventilation air is latent heat (humidity). To reduce this part of the building's heat gain, provide the HVAC designer with a budget which allows the ventilation air to be reduced in spaces which are unoccupied.

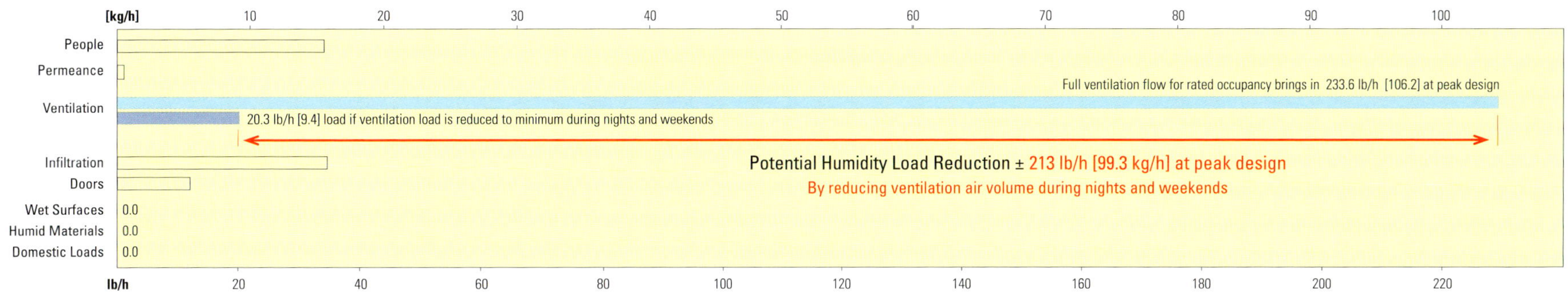

about 410 l/s, compared to the unoccupied ventilation rate of only 30 l/s. In other words, a 13-fold reduction in the heat and humidity load when the ventilation air to that classroom can be modulated down to its minimum.]

However, such modulation is not achievable with low-budget HVAC systems. Each space needs some form of occupancy sensor, plus a controller and damper which meters the right amount of ventilation air into that space. The most direct means of measuring the true ventilation requirement is a carbon dioxide sensor. These basically measure human respiration and metabolism. On a rise in CO_2 concentration, the system will send more than the minimum ventilation air to that space.

The extra cost is not just in sensors and controls, but also in the duct arrangement which separates the ventilation air from the rest of the supply air. There are many clever ways that the HVAC designer can accomplish this dedicated and defined ventilation air, but they all cost more money than the typical low-budget, highly wasteful approach of "constant-minimum-ventilation-to-the-whole-building."

The main point for the owner and for the architectural designer is that, after saving all that money by removing the wasteful glazing, the resulting savings can further reduce heat gain by giving the HVAC designer the money to provide demand-controlled ventilation.

Some green building rating systems will then give the building extra credit for superior ventilation at the design stage. But the real benefits are much more long-lasting. The occupants will enjoy better indoor air quality, the owners will spend much less money on HVAC operation, and the risk of mold in the building will be greatly reduced, as explained in detail in Chapter 5.

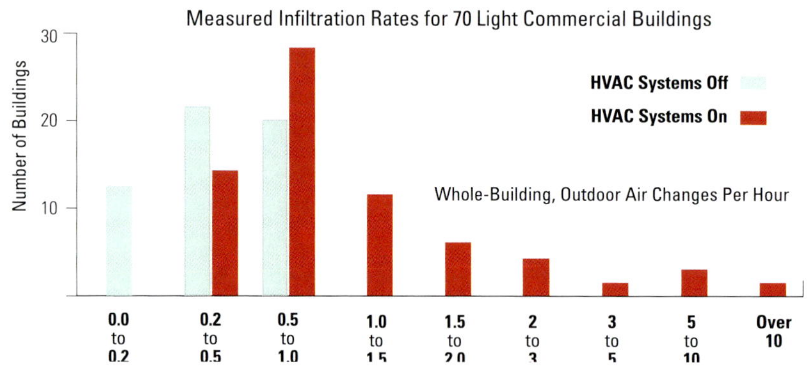

Fig. 9.13 Leaky air systems pull hot and humid air into the building
To avoid this heat gain, provide the HVAC designer with a budget large enough to eliminate the use of building cavities as supply and return plenums.

Allow enough ceiling height and enough money for ducted supply and return air

The last suggestion for owners and architectural designers concerns ceiling height. One could reasonably ask, "How could the ceiling height and ducted supply and returns possibly affect the heat gain of a building?" Well, it's all about preventing air leaks.

Moving air through tight duct work, rather than dumping air into building cavities reduces the amount of air leakage into and out of the building enclosure. HVAC fans create big air pressure differences. In below-floor supply and above-ceiling return air plenums, those large pressure differences are carried out to the exterior walls. As the fans pull air from the return plenums, small gaps in the exterior wall joints can allow hot and humid air into the building, where it becomes a cooling and dehumidification load. And if supply air is pushed into a leaky plenum under the floor (a currently popular but frequently problematic design), then as air escapes out of the gaps in the exterior wall, it must be replaced by an equal amount of makeup air, which generates extra cooling and dehumidification loads.

The typical whole-building air leakage numbers are not small. Figure 9.13 shows the difference in air exchange rates in 70 light commercial buildings when the systems are off (only wind and stack pressures driving air exchange), compared to the air exchange rates when those leaky air systems are turned on. The shift to the right shows the vast increase in outdoor air exchange which comes from leaking duct work, and especially from leaky return air plenums.[8]

It's basically impractical to seal up supply and return air plenums as tightly as metal duct work to avoid this huge increase in cooling and dehumidification loads. Just as soon as the plenum is sealed up—along comes the telephone service guy, or the cable guy, or the plumber, or all three, and then wham—another set of air leaks in that plenum wall. In a recent survey of seven high-budget Federal buildings with underfloor supply air distribution plenums, the U.S. General Services Administration measured air leakage rates of 40 to 100% of the total design air flow—after all the air sealing was complete.[9] In other words, those systems had to somehow come up with 40 to 100% extra supply air to meet the design HVAC loads.

Certainly, the HVAC designer would prefer to avoid this lost capacity and increased load. But if the owner and architectural designer do not allow enough space between floors for supply and return duct work to be fit between the structure (and around plumbing, wiring, fire protection and the attendant support brackets), the designer will be forced into using return air plenums. He or she might then specify in a stern, no-nonsense voice: "All plenums shall be sealed up air tight using spray-applied fire sealant after all carpentry, electrical work, communications cabling, security wiring and plumbing is complete..." And that specification may even get into the General Contractor's scope of work. But do you *really* think that air sealing will actually happen? We don't, either. The evidence gathered from field investigations of building-related problems supports that skeptical view.

So that's why the owner and architectural designer should provide the HVAC designer with enough space between floors, and enough money for air-tight duct connections. That way, the heat and humidity loads in your building will be much less than in typical buildings. Your building will be more comfortable, it will cost less to operate, and it will have a reduced mold risk compared to typical air conditioned buildings in hot and humid climates. Now, some suggestions for the HVAC design.

HVAC Designer Decisions

After the architecture has established the baseline cooling loads, the HVAC designer can consider some suggestions for keeping heat out of the building through clever engineering.

Engineers have sometimes been described as those who "... can do for a dollar what any fool can do for ten." Without comment on the related problem of underpriced engineering services, that thrifty bias guides the suggestions on the HVAC side. They are arranged in order, beginning with the lowest-cost, highest value ways to keep heat out of a building.

Seal up all air-side joints and connections

When the connections in air systems are not air tight, the suction and positive pressure generated by the fans is transferred to the building cavities, and then to the exterior walls. Since walls and wall joints are seldom air tight, outdoor air is pulled and pushed through the joints by the pressures created by the HVAC fans. In light commercial buildings, the amount of HVAC-driven, outdoor air exchange is really astonishing. Figure 9.13 showed the difference in outdoor air exchanges rates in 70 buildings with the systems on and off.[8] When the systems are turned on, the air exchange rate skyrockets. This is because in the past, most HVAC designers and most HVAC installers have not understood how important it is to seal up all the duct connections.

The lowest-cost, highest value way to keep extra heat out of the building through HVAC design is to simply specify that all duct joints, and most especially all duct *connections* to any box containing a fan, must be sealed, using mastic.

All connections need to be sealed, including the *connections at VAV boxes, filter boxes, cooling and heating coil housings, PTAC cabinets and all grills, registers and diffusers.* That seal-with-mastic specification also includes all joints and connections in *exhaust air ducting*, such as that from bathrooms, showers or kitchens.

This suggestion should please the most thrifty owners and HVAC designers. According to sheet metal contractors, sealing up the connections with mastic will probably add only 3 to 5% to the cost of the duct installation. And tight connections save a great deal of fan energy, as well as reducing the amount of heat and humidity which the HVAC fans would otherwise pull into the building. The energy losses of leaking duct connections account for about 30 to 40% of the total annual cost of operating the HVAC system.[10] Not only that, but those leaking connections are often a major reason for mold problems in buildings in hot and humid climates, as explained in detail in Chapter 5.

So if a hyper-thrifty owner or architect is concerned with the cost of tightly-sealed air connections, they can consider this question: Is there any less-costly way to reduce HVAC-driven mold risk while saving 30 to 40% of annual HVAC operating costs? Sealing connections is very cost-effective, and increasingly, it is required by energy codes.

Don't use building cavities to carry supply or return air

Another description for a ceiling return plenum or for an underfloor supply air plenum is "a very leaky air duct."[8,11] To keep extra heat and humidity out of the building, don't use building cavities as supply or return air plenums.

If the owner and architectural designer have not provided the money and the space needed for hard-connected, tightly-sealed ducts, it might be in everybody's best interest for the HVAC designer to point out that the building will pull in more heat and humidity than necessary, and that the risk of mold will also be higher than necessary.

If that conversation does not obtain the space and the budget for sealed duct work, then the prudent HVAC designer will make sure the building is equipped with more-than-normal dehumidification capacity, especially for operation during the part-sensible-load hours when humidity is at its peak. It might also be useful to note for the record the concerns about lost AC capacity and increased mold, and to provide the architectural designer with a specification to seal all joints and penetrations of the plenums with fire-rated sealant so that they are air-tight. (One can hope... but see Figures 9.14 and 9.15.)

Fig. 9.14 Air-tight plenums?... not for long

Underfloor supply air plenums and above-ceiling return air plenums are very difficult to seal air-tight, especially over time. Instead of relying on leaky building cavities, use air-tight, sealed ducts and mastic-sealed duct connections.

This is not to suggest that hundreds of thousands of buildings with return and supply air plenums are not "operating successfully" all over the world. But, success is a relative concept. The air that leaks out of and into those plenums is a huge energy waster and a mold risk. Adding "high-efficiency" cooling equipment to a building with leaky supply and return air plenums is basically like putting lipstick on a pig.

Fig. 9.15

Air-tight plenums - Attractive in theory, difficult in practice

With so many different trades working in building cavities, it is really difficult to ensure that all gaps, holes and joints are sealed air tight. These photos show examples of underfloor supply air plenums that (theoretically) had been "sealed up, air-tight."

Fig. 9.16

Reduce ventilation air flow when rooms are unoccupied

Outdoor air is hot and humid, all year long. It's a very large load. So, reduce the ventilation air whenever the building is not fully occupied.

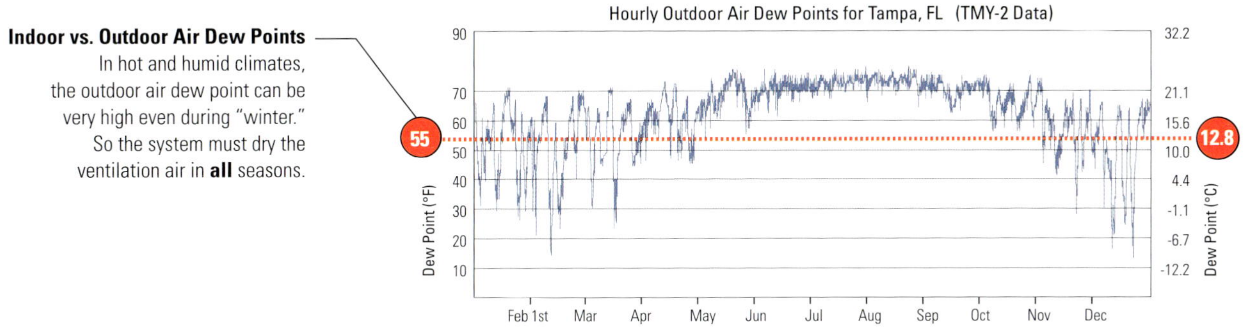

Install demand-controlled ventilation

Outdoor air is hot and humid, and it costs a lot of money to clean it and dry it out. Don't bring it in until you need it. And when you have to bring in ventilation air, don't bring in any more air than you really need for the number of people actually occupying the building.

Easy to say—but difficult and expensive to do. But demand-controlled ventilation is especially worth doing in hot and humid climates because the cost of adequate dehumidification is so high for so many hours per year. In moderate climates, a constant volume of ventilation air is not quite as expensive, because often that air is reducing rather than increasing the AC load. Not so in hot and humid climates. There is nearly always a dehumidification load associated with ventilation air, for nearly all the hours each year.

Note the graphic in Figure 9.16, which shows the hourly dew points for a typical year in Tampa, FL. Note that even during "winter" months, the outdoor dew point is far above the indoor dew point.

There are several ways to avoid excess ventilation without the risks of inadequate ventilation. More detailed suggestions for different ways to modulate ventilation air in response to vaiable occupancy are described in Chapter 3 (Managing Ventilation Air) and in Chapter 15 (Designing Ventilation Air Systems).

The main point is that demand-controlled ventilation is more cost-effective in hot and humid climates than in moderate climates because the loads are higher. And especially in hot and humid climates, less ventilation humidity also means less mold risk for the building. So demand-controlled ventilation deserves a careful look from the HVAC designer, especially for public buildings like schools and courthouses. These have highly-variable occupancy, and in the case of schools, long periods of little or no occupancy when the AC system must still operate. Constant-volume ventilation makes no sense for such buildings.

Don't let air economizers fill the building with humid air

In hot and humid climates, using outdoor air for "free cooling" usually results in a higher, not a lower load for the building.

In hot and humid climates, the outdoor air is nearly always more humid than what you'll want indoors. So for most of the hours in a year, even when the temperature outdoors is below the indoor temperature, the ventilation air will still need to be dried.

Figure 9.17 shows the ventilation load indices (VLI) for several different U.S. locations.[12] The VLI is the sum of the energy needed to bring one cfm [or one l/s] from the outdoor air conditions down to neutral indoor air conditions, over all 8760 hours in the year. The VLI has two components—the sensible load and the latent load imposed by that one cfm of ventilation air. Both of these annual loads are expressed in ton-hours per cfm per year. [kW per l/s per year].

Note how the ventilation air's latent load—its humidity—is far greater on an annual basis than its sensible cooling load. That's a reminder that an air-side economizer is not usually economical in

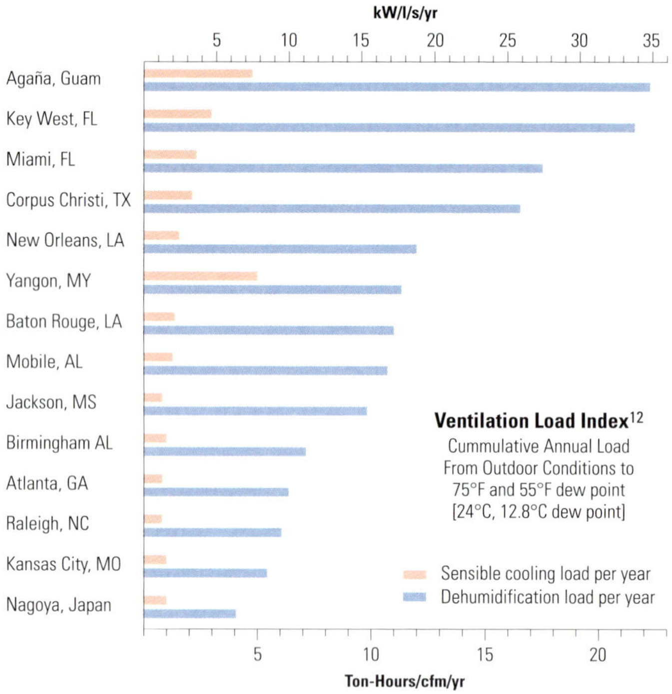

Fig. 9.17

Control air-side economizers based on both dew point and dry bulb temperatures

In ventilation air, the annual dehumidification load is far larger than the sensible cooling load. Therefore, any air-side economizer must be controlled by the outdoor air dew point in addition to the outdoor air temperature.

most hot and humid climates. It costs a great deal to remove that latent load, even when the indoor sensible load is reduced by the economizer air.

If your site-specific analysis shows that an outdoor air economizer cycle will indeed reduce the total annual loads, it's important that the economizer be controlled not only by the outdoor dry bulb temperature, but also by its dew point. If the outdoor dew point is above the target indoor dew point (usually 55°F [12.8°C]), then the economizer should not flood the building with humid outdoor air, even if the outdoor air's dry bulb temperature appears attractive.

When the indoor dew point is too high, the occupants crank down the thermostat setting in a desperate search for better comfort, leading to high energy costs even when outdoor temperatures are moderate. So avoid the use of air-side economizers, unless these are controlled by dew point, as well as by dry bulb temperatures.

Use exhaust air to precool and predry ventilation air

When the building exhausts large amounts of cool and dry air, it makes sense to use that air to pre-dry and to precool the incoming ventilation air, using an enthalpy heat exchanger. These devices greatly reduce the loads on buildings in hot and humid climates.

In many cases, adding an enthalpy heat exchanger actually reduces the net installed cost of the cooling and dehumidification systems. Plus, the operating cost of the ventilation air is much lower for the entire life of the system. Enthalpy heat exchangers in hot and humid climates has been called "the closest thing to a free lunch in HVAC engineering."

To take advantage of these big benefits without the downside, just keep in mind three cautions. First, recognize that the effectiveness of ventilation air pretreatment depends on the volume, the temperature and the dryness of the exhaust air. So try to collect as much clean exhaust air as possible and bring it back to the heat exchanger before it leaves the building. Second, an enthalpy heat exchanger cannot dry the incoming ventilation air unless the exhaust air is also dry. In other words, the system still needs effective indoor dehumidification even when the outdoor temperature is low, when the cooling system alone may not be operating long enough to dry effectively.

Third, keep in mind that the heat exchanger presents a significant resistance to air flow, on both the exhaust and ventilation air streams. For many hours each year, even in hot and humid climates, the outdoor temperature will be low enough that one does not want to *heat* that incoming air with the warmer exhaust. During those hours (sometimes thousands of hours each year depending on the climate) it makes sense to bypass air around the heat exchanger. Such a bypass avoids the expense of the fan horsepower needed to push the air through the heat exchanger.

Finally, remember that if the design follows ASHRAE Standard 90.1-2004, it's a mandatory requirement to recover energy from someplace in the building when an individual fan system has a supply air flow over 5,000 cfm, and when the *outdoor air portion* of that total flow is more than 70%. So when the architectural design allows the exhaust and ventilation air streams to come close together, an exhaust air heat exchanger is one good way to meet this requirement of ASHRAE standard 90.1.

Keep the indoor dew point low, allowing warmer indoor temperatures

When it's hot outdoors, the colder the air is kept inside the building, the greater is the heat flow through its windows, walls and roof. So to reduce the amount of incoming heat, allow the indoor air temperature to rise higher.

In some parts of the world, energy use laws prohibit cooling the indoor air to the levels which are quite common in North American buildings. In Japan for example, office buildings in Tokyo are seldom cooled below 81°F [27°C], because lower temperatures are considered quite wasteful of energy.[12] In contrast, air conditioned buildings in North America are routinely cooled to 75°F [23.9°C]. Indeed, strange as it may seem to those who live in other parts of the world, U.S. buildings are often chilled down to 72°F [22.2°C] and sometimes even lower.

One of the many reasons for such deep cooling is the occupants' desperate attempts to achieve comfort when indoor humidity is too high. If the only control you have is the thermostat, then dropping the temperature set point will be the quickest way to improve comfort when the dew point is too high. Chapter 2 explains the interacting variables of human thermal comfort in more detail. But for purposes of this chapter, it's enough to note that when the indoor dew point is kept low (below 55°F [12.8°C]), even people accustomed to North American air conditioning levels are often willing to let the thermostat set point rise to 79°F [26°C] before comfort complaints are registered.[14,15]

So to reduce the amount of heat that gets into the building, keep the dew point low and then let the dry bulb temperature float upwards, until it's just below the temperature at which occupants notice the temperature. As explained in chapter 2, each building will be different, and each group of occupants will respond differently to temperature and dew point levels. But as a starting point, try keeping the dew point below 55°F [12.8°C] and setting the dry bulb temperature at 79°F [26°C]. Field-measured data suggests that these levels can save about 15% of annual cooling costs.[14,15]

References

1. Most architectural designers will recognize these quotes, which describe the design aesthetic of many famous modernist architects active during the early and middle of the 20th century. Notable for elegantly simple buildings, moderist design has been a powerful inspiration for the architectural assumptions and therefore the design preferences of owners. Less helpfully however, these buildings were often a thermal disgrace, reflecting the astonishingly low energy costs in the US during that short period in history. Many were built with uninsulated, highly conductive steel frames, infilled with huge sheets of glass. HVAC designers could wish that *currently* famous architectural designers would follow the guideline that "less is more" with respect to glass. Creative designs which use *very little* glass would provide future generations of designers and owners with a more sustainable visual inspiration than the current wasteful fashion preference for "all glass, all the time."

2. Gronbeck, Christopher *Window Heat Gain Calculator* 2007. http://www.susdesign.com/windowheatgain/
3. Turner, Cathy and Frankel, Mark. 2008. *Energy Performance of LEED© for New Construction Buildings*. - March 4th, 2008. New Buildings Institute, Vancouver, WA www.newbuildings.org
4. Lstiburek, Joseph. 2007. "A Bridge Too Far" ASHRAE Journal, October 2007 pp. 64-68.
5. This analogy comes from John Straube, professor of civil engineering and building science at Waterloo University, in Waterloo, Ontario, Canada—a nation which, on the whole, has been quite conscious of the importance of insulation.
6. Cummings, James. Private communication. Project leader, Florida Solar Energy Center, Cocoa, FL
7. ASHRAE Standard 62.1-2007 - Ventilation for Acceptable Indoor Air Quality. ASHRAE, Atlanta, GA www.ashrae.org
8. Cummings, James; Withers, Charles; Moyer, Neil; Fairey, Philip; McKendry, Bruce. *Uncontrolled Air Flow in Non-Residential Buildings*. 1996. Final Report, FSCEC-CR-878-96, 1996. Florida Solar Energy Center, 1679 Clearlake Rd., Cocoa, FL. 32922
9. Holley, William, Wood, James and Gupta, Vijay. 2006. *Measured air leakage rates in systems with underfloor supply air distribution plenums.* Unpublished reports of tests conducted in 11 Federal buildings, with a combined gross floor area of 8 million ft^2 [750,000 m^2]. Office of the Chief Architect, U. S. General Services Administration, Washington, DC. Examples included:
   - U.S. Courthouse 1 - 70% air leakage
   - U.S. Courthouse 2 - 43% leakage
   - U.S. Courthouse 3 - 34 to 68% initial leakage and 26 to 59% leakage after remediation
   - Census Bureau office building - 60% leakage
   - NOAA Office building - 40% air leakage
   - Federal office building: Zone 1, 200% leakage; Zone 2, 45%
10. Delp, William; Woody, Nance; Matson, E.; Tschudy, Eric; Modera, Mark & Diamond, Richard. *Field Investigation of Duct System Performance in California Light Commercial Buildings.* 1998. Report LBNL #40102, Building Technologies Program, Lawrence Berkeley National Laboratory, Berkeley, CA
11. Henderson, Hugh; Cummings, James; Zhang, Jian Sun; Brennan, Terry. *Mitigating The Impacts of Uncontrolled Air Flow on Indoor Environmental Quality and Energy Demand in Non-Residential Buildings*. 2007. Final Report - NYSERDA Project # 6770. New York State Energy Research & Development Authority, 17 Columbia Circle, Albany, NY 12203-6399
12. Harriman, Lewis G. Kosar, Douglas and Plager, Dean. 1997. *Dehumidification and Cooling Loads from Ventilation Air.* ASHRAE Journal, November, 1997 pp.37-45. ASHRAE, Atlanta, GA. www.ashrae.org
13. Moffett, Sebastian. "Japan Sweats it Out as it Wages War on Air Conditioning." *Wall Street Journal*, Sept. 11th, 2007.
14. Spears, John; Judge, James. "Gas-Fired Desiccant System for Retail Super Center" 1997. ASHRAE Journal, October 1997 pp.65-69.
15. Fischer, John; Bayer, Charlene. "Failing Grade for Many Schools - Report Card on Humidity Control" ASHRAE Journal, May 2003. pp.30-37.

Image Credits

Fig. 9.4. Tips for Daylighting. O'Connor, Lee, Rubenstein & Selkowitz. U.S. Department of Energy - Lawrence Berkeley Laboratory, Berkeley, CA

Fig. 9.5 Tips for Daylighting

Fig. 9.8 Building Science Inc. www.BuildingScience.com

Fig. 9.9 Journal of Light Construction www.JLConline.com

Fig. 9.10 Building Science Inc. www.BuildingScience.com

Fig. 9.11 Building Science Inc. www.BuildingScience.com

Fig. 9.15 Courtesy of the U.S. General Services Administration, Office of the Chief Architect, and James Woods, Ph.D, P.E.

Chapter 10

Architectural Lessons From Tropical Storms

By Joseph Lstiburek

Fig. 10.1 Storms happen often in hot and humid climates

For example, Hurricane Katrina hit the U.S. Gulf Coast in 2005, followed by Hurricane Ike in 2008. Tropical storms teach painful lessons. History can guide architectural decisions for buildings in hot and humid climates—unless the owner or architect prefer to have the next storm teach the same lessons one more time.

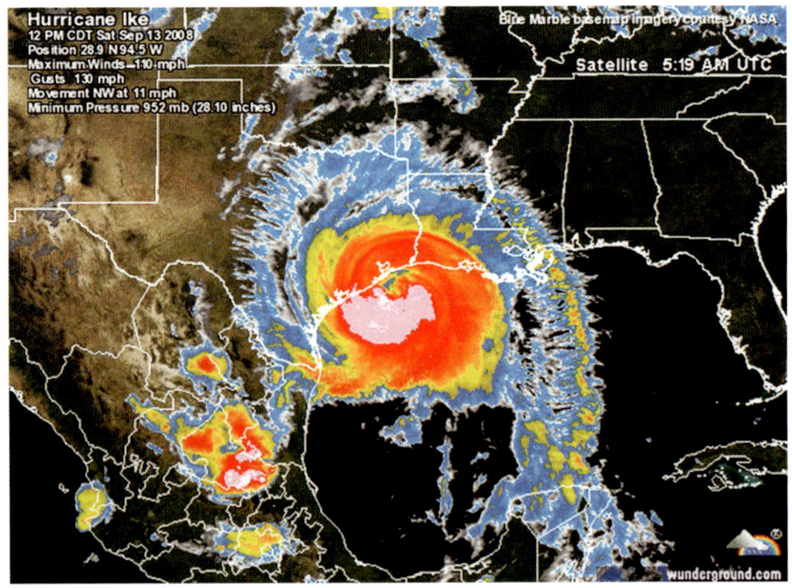

Tropical Storms Are Compelling Teachers

We learn our most enduring lessons from disasters. In North America, Hurricane Andrew (1992) taught us about wind. Hurricanes Charley, Frances and Jeanne (2004) taught us about rain. The Red River of the North Basin taught us about floods (1997). Hurricanes Katrina (2005) and Ike (2008) had it all: wind, rain and flood.

We tend to build and rebuild in pleasant places—that periodically get flattened and flooded by hurricanes, typhoons and cyclones. This is what we do, for better or worse. So how should we rebuild in those areas? And when we build entirely new buildings on sites that we *know* will see storms over the useful life of those buildings, what measures should the owner insist that the designers include in the plan?

We often design and build to resist wind, rain and flood as individual forces. It's much less common to design and build in a way which resists all three forces simultaneously. But the lesson of Hurricane Katrina was particularly harsh. If it was not obvious before 2005, it is certainly clear today that structures built in areas prone to tropical storms must address wind, rain and flood—all at the same time.

Fig. 10.2 Coastal buildings
Raising buildings above the expected storm surge is the foundation of a rational architectural style for buildings at the ocean shore in hot and humid climates. These examples are located in Charleston, SC and in the Florida Panhandle. Note the location of the AC condenser in the lower photo.

Resisting wind and rain

To address wind and rain, we can look to Florida for good advice. When you are hit by hurricanes on a regular basis, you tend to learn a thing or two. The Florida Building Code underwent significant revision after Hurricane Andrew. The success of those changes was apparent during the four hurricanes that hit Florida during August and September 2004. During those storms, *no one died in any structure built under the revised code.*

Beyond that impressive life safety accomplishment, structural damage due to wind was minor in buildings built under the new code. And the Florida code has been further improved as a result of the 2004 hurricane season. Excellent specific guidance is readily available from those Florida experiences.[1,2]

On the other hand, in an irony that is best appreciated by architects, engineers and contractors, the success of the code changes in Florida regarding wind resulted in significant unhappiness with *rain* entry. When you are missing a roof and an entire façade, you're not likely to complain about rain leaks. New Florida buildings did not blow away. Roofs and façades stayed intact during the 2004 hurricane season. But they did leak rainwater.[3]

Preventing rainwater entry is rather straightforward[4]—but only if design decisions are made to address it. As a result of the 2004 hurricane season, control of rainwater entry is now a Florida priority. It should be a priority everywhere it rains. In addition to the information provided by Reference 10.4, some specific suggestions are provided here in this book, in Chapter 8 (Keeping Water Out of Buildings).

Resisting storm surges and floods

To address floods, we should look to common sense. Specifically:

- Elevate the building to minimize its contact with flood water and any coastal storm surge;
- Build with materials that tolerate wetting; and
- Design assemblies to easily dry when they become wet.

The U.S. Federal Emergency Management Agency (FEMA) has all of the basics correct in their guidance, and has had them correct for a long time.[5,6,7]

The obvious point is to build above the ground level—as high as possible and practical. Near the edge of an ocean, in areas such as Charleston, S.C., and Florida's panhandle, the first level in new buildings is often used for parking, storage and building access. Utilities, services and equipment also are elevated (Figure 10.2). Raising the bulk of the building above the predictable storm-surge levels should be the fundamental basis of design in areas which are at high risk for coastal flooding.

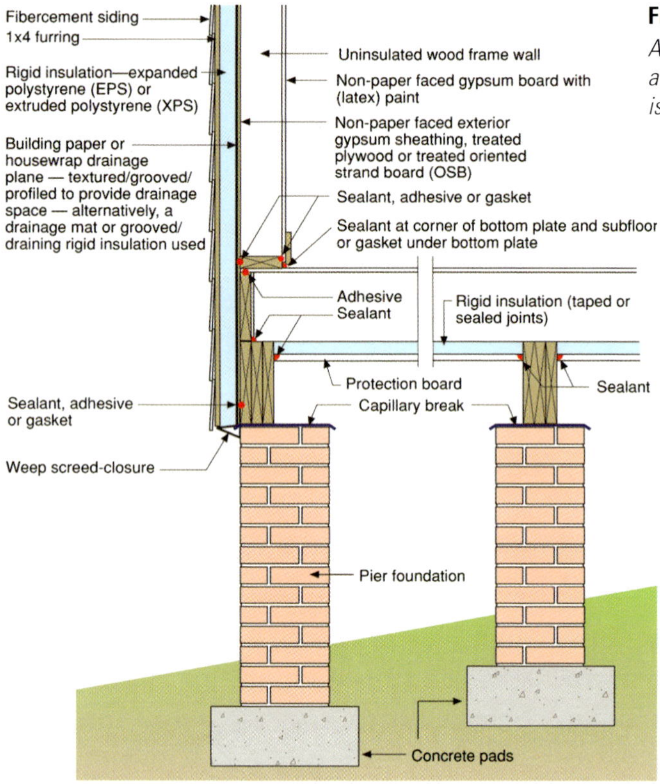

Fig. 10.3 Pier foundation

Away from the coastal storm surge, but still in a zone with a high probability of flooding, pier foundations are useful. Note how the stud wall cavities are kept clear, for easier drying after and moisture event. This is possible because the insulation is outboard of the sheathing rather than inside the stud cavity.

Also in Florida—in a modern adaptation of medieval design practices—residential buildings are constructed from masonry at the lower level and wood frame on the level above. In medieval times the lower level was built out of strong materials to resist marauding invaders and the usual intoxicated noblemen and cranky neighbors. In modern times, in areas prone to tropical storms, the lower level is built more durably to resist marauding floods and the usual surface water from storm and rainwater runoff.

Materials and assemblies which tolerate water

The most water-sensitive building materials in widespread use today are paper-faced gypsum board and fiberglass batt insulation. When

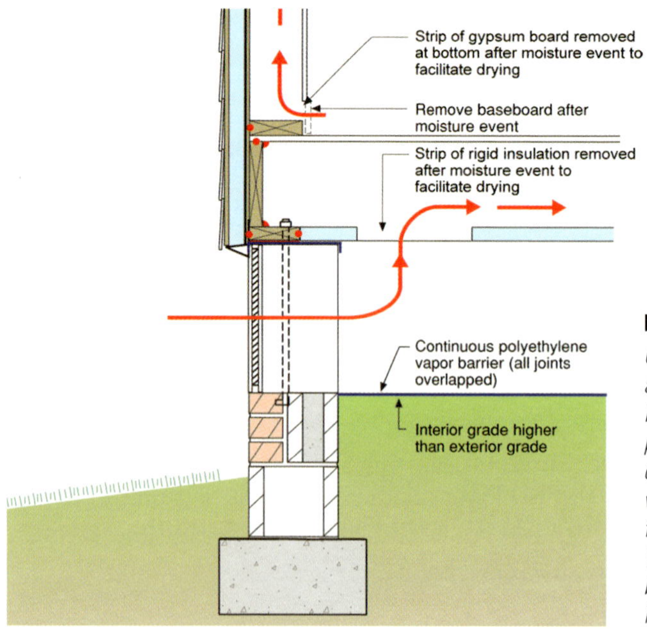

Fig. 10.4 Raised crawl space

Crawl spaces should always be above grade, in any climate. But this detail is especially important in areas prone to flooding. This approach limits the potential for water intrusion, and allows easier drainage if the worst should happen. The flood vent must be essentially air-tight, until an actual flood occurs. Crawl spaces should not be vented. They should be treated basically like occupied basements, for many reasons which are discussed in detail in Chapter 8 .

Further inland, pier foundations and elevated crawlspaces can be appropriate, because the risk of coastal flooding is reduced. (Figures 10.3 and 10.4) Inland, the more common risks are from surface water runoff. Slabs-on-grade should be avoided in flood zones. If slabs are used, they should be constructed as raised slabs (Figures 10.5 and 10.6).

Additionally, we should use water-resistant and water-tolerant materials. Again, FEMA guidance has it right.[8,9] Recent work has demonstrated the effectiveness of constructing assemblies from water-resistant materials.[10] Not surprisingly, in North America, Florida leads the way. Most commercial buildings in Florida are constructed from masonry and concrete (Figures 10.7 and 10.8).

Fig. 10.5 Raised slab - Diagram

Slabs-on-grade offer no protection against floods and storm water runoff. But raising the slab, as shown here, keeps the construction economical and avoids the often-problematic crawl space. Note that the baseboard can be removed, to allow efficient and rapid wall drying in the event of a flood.

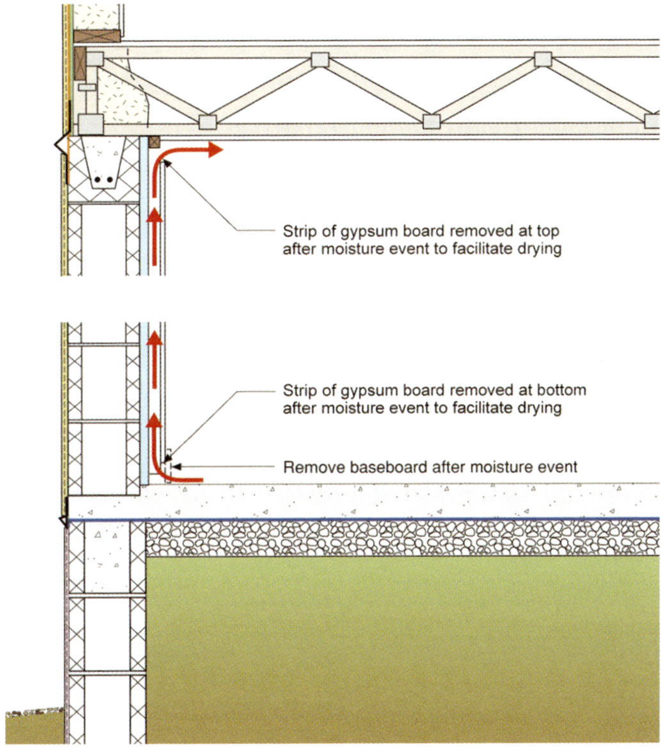

Fig. 10.6 Raised slab - Photo

Fig. 10.7
Masonry can be water-tolerant

Masonry resists floods and storms very well. But it's critical to design the walls so they dry out, rather than trap water. See figure 10.8 for useful suggestions.

soaked, paper is easily digested by mold and bacteria. And when glass fiber batt insulation gets soaked inside building cavities, it never really dries out. Soaked insulation serves as a reservoir, providing the moisture which leads to microbial growth in flooded buildings, even after the visible indoor surfaces appear dry.

Gypsum board requires more care in tropical storm zones

In typical North American design practice, it's difficult to avoid paper and glass fiber batts. The entire interior of most buildings is typically lined with paper-faced gypsum board. Shaft walls, utility and service walls are wrapped with paper-faced gypsum board. And in many commercial noncombustible building assemblies, the exterior is also wrapped with paper-faced gypsum board. We are building paper buildings. Especially for areas prone to tropical storms, this is a problem.

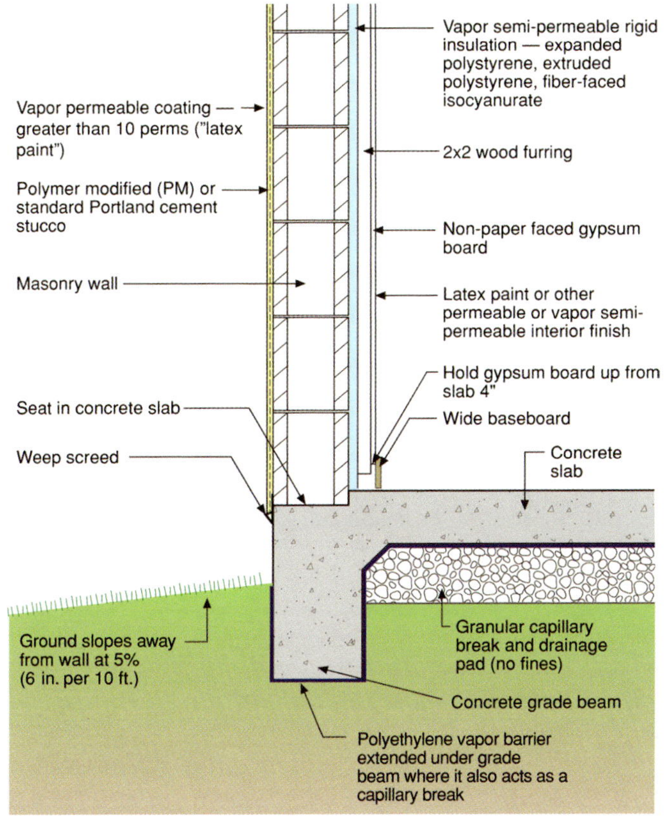

Fig. 10.8 Masonry wall details which avoid trapping water

Note the seat in the foundation, which acts like flashing. Then the weep screed allows trapped water to flow out of the block. Indoors, the wall board can dry, because there's an air gap behind it and its wall covering is vapor-permeable.

The simple answer is; don't use untreated paper-faced gypsum board. Instead, use exterior-grade, glass-fiber-faced gypsum board on the exterior (Figure 10.9). And on the interior, use interior-grade glass-fiber-faced gypsum board. For the more adventurous designer, paper-faced, interior-grade gypsum board with antimicrobial treatments has recently become available. These products promise resistance to microbial growth for longer periods of wetting than untreated paper-faced board.

Fig. 10.9 Moisture-tolerant exterior gypsum board

The yellow face is made of glass fiber cloth, impregnated with resin. This is much more tolerant of moisture than conventional paper-faced gypsum board. Also note that the stud cavities are not insulated. That's because the insulation will go on the outside of the moisture-tolerant gypsum board—the best location for insulation in a steel-stud exterior wall.

But in all cases, keep the lower edge of interior gypsum board up above the floor level. Create a gap, which serves as a capillary break to prevent the upward flow of moisture into the walls from damp floors. That gap at the floor should be sealed with fire-rated material, so the walls will still keep fire from moving from room to room, and so the air at that gap does not transmit noise between rooms.

Insulation and exterior cladding choices

The simplest answer to the risks of soaked glass fiber batt insulation is—don't use it. In commercial steel stud assemblies, do not insulate the cavity with loose batts, or anything else. Instead, install a different form of insulation on the exterior of the assembly. Then, design the cavity to be ventilated after a moisture event to facilitate drying (Figures 10.10 and 10.11). The insulation on the exterior can be semi-rigid fiberglass, foam plastic boards or rock wool.

Insulating steel stud cavities has always been a bad idea. The insulated cavity will slow down the heat gain a little bit, but the steel studs will conduct a great deal of heat into the building. An R-19 fiberglass batt installed in a 5.5 in. [140 mm] steel-frame wall yields an effective thermal resistance of ***approximately R-7*** when insulation is installed "with good workmanship" (using the isothermal planes method described in Chapter 25 of the *ASHRAE Handbook—Fundamentals*). But how often do we see "good workmanship" to that standard, which in any case would only reach R-7? In real-world installations the enclosure leaks air, reducing the composite R-value still further. Also, the batts are usually compressed by wires and pipes, and the batts often leave air gaps at the top, sides and bottom of the cavities. Given the real-world shortcomings of batts inside steel stud cavities, why even bother insulating? (Also consider that ASHRAE Standard 90.1 calls for R-19 for walls in hot and humid climates, to minimize energy use.)

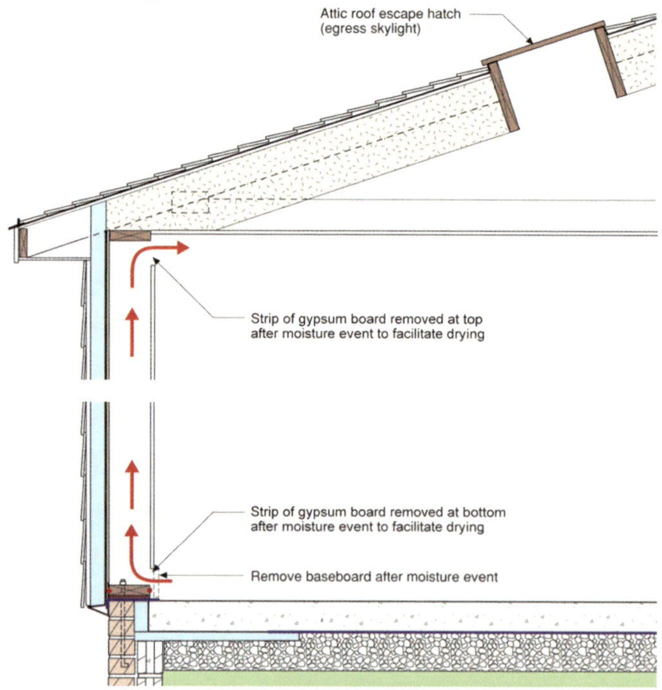

Fig. 10.10
Residential walls which dry easily

With the insulation on the outdoor side of the sheathing, the stud cavities can pass air freely for drying, after the baseboard is removed and after temporary openings are made at the ceiling.

Fig. 10.11

Commercial walls which dry easily

Again, with the insulation on the outdoor side of the sheathing, the stud cavities can pass air freely for drying, after the baseboard is removed and after temporary openings are made at the ceiling.

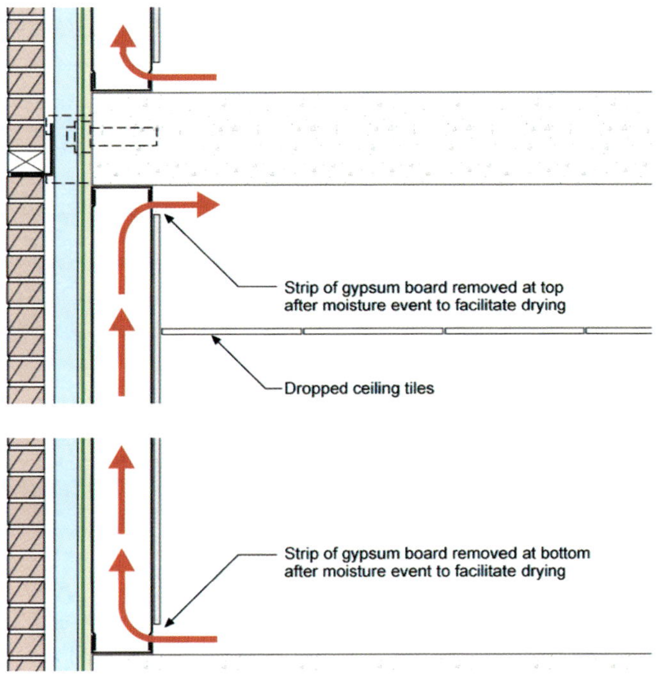

One can almost hear the glass fiber industry and the steel industry grinding their teeth and gearing up for combat as they read this chapter. They should relax. The aftermath of tropical storms strongly favors steel and glass fiber insulation—but in different forms. Steel framing has always been a problem due to thermal bridging. On the other hand, when *all of the insulation is on the exterior* and when the walls have a drainage plane, steel framing has advantages over wood after floods. For one, rust is less of a problem than are mold and rot. Even better, insects don't eat steel and it doesn't burn. What's not to like about steel when the cavity is dry and empty, and the insulation is on the outside?

As for the glass fiber insulation industry, *semi-rigid glass fiber board* insulation has advantages over foamed plastic insulation. It is extremely vapor-permeable, so it will dry well when it is located on the exterior of the sheathing, in a vented cavity. Also like steel, it does not burn. One might even expect that more money can be made selling semi-rigid fiberglass board for exterior insulation than can be made selling low-density fiberglass batt insulation for stud cavities. Glass fiber board exterior insulation provides a major performance improvement compared to past practices, and it is flood-resistant.

Unlike steel-framed buildings, for masonry mass wall assemblies there are two choices: exterior insulation and finish systems (EIFS), or interior insulation, using moisture-tolerant, semi-vapor-permeable rigid foam insulation applied to the indoor surfaces of the exterior walls (Figure 10.12).

In areas subject to periodic tropical storms, it's best to avoid wood and wood-based cladding and trim materials. Better choices include fiber cement and plastic composite materials for exterior cladding and trim. But regardless of the material, all exterior cladding and trim should be back-vented so it can dry, and so that water which gets under the siding cannot jump that vent-gap to soak the sheathing. If

Fig. 10.12 Interior extruded-board insulation

The insulation resists moisture absorption in a flood. And the furring strips provide a clear air passage for drying the interior wall board from behind, after the baseboard is removed. Note the spray-type insulation in the attic, which allows useful functions up there, such as running duct work.

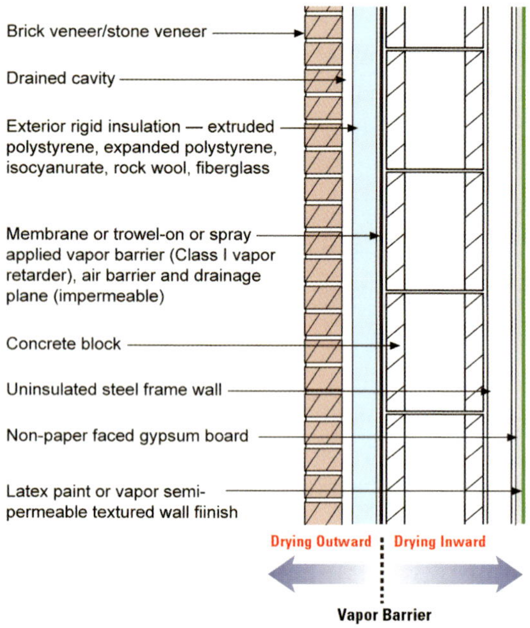

Fig. 10.13
Wall #1 - Masonry with brick veneer

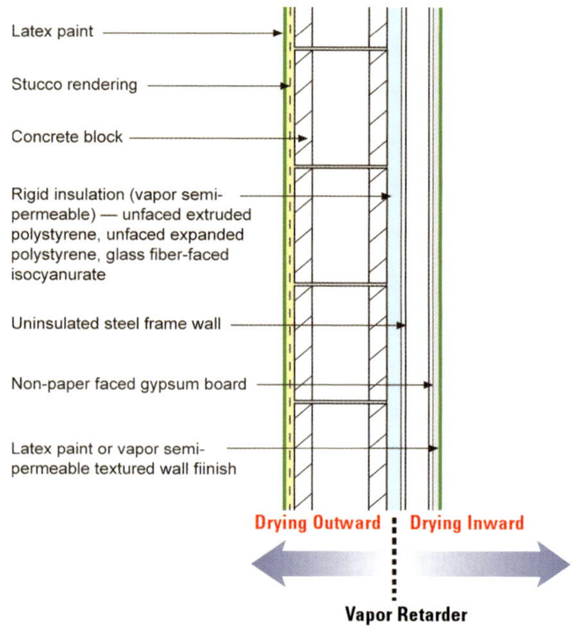

Fig. 10.14
Wall #2 - Masonry covered by stucco

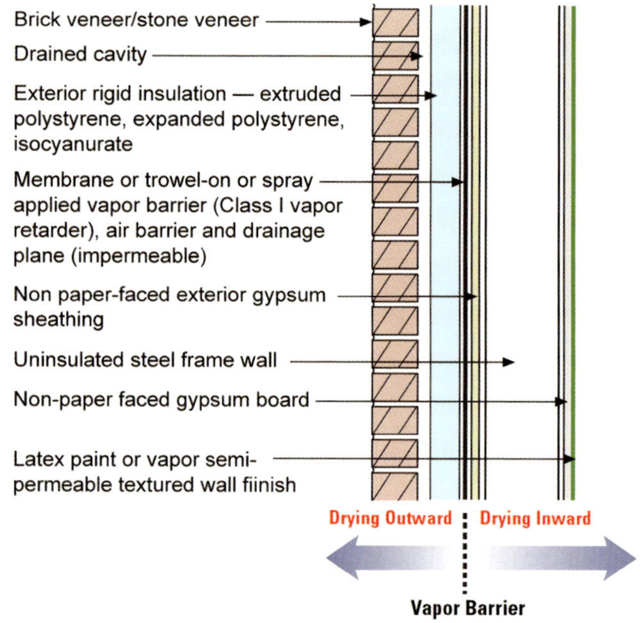

Fig. 10.15
Wall #3 Steel studs with brick veneer

wood trim and cladding is used (grudgingly), it should be coated on all six sides prior to installation to reduce water absorption.

Assemblies which dry easily

Nothing dries like airflow, especially when very little electrical power is available after the storm wipes out the utilities. Cavities should be designed to be easily vented after a flood. By cutting holes at the tops and bottom of each stud bay, the walls can begin drying through natural convection, until disaster drying professionals arrive to dry the building using portable equipment.[11] To allow both natural and forced drying, interior cavities should not contain absorptive insulation—or indeed any insulation at all. Keep the insulation on the outdoor or indoor sides of the structure, in order to keep the cavities free of obstructions. Then, design the cavities to allow temporary vent openings to be made easily at both the tops and bottoms, to allow a flow of drying air.

Note, of course, that permanently-open, interior vented cavities can transfer fire. Therefore, cavities should be sealed and compartmentalized under normal building operation. They should only be opened temporarily after a moisture event, to speed drying.

Finally, as described in more detail in Chapters 5 and 8, the wall materials and the wall coverings should be vapor-permeable in addition to being non-water-sensitive. Such porous materials facilitate constant drying by diffusion, from damp cavities into the drier occupied (air conditioned) spaces.

Examples of six flood-tolerant-easily-dried exterior walls are shown in Figure 10.13 through 10.18. Note that none of them have stud-cavity insulation. All are insulated on the exterior or the interior of the structure, so that the insides of the walls can be dried after a flood without major destruction or reconstruction. These alternatives are economical, and they have been proven in widespread use. They

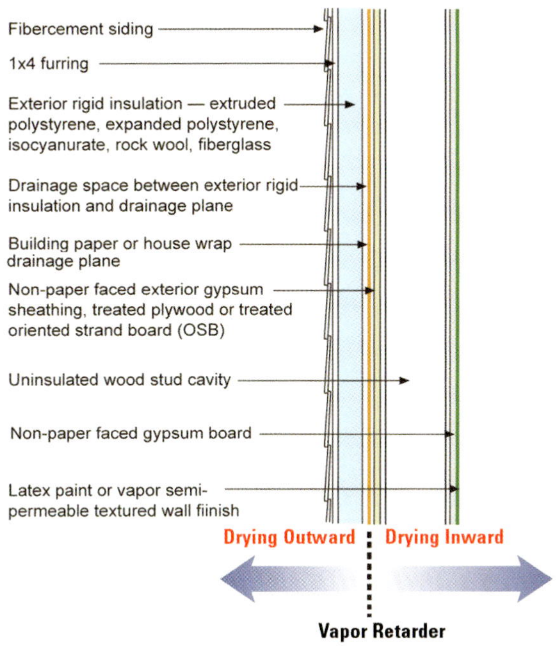

Fig. 10.16
Wall #4 Wood studs with fiber-cement siding

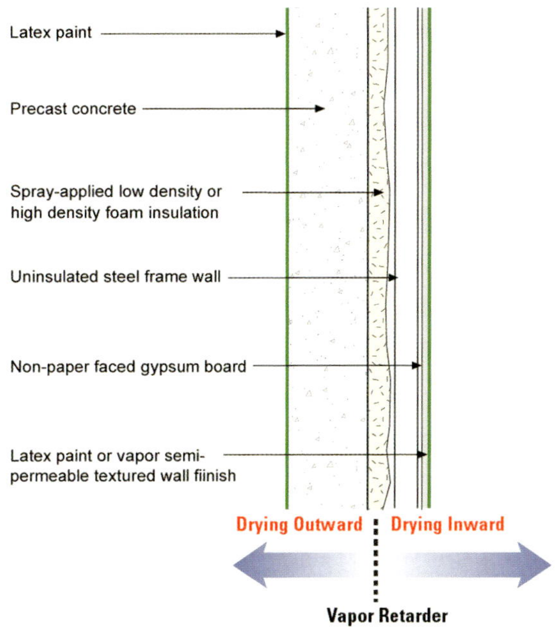

Fig. 10.17
Wall #5 Precast concrete with spray-foam insulation

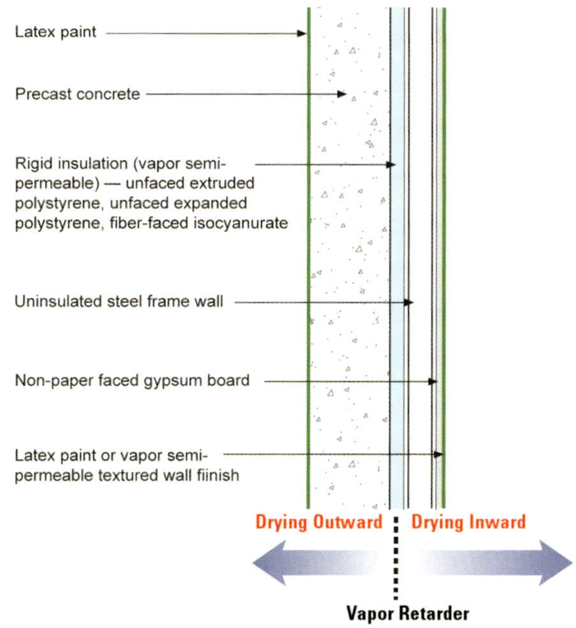

Fig. 10.18
Wall #6 Precast concrete with extruded board insulation

are also better walls than usual from an energy perspective. Such buildings can be made more air-tight than typical buildings, and their insulation is continuous, without thermal bridges and voids. So there's less hot and humid air infiltration into the building, and less conductive heat transfer. Finally, because these walls resist soaking and can dry more easily if water does intrude, they reduce the risks of mold growth—always an issue in buildings in hot and humid climates.

Summary

Buildings subject to tropical storms will have to address three principle disaster functions simultaneously: wind, rain and flood. The good news is that we have experience dealing with each of these disaster functions individually. It should be straightforward to integrate them. In storm-prone areas, design and construct buildings using these steps, in the following order of priority:

- Keep them from blowing away;
- Keep the rain out;
- Elevate the structures above the flood plane;
- Build with materials that tolerate soaking; and
- Design assemblies to easily dry when they become wet.

In fact, these steps apply to all structures no matter where they are built. But in areas subject to tropical storms, the consequences of failing to follow any one of them will come sooner and more frequently, and will cost more to fix.

References

1. Institute for Business and Home Safety. 2005. Fortified Builder's Guide. www.ibhs.org/publications/downloads/468.pdf.
2. 2004 Florida Building Code. Florida Department of Community Affairs.
3. Lstiburek, J.W. 2005. Rainwater Management Performance of Newly Constructed Residential Building Enclosures During August and September 2004. Home Builders Association of Metro Orlando and the Florida Home Builders Association Report. www.buildingscience.com/resources/walls/rainwater_management.pdf.
4. Lstiburek, J.W. 2004. Water Management Guide. Westford, Mass.: Building Science Press. (www.BuildingSciencePress.com)
5. Federal Emergency Management Agency. 1996. Elevated Residential Structures. FEMA-54. www.fema.gov/pdf/hazards/floods/fema54.pdf.
6. Federal Emergency Management Agency. 2000. Coastal Construction Manual, 3rd ed. FEMA-55.
7. Federal Emergency Management Agency. 2001. Crawlspace Construction for Buildings Located in Special Flood Hazard Areas. FIA-TB-11-01. www.fema.gov/pdf/fima/tb1101.pdf.
8. Federal Emergency Management Agency. 1993. Wet Floodproofing Requirements for Structures Located in Special Flood Hazard Areas. FIA-TB-7-93. www.fema.gov/pdf/fima/job14.pdf.
9. FEMA, 1993. Flood-Resistant Materials Requirements for Buildings Located in Special Flood Hazard Areas. FIA-TB-2-93. www.fema.gov/pdf/fima/job4.pdf.
10. Aglan, H., R. Wendt, and S. Livengood. 2004. Field Testing of Energy-Efficient Flood-Damage-Resistant Residential Envelope Systems. Oak Ridge National Laboratory ORNL/TM-2005/34
11. Harriman, Lewis G. III., Schnell, Donald and Fowler, Mark; "Preventing mold by keeping new construction dry." *ASHRAE Journal,* September 2002. pp 28-34

Image Credits

Fig. 10.1 - National Weather Service, U.S. National Oceanographic and Atmospheric Administration, Silver Spring, MD. (www.weather.gov) and the Weather Underground, www.weatherunderground.com

Figures 10.2 - 10.18 were provided courtesy of Building Science Corporation, Westford, MA (www.BuildingScience.com)

Chapter 11
Estimating Dehumidification Loads

By Lew Harriman

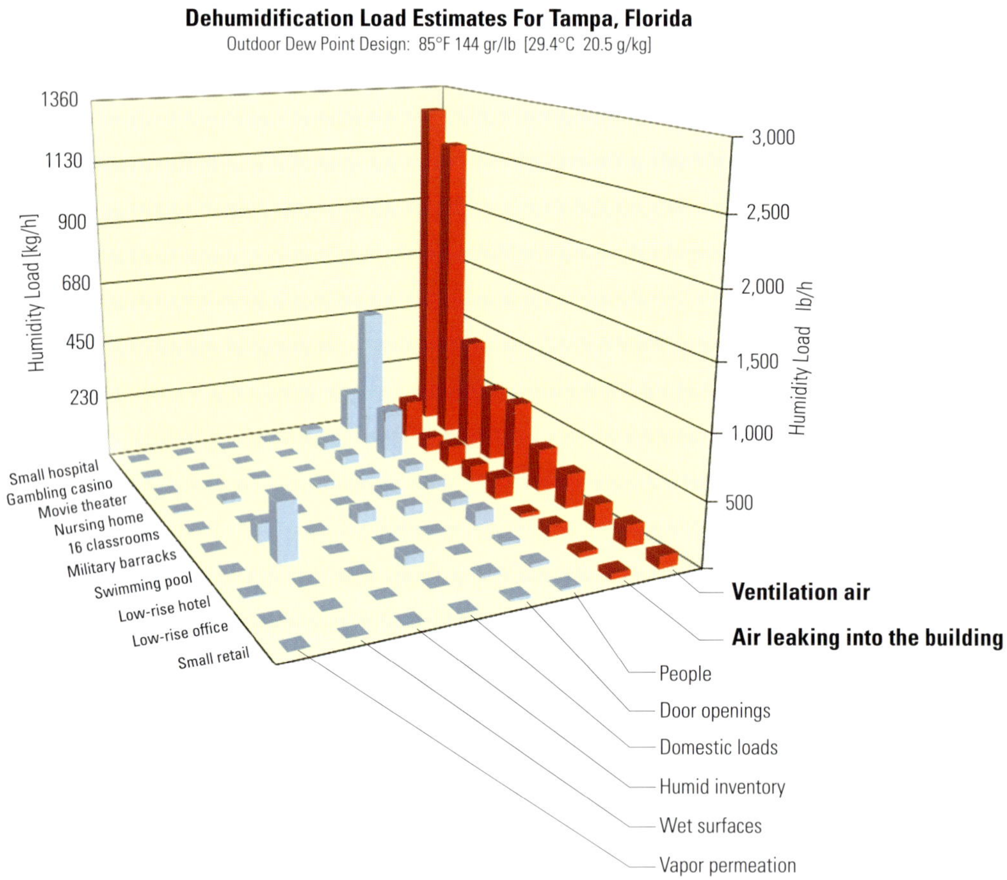

Fig. 11.1 What's important—and what's not

These load estimates show that the ventilation air and the air which leaks into the building (infiltration) are the largest dehumidification loads in nearly all building types.

The wise designer spends time to carefully quantify those two loads first. Quantifying these loads will require a conversation with the owner about the number of people who will occupy the building, and about the importance of continuous air barriers.

Dehumidification (DH) Loads

In hot and humid climates, the DH loads are not only high, they also remain high for much of the year. As the reader can see from figure 11.1, the largest DH loads are in the ventilation air, the infiltration air and the door openings. This is true in nearly all building types. Provided that those three load elements are well-estimated, the design can proceed on a firm foundation.

When the owner wants humidity control rather than just humidity moderation, a well-considered load estimate will be an essential first step in the HVAC design. Making the DH load estimate a collaborative effort is a useful opportunity to clarify the owner's expectations compared to his or her construction budget.

Any humidity load estimate is an agreement on shared assumptions between the owner and the HVAC designer. So to avoid needless costs and rework of the HVAC design, it's important that both parties understand and agree on three critical assumptions before the calculations begin.

The Estimate Begins With Owner's Decisions

The owner makes three decisions which guide all the rest:

- What is the target maximum indoor dew point? (55°F [12.8°C]—or some other value).
- During how many hours in a typical year will it be acceptable to have a greater risk of being above the desired indoor dew point? (35, 88 or 175 hours).
- How *many* people will probably occupy the building, and *when* are they likely to occupy *which parts* of that building?

After the owner has decided these issues—usually with the assistance and guidance of the HVAC designer the DH load estimate can begin.

Step 1 - Selecting the outdoor design condition

Peak dehumidification loads occur when the outdoor dew point is at its highest point for the year—*not* when the outdoor dry bulb temperature is at its peak.

Fig. 11.2 The importance of using peak outdoor dew point for dehumidification load calculations

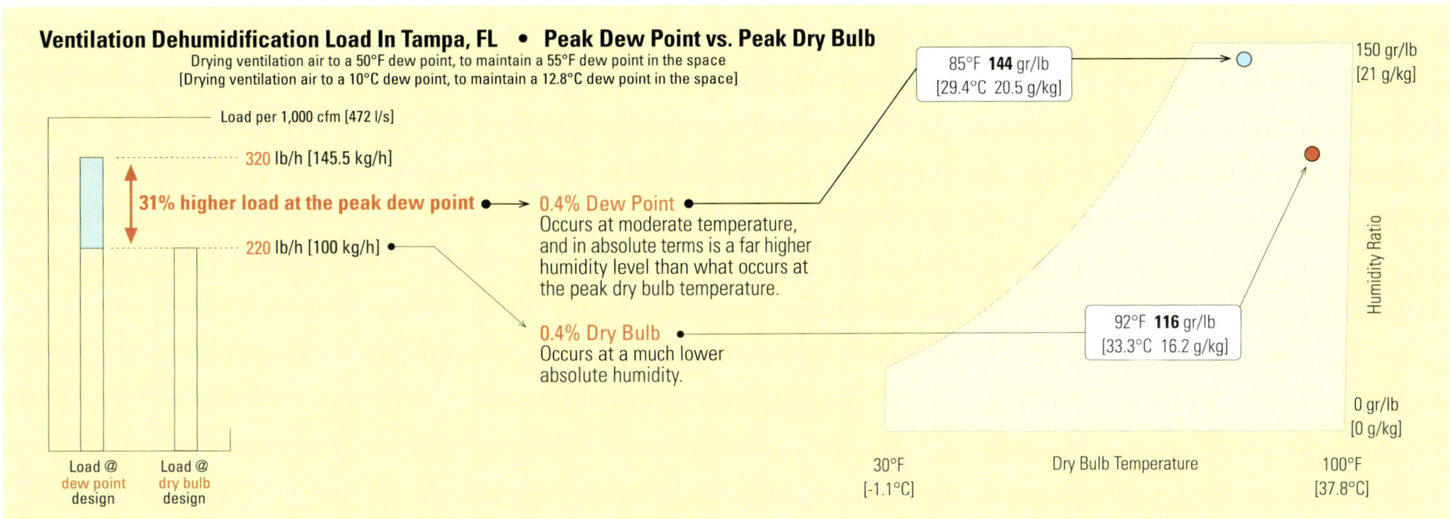

Outdoor humidity is 30 to 35% higher at the peak dew point condition compared to the humidity load at the peak dry bulb condition. Figure 11.2 shows the difference between indoor and outdoor humidity levels at both the peak dew point and peak dry bulb conditions for Tampa, FL. To avoid major shortcomings in the design, *make sure to use the peak dew point values* for DH load calculations.

Next, the designer must know *which of three* peak dew point values the owner prefers him to use in the calculations. The *ASHRAE Handbook—Fundamentals* provides peak dew point values which are only likely to be exceeded for 35, 88 or 175 hours during a typical year. (These are the 0.4%, 1% and 2% annual values, respectively.) This is really an owner's decision, even though few owners are in a position to make that judgement without advice from the HVAC designer.

It comes down to these questions: What are the consequences of having loads which might be above the expected design value? Does any catastrophe occur if the indoor dew point is higher than expected for a total of 88 hours in a typical year? Or—if occupants will simply be a bit less comfortable for 88 hours—will their reduced comfort have important economic or safety consequences for the owner? Only the owner can decide these questions. The more extreme the outdoor design condition, the more expensive the dehumidification equipment will be.

Probably, the owner will have less tolerance for above-spec indoor humidity in a critical-care surgical suite compared to a few hours of high humidity in a quick-service restaurant. And the tolerance will probably be more critical for an industrial process involving explosives manufacturing than for an enclosed swimming pool in which swimmers expect to enjoy splashing water on each other.

So it's up to the owner: How many hours "in a typical year" is the owner willing to tolerate the risk of humidity loads above the design values?

For most commercial and institutional occupancies (other than museums, archives or hospitals), when the indoor dew point is slightly higher-than-usual for a few hours a year, there will be no major problems. So for the majority of applications, either the 1% (88 hrs) or 2% dew point (175 hrs) is usually the most economically-practical choice. The more extreme 0.4% value (35 hrs/yr) is more frequently used for industrial or medical applications.

But in all cases, it's useful for the owner to keep in mind that weather varies, and that *any* design still assumes *there will be some number of hours above the calculated peak DH loads*. So unless the budget is unlimited, the owner's indoor humidity expectations should be set according to the number of hours each year which he expects to be above the design values.

For peak dew point values the designer can consult the climatic design chapter of the current ASHRAE Handbook—Fundamentals. The peak dew point values have been included in that reference since the 1997 edition. (Climatic data in *earlier editions do not* contain the peak dew point values, and therefore should not be used for DH load calculations) The digital edition of the Fundamentals volume currently contains peak dew point values for more than 5,000 locations worldwide.

Finally, keep in mind that the ASHRAE design values are based on 30 years of weather. Those values are essentially averages of extremes. This means that during some years—the years which are "less typical"—the number of hours above and below the ASHRAE design values will be different than the number of hours during typical years.

There are no guarantees with future weather. Meteorologists believe that future weather will be similar to weather patterns of the past. But it's understood by all technical professionals that weather can never be *identical* to the "typical" years that ASHRAE used for establishing design values.

Fig. 11.3 Ventilation is for people
*When the space is **not** occupied, the enormous humidity load on the building can be reduced by reducing the volume of ventilation air.*

Step 2 - Selecting the target maximum indoor dew point

In this book, we have consistently suggested that a 55°F dew point (65 gr/lb) [12.8°C dew point, 9.24 g/kg)], is a prudent upper limit to:

- Reduce mold risk in hot and humid climates.
- Allow the cooling system to respond quickly, for better comfort more quickly and therefore lower energy cost.

- Allow the dry bulb temperature set point to rise above historically cold North American set points, for better-than-typical energy efficiency.

It may be useful to understand the logic behind this suggested upper limit. With respect to mold risk, most air conditioning systems don't create surface temperatures which are much lower than 65°F [15.6°C]. A dew point of 55°F, together with a surface temperature of 65°F, means the *surface* relative humidity won't be greater than 70%rh—which, when the air and material are at equilibrium, is 10% below the moisture content which is likely to support mold growth in most materials.[1,2]

[A dew point of 12.8°C, together with a surface temperature of 18.3°C means the *surface* relative humidity won't be greater than 70%rh—which, when the air and material are at equilibrium, is 10% below the moisture content likely to support mold growth in most materials.[1,2]]

With respect to cooling system responsiveness, when the indoor dew point is kept below 55°F [12.8°C], the cooling coil will cool the return air more quickly than if it had to first condense large amounts of moisture out of that air. So the time needed to cool the building down to the thermostat set point is reduced. The system responds more quickly to increased sensible heat loads, and more quickly to changes in the thermostat's set point.

And finally, keeping the dew point below 55°F [12.8°C] allows the indoor dry bulb temperature to rise as high as 78° or 79°F [25° or 26°C] or perhaps even higher, before the occupants become uncomfortable. Then (all other things being equal) with the higher indoor temperature, less energy is required for cooling.

That said, it is important to understand that this suggestion comes from the experience of the authors and many of the members of the Project Monitoring Committee for this book, rather than from the public, consensus-based process required for establishing an ASHRAE Standard. In other words, there is room for disagreement between experts about the appropriate, prudent and economically-optimal maximum indoor dew point for buildings. Some will prefer a *lower* limit for better comfort results and reduced mold risks. Others may believe that *higher* limits do not raise mold risk, energy use or discomfort to unacceptable levels.

For more detailed reference material to guide the selection of the indoor dew point limit for load calculations, ASHRAE has published a separate book: *The Humidity Control Design Guide*.[3]

Again, in the final analysis the upper limit for the indoor dew point is a decision which must be made, or at least validated, *by the owner*. Lower dew points mean lower risks, but also higher initial equipment costs. Higher dew points save construction costs, but they increase cooling costs and increase mold risks over the life of the building.

Our suggestion is clear: use a 55°F dew point [12.8°C] as the target indoor maximum and therefore as the basis of the dehumidification load calculations. But this is a suggestion—not an ASHRAE standard. Until ASHRAE or local code authorities establish firmer upper dew point limits, the owner must make his own decision based on his own experience, needs and preferences, no doubt with advice from the HVAC designer.

Step 3 - Quantifying & locating the people in the building

The largest humidity load will be carried into the building by the ventilation air. *The amount of ventilation air depends primarily on the number of people who occupy the building.*[4] The owner has a clearer idea of how many people will occupy the building than does the HVAC designer. The owner also has a better understanding of where those people will be at any given time within the building.

Occupancy information is very important for the HVAC designer to understand before load calculations begin. Over-ventilation based on assumptions of the "worst-case" occupancy guesstimates are frequently responsible for poorly-performing systems and/or dehumidification components which are needlessly large and costly.

The ventilation humidity load is huge and therefore very costly. Don't guess at it.

See how easy it is to print that advice in a book ...compared to how difficult it is to get accurate occupancy estimates from owners in the real world?

Yes, it's very difficult for the owner to know how many people will really occupy all spaces of the building over its lifetime. Consequently both owner and HVAC designer sometimes just make a "safe" guess—assuming the maximum allowable occupancy for all spaces at the same moment, when the outdoor dew point is at its peak. That "safe" guess may sometimes be inevitable. But keep in mind the huge penalty paid in construction cost, operating energy and discomfort based on such an unlikely, worst-possible case.

The better approach is to have a considered discussion with the owner, allowing the HVAC designer to make an educated guess at a population diversity factor—the probable mixed use maximum occupancy of all spaces served by each separate system, when the outdoor dew point is at its peak. Then use that mixed-use, system-specific maximum population for the estimate of the ventilation air requirement, and therefore for the DH load calculation for each system.

Later, during the actual design of the systems it may be logical to serve all systems or all spaces with a central source of pre-dried ventilation air. But making the ventilation DH load calculation system-by-system helps avoid poor cooling and humidity control performance when the building is *not* served by a continuous source of dry ventilation air. As long as the designer "sees" the load, he or she can design the systems to remove it. But if the DH load is not calculated by the designer, it's likely to be overlooked in the system design, with the usual result being cold, damp spaces and high energy costs.

Calculating the DH load from ventilation air in hot and humid climates using the peak dew point outdoor design condition is sometimes a traumatic but educational experience for designers who are more familiar with other climates, and for those who are used to assuming, incorrectly, that the peak DH load occurs at the peak outdoor dry bulb condition. At first, it's difficult to believe the DH load could be so large.

Variable-volume and/or intermittent ventilation are ways to reduce the DH load. The actual occupancy of buildings is rarely if ever at its maximum allowed by law. So don't ventilate as if it were. Vary the ventilation air by the design of the system, or ventilate intermittently to keep pollutant concentration within prudent limits.

ASHRAE Standard 62.1 (Ventilation for acceptable indoor air quality) contains detailed guidance for calculating the ventilation air requirements of commercial, multi-family residential and public buildings. But of course the local ventilation codes will also govern the designer's decisions. In some cases, local codes will call for greater than ASHRAE 62.1-2007 recommended air flows. (Northern Europe, in particular, has more strict requirements for ventilation.) And in other locations throughout the developing world, ventilation may not yet be a code requirement.

To estimate the appropriate ventilation air flow in the absence of local code guidance, the authors suggest that ASHRAE Standard 62.1 is a good choice, because it represents the result of a rigorous international consensus process. Some of the recommended air flows of table 6-1 of the 2007 edition of that standard are shown at the end of this chapter. But for full details and more occupancy types, we strongly recommend consulting the current edition of the standard, which is updated more frequently than this volume.

After Owner's Decisions, Engineering Judgement and Calculations Begin

In steps one, two and three, the owner has decided the design limits, so calculations can now begin. The calculations are arranged in order from most to least effect on the total DH load.

Step 4 - Estimating the ventilation & makeup air load

Since the largest DH load comes from ventilation air it makes sense to begin by estimating that load.

The recommendations of ASHRAE Std 62.1 can be used to estimate how much ventilation air is needed in which zones. Table 6-1, located at the end of this chapter, shows a subset of the recommendations of Std 62.1-2007. Note that the standard calls for a certain amount of air per person, but also an additional amount of air to dilute the concentration of pollutants generated by the materials and equipment inside the building—carpets, fabrics, cleaning liquids, copier and laser-printer emissions, and so forth. The additional air for dilution of building pollutants is determined by the amount of occupied space: a certain number of cfm/ft^2 [or l/s/m^2].

Based on the sum of the zone-level ventilation requirements, ASHRAE Standard 62.1 provides procedures for calculating the required system-level intake air flow, which will depend on the type of system selected.

Add outdoor air for pressurization

To the sum of the air flows needed for people and building-generated pollutants, add the amount of air needed to keep the building under a slight positive air pressure. That amount will depend on two factors: the air leakiness of the construction and the amount of air exhausted from toilet exhausts, kitchen exhaust hoods and any other exhaust fans.

There is no easy way to be certain of how much air will be needed to keep the building under a slight positive air pressure. It's a complicated problem, with several impossible-to-quantify parameters when the building does not yet exist. Perhaps for that reason, ASHRAE has no useful guidance on this subject, important though it might be.

In the past, industry practice has been to sum the *exhaust* air flows. Then to that total, add between 5% and 10% of the total *supply* air flow and verify that the incoming outdoor air is at least that amount. That way, in theory the building should stay under a slight positive air pressure, reducing the amount of humid air that leaks in through the building enclosure.

That practice is almost certainly inadequate in many buildings, but is overkill for others. It all depends on how air-tight the building is built, how air-tight it stays over time, whether the duct connections are sealed and how much internal air pressure is really necessary to minimize air infiltration.

Although the practice of adding an excess of outdoor air in the amount between 5% and 10% of the supply air is anecdotal rather than supported by research, to experienced practitioners it seems on balance to be prudent practice. While the 5-10% practice does not, therefore rise to the level of a recommendation, the authors acknowledge its common use in HVAC designs, and submit that observation to aid the readers' engineering judgement.

The goal is to have a small amount of *dry indoor air leaking out* of the cracks and joints of the building enclosure *rather than allowing humid outdoor air to leak in* through those same cracks and joints.

Balance exhaust flows with dry ventilation air

In buildings without food preparation, often the ventilation air requirement plus the makeup air for toilet exhausts will total more than enough to provide excess air for pressurization. In contrast, buildings which *do* have kitchen exhausts, fried food exhausts, swimming pool exhausts or research lab hoods often suffer from negative internal air pressure (suction of humid outdoor air through the building enclosure). Often, these problems occur because the kitchen exhaust systems and the HVAC systems are designed by different people.

Ventilation Air Load Per Hour (lb or kg)

Density of Standard Air
I-P d = 0.075 lb/ft³
SI d = 0.072 kg/l/s

$$W_V = Q \times (60 \times d) \times (M_1 - M_2)$$

- Q: Ventilation air flow (cfm or l/s)
- d: Air density (lb/ft³/min or kg/l/s)
- M_1: Outdoor humidity ratio (lb/lb or kg/kg)
- M_2: Indoor humidity ratio (lb/lb or kg/kg)

Fig. 11.4 Ventilation air - Dehumidification load equation

Kitchen and lab exhaust hoods and "exhaust ventilation" fans are notorious for not being balanced by dry makeup air.

To avoid problems, the HVAC designer must take care to locate and quantify *all* the exhausts from the building, even if these are not under his or her direct control. Then, the HVAC designer must either provide the extra ventilation air, or make certain the other responsible parties have balanced that exhaust flow with dry make up air. (Not just "tempered" air—*dry* air is what is needed in a hot and humid climate when the building is humidity controlled).

Net load from incoming outdoor air

After all the outdoor air flows have been totaled for diluting contaminants from people and the building, and for balancing the exhausts and providing positive internal pressurization, the designer can use the equation in Figure 11.4 to calculate how much water vapor will need to be removed from that incoming air.

Step 5 - Estimating the infiltration load

All buildings leak.[5] Air entering through those leaks carries humidity which must be removed by the dehumidification equipment.

The amount of air leaking into the buildings depends on the *local* pressure difference between indoors and outdoors. More air leaks in when wind blows against a wall, as shown in Figure 11.5. Low pressure on the leeward side of a building pulls air out of the wall which is replaced by air leaking in on the windward side.

Leakage varies greatly from one moment to the next as wind pressure changes. It also varies greatly between one part of the building wall and another, creating inward leakage in one part, and outward leakage nearby.

Because of these complex phenomena, infiltration is notoriously difficult to predict with any accuracy. Although many attempts have been made to correlate the Architect's design and the building code requirements with measured leakage, the results are random.

For example, Canadian building codes are considerably more strict regarding building tightness than requirements in most other countries, and contractors are quite experienced with these strict requirements. But one study of high-rise apartment buildings in the prairie region showed that while codes call for leakage of less than 0.02 cfm · ft² [0.10 l/s · m²], measured results showed actual leakage rates between 0.39 and 0.60 cfm · ft² [2.06 and 3.15 l/s · m²]. In other words, 200 to 315% more than the expected infiltration. Similarly, measured leakage for 70 commercial buildings in Florida showed conclusively that North American commercial buildings leak a truly astonishing amount of air—in some cases over 10 air changes per hour.[6]

Figure 11.6 shows typical leakage rates for exterior walls of commercial buildings. These values are extracted from Chapter 27 of the 2005 *ASHRAE Handbook—Fundamentals*.[9] Typical low-rise, low-cost commercial buildings have either average or leaky walls. Institutional and high-rise buildings may also have leaky walls, but are more likely to have either average or tight wall construction.

Most technical professionals are understandably uncomfortable using vague terms like "tight" and "leaky" and "typical low-rise buildings" to select values to enter into load calculations. Experts in DH load calculations are equally uncomfortable. But the only alternative is to perform a whole-building leak test after

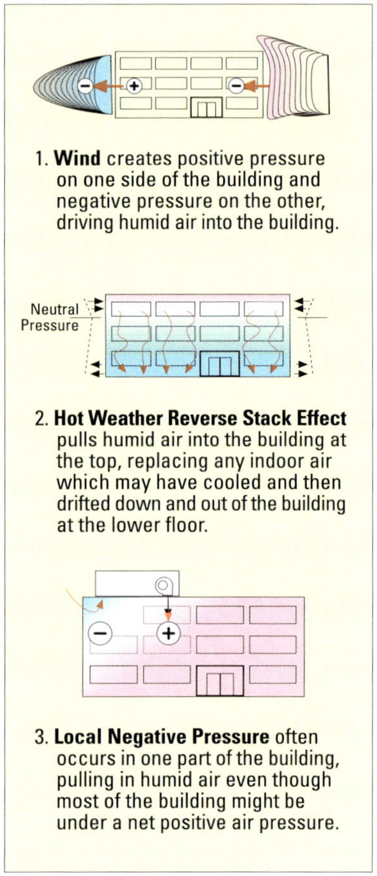

Fig. 11.5
Air pressures drive infiltration

Exterior Leak Rate - Under Pressure

Wall Construction	cfm • ft²	l/s • m²
Tight	0.1	0.004
Average	0.3	0.013
Loose	0.6	0.026

Leakage rates are based on field measurements for North American commercial buildings at a pressure difference of 0.3" [75Pa]

Fig. 11.6
Pressure-driven humid air infiltration rates

Fig. 11.7
Infiltration air - Humidity load equation

Infiltration Load Per Hour (lb or kg)

Density of Standard Air
I-P d = 0.075 lb/ft³
SI d = 0.072 kg/l

- A: Wall surface exposed to wind (ft² or m²)
- I_r: Infiltration rate (cfm/ft² or l/s/m²)
- d: Air density (lb/ft³ or kg/l)
- M_1: Outdoor humidity ratio (lb/lb or kg/kg)
- M_2: Indoor humidity ratio (lb/lb or kg/kg)

$$W_i = A \times I_r \times (60 \times d) \times (M_1 - M_2)$$

the structure is complete—an expensive and complex practice, and one which is not possible in the design stage.

In the past, this discomfort has led even highly-skilled professionals to simply *ignore* the entire issue of air leakage, frequently with the rationale that "if ventilation air flow is greater than the exhaust air flow then any leakage will be from inside to outdoors, eliminating the leakage moisture load." Unfortunately, not a single field investigation validates that optimistic assumption. In fact, quite the reverse.[5,6,7] All buildings pull in *some* outdoor air at some times at some points—even when the *overall average* internal air pressure is positive.

Unless the designer has reason to believe the building will not leak—making it unique among buildings investigated throughout the world to date—engineering judgement must be applied and a decision made as to leakage rates. A leakage rate for the walls exposed to the prevailing winds can be entered into the equation shown in Figure 11.7.

As a practical matter, very few commercial buildings are controlled to extremely low humidity levels, and even fewer are defined well enough at the design stage to merit a time-consuming component-level DH load estimate. However, when the designer needs to estimate infiltration more precisely, and when great detail is available concerning the construction of the humidity-controlled space, Chapter 25 of the *ASHRAE Handbook—Fundamentals* (1997) contained leakage rates for individual components such as different types of doors, windows, electrical outlets, fireplaces and surface-mounted lights. These values are useful when calculating loads in very well-defined small spaces that must be maintained at extremely low dew points.

Rain-soaked masonry walls

Investigations into moisture problems in humid climates have noted that porous, rain-soaked bricks or concrete blocks add moisture to any infiltration air as it traverses the wall. In most cases, the architectural designer will be aware of the need to use water and air barriers on the indoor side of such materials to seal the wall against moisture and humid air leaks.

But if such barriers are not in place for any reason, the humidity control designer can estimate the enormous increase in the moisture load by first estimating the surface temperature of the exterior wall after a rainstorm. Assume that any air entering through that wall will be saturated at the exterior surface temperature, and then recalculate the moisture load from infiltration using the equation in Figure 11.7. The resulting load will be fearsomely large, which should prompt the architectural designer and contractor to make certain that a continuous water and air barrier will be in place behind the brick or over the masonry block to limit humid air infiltration.

Leaking return air ducts

In most buildings, return air duct work passes through conditioned spaces. But where ducts pass through *unconditioned* spaces—such as attics and crawl spaces— any humid air leakage into the duct adds a dehumidification load to the system. Good practice suggests that the HVAC designer should specify that all return and supply duct work in dehumidified buildings shall be sealed on all longitudinal and transverse joints, and sealed to the inlets of all air handlers, cooling coils, VAV boxes, supply diffusers, return grills and filter housings.

On the other hand, not all systems are installed with good practices. Also, older buildings sometimes need humidity control added long after initial construction. When for any reason ducts have not been sealed, and when the return duct work passes through humid

Probable Return Duct Infiltration

Return Duct Pressure		cfm • 100 ft² Lab vs. Field	l/s • m² Lab vs. Field
-0.1 "	[-25Pa]	11 vs. 770	0.0042 vs. 0.3376
-0.25 "	[- 62Pa]	20 vs. 1400	0.0088 vs. 0.6138
-0.5 "	[-125Pa]	31 vs. 2170	0.0136 vs. 0.9513
-1.0"	[-249 Pa]	48 vs. 3360	0.0210 vs. 1.4730

Leakage rates are based on the exterior surface area of the ducts, not the occupied floor area of the building.

Lab-tested leakage rates of partly-sealed duct sections are found in the SMACNA Duct Inspection Guide (Reference 8)

Field leakage rates are based on FSEC report of measured return air leakage in light commercial buildings in Florida (Reference 6). The field-measured leakage rates in return systems of 46 existing buildings averaged 70 times greater than the SMACNA-based expectation.

Fig. 11.8
Probable infiltration rates into return ducts

air, Figure 11.8 provides estimates of air infiltration rates at different suction pressures.[6,8] These allow the designer to estimate the dehumidification load from return air duct leaks, using the equation in Figure 11.7.

Step 6 - Estimating the load from people

People release water vapor as they breathe, and they evaporate moisture as they perspire. Also, their clothing absorbs moisture outdoors from both humid air and rain. That moisture becomes a dehumidification load inside the building as the clothing dries.

Respiration and perspiration

Each time someone breathes out, they release a lung full of air at a temperature of 98.6°F and 283 grains per pound—nearly saturated air at normal body temperature. [37°C, 40.4 g/kg]

The number of breaths per hour depends on the person's physical activity. More active people will breathe more frequently and more deeply. Figure 11.9 provides typical moisture release rates for respiration and perspiration based on different levels of physical activity. The

Activity	Typical of	lb/h	kg/h
Seated, at rest	Theater patron	0.10	0.05
Seated, very light work	Hotel or restaurant patron	0.15	0.07
Seated, moderately active	Offices, retail cashier	0.19	0.09
Standing, light work, walking	Offices, retail patron	0.19	0.09
Walking, standing	Offices, retail floor clerk	0.24	0.11
Seated, light work	Electronic assemblers	0.45	0.20
Moderate dancing	Dancing, nursing care	0.52	0.24
Walking briskly with loads	Restaurant servers	0.60	0.27
Light exercise	Bowling, slow treadmill	0.83	0.38
Heavy work with lifting	Factory, health club machines	0.92	0.42
Athletics	Basketball, heavy excercise	1.04	0.47

Fig. 11.9
Humidity loads from occupants

equation shown in Figure 11.10 can be used to total the respiration and perspiration loads from long-term occupants of the space.

Moisture desorbed from clothing

In retail stores, restaurants, theaters, schools and other high-traffic, high-density buildings, the moisture collected outdoors by clothing can be a significant load. That moisture is released by evaporation when the person walks into a humidity-controlled building. The load depends on the amount of desorbed water vapor, the moisture level indoors and the duration of the visit. The lower the indoor humidity ratio, the more moisture will be released, and most of the clothing moisture load is released during the first 10 minutes.

Respiration & Perspiration Load Per Hour (lb or kg)

- n_1: Group @ activity level 1 (Number of people)
- rp_1: Moisture release per person (Grains or grams per hr)
- n_2: Group @ activity level 2 (Number of people)
- rp_2: Moisture release per person (Grains or grams per hr)

$$W_p = (n_1 \times rp_1) + (n_2 \times rp_2) + \ldots$$

Fig. 11.10
Occupants - Humidity load equation

Fig. 11.11
Humidity loads from visitors

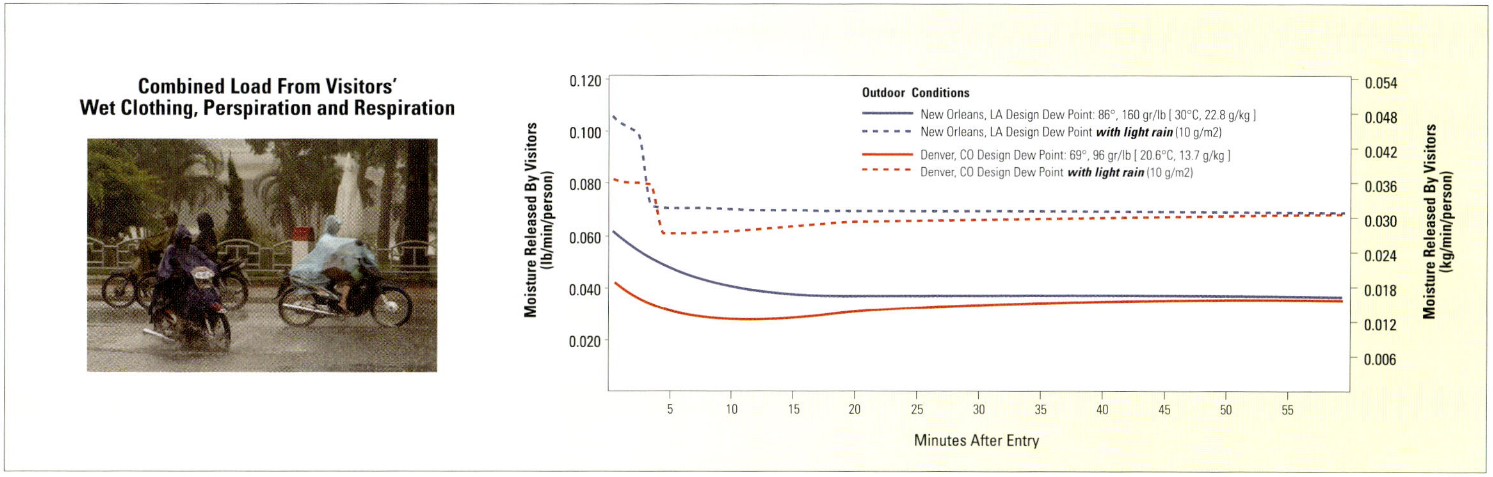

Fig. 11.12
Visitors - Humidity load equation

Visitor Load Per Hour (lb or kg)

$$W_V = n \times \frac{t}{60} \times \frac{(R_{r1} - R_{r2})}{2}$$

- n — Visitors (Number of people per peak hour)
- t — Average length of visit (Minutes)
- R_{r1} — Initial moisture release rate (lbs/h or kg/h)
- R_{r2} — End-of-visit moisture release rate (lb/h or kg/h)

Figure 11.11 shows the load released by adults dressed in cotton sweat clothes as they enter a humidity-controlled building after walking outdoors in design moisture conditions for Denver, CO (a dry climate) compared to New Orleans, LA (a humid climate).

The designer can decide which moisture release rate to assume by asking the owner to estimate the average length of a typical visit to the building, and by taking the average of the initial and final release rates for a visit of that duration. For example, if an average "take-out" patron in a quick-service restaurant in Tampa, FL usually stays in the building for 5 minutes, the initial release rate is appropriate to use in the calculation. But if a patron in a family restaurant stays for 45 minutes, the designer could use the average of the initial and the 45-minute release rates in the equation. If several different "visit lengths" are typical—as in a restaurant with both take-out and sit-down meals—then the designer might choose to solve the equation separately for each category of visitors and combine the results to obtain the total clothing desorption moisture load.[10]

Step 7 - Estimating the load from door openings

As exterior doors open and close, they allow humid air to enter the building. The amount of air depends on the open area of the door, how long the door remains open, and the air pressure difference between indoors and outdoors.

Also, for tall doors the temperature difference across the wall leads to pressure differences caused by the different buoyancy of the air masses on each side of the door. This is the most important contributor to leakage through doors in cold storage buildings, where doors are tall, and where temperature differences are large during humid months.

For low-rise commercial buildings and for typical personnel doors, the height of the door is not as important as the exterior wind pressure exerted when the door is open. When the wind blows against

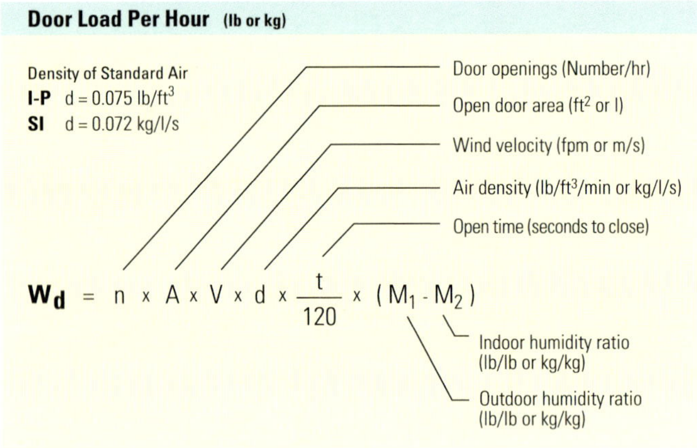

Fig. 11.13 Door openings - Humidity load equation

the exterior wall, it forces humid air into the building every time the doors on that wall open. The faster the wind, the more humid air enters while the door is open.

The extreme wind speeds are presented in the climatic design information chapter of the *ASHRAE Handbook—Fundamentals*. However, extreme winds usually occur when the weather is changing rapidly or during storms. Those may not be the periods when a commercial building has the most door openings. Some designers prefer to use the average annual wind velocity instead of one of the three extremes. The average annual wind velocity can be obtained from the local airport weather station, or from the U.S. National Climatic Data Center in Asheville, NC. (http://www.ncdc.noaa.gov/) That organization records hourly weather data and summaries for over 10,000 sites worldwide. Annual average wind speeds are often between 25 and 50% of the 5% extreme velocity shown in the *ASHRAE Handbook—Fundamentals*.

After the designer decides which wind velocity to use and after the owner estimates the door traffic, the designer can estimate the door moisture load using the equation in figure 11.13.

Vestibules or air locks

A vestibule or "air lock" is a small chamber between two doors—a common feature of exterior entryways into commercial buildings. Since the entering person does not open both doors at the same moment, wind cannot carry humid air into the building freely—the air halts in the air lock, and only a small amount drifts into the building when the interior door is opened. For doors which must be opened manually, the designer may assume that some fraction of the air in the vestibule is moved into the building with each door opening, and that the moisture content of that air is the average of the indoor and outdoor humidity ratios. The equation in Figure 11.15 allows an estimate of the DH load from vestibules with manual doors.

Sometimes, vestibule doors open automatically, as in supermarkets and any building in which customer convenience is more important than limiting dehumidification loads. With automatic operation, both doors may be open at the same time, but the net open time to the weather is considerably less than if there were only a single door between the weather and the indoors. For automatic doors on vestibules, the designer can estimate a shorter open time, using the equation for doors.

Wind Speed Conversion		
mph	**fpm**	**m/s**
1	88	0.4
2	176	0.9
3	264	1.3
4	352	1.8
5	440	2.2
6	528	2.7
7	616	3.1
8	704	3.6
9	792	4.0
10	880	4.4
11	968	4.9
12	1056	5.3
13	1144	5.8
14	1232	6.3
15	1320	6.6
16	1408	7.1
17	1496	7.6
18	1584	8.0
19	1672	8.5
20	1760	8.9

Fig. 11.14 Wind speed conversion

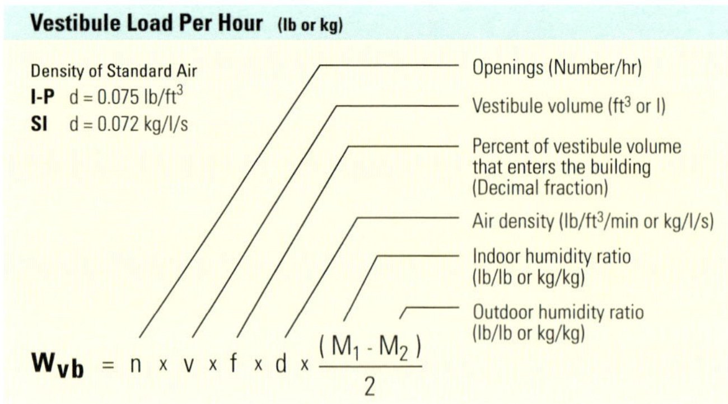

Fig. 11.15 Vestibule openings Humidity load equation

Door curtains using air or plastic strips

In truck loading docks, "air curtains" are sometimes used to limit the amount of air drifting into the building. An air curtain forces a high-velocity jet of indoor air across the opening. The effectiveness of that air as a barrier depends on its momentum—the product of its mass times its velocity. More air at a higher speed will be a more effective barrier to wind blowing against the opening.

For dehumidification load estimates through air curtains, the designer should consult the manufacturer of the device. These firms can sometimes provide estimates of outdoor air leakage when wind is blowing against the side of the opening at the design velocity. But usually, air curtains are not intended to resist large wind pressures. They function best when the wind is at levels well below design extremes. Leakage at design or even average outdoor wind velocities is likely to be quite large. Buildings which need humidity control seldom use air curtains, because their infiltration load is so large that it greatly increases the size and cost of dehumidification components.

Door curtains made of overlapping plastic strips are often used in warehouses to limit air movement through doorways otherwise open to unconditioned spaces. Estimating the continuous humid air infiltration through these curtains has not yet been the subject of research. The designer is left to make an educated estimate.

If the end of the strips hang above the floor, the designer can estimate the open area, and assume that the humid air moves through that space continuously. If the ends of the strips rest on the floor and bend, leaving openings between strips near the bottom of the curtain, the designer can estimate the amount of open area as a percentage of the door space, and calculate infiltration through that area based on some assumed velocity.

If the doorway is on an exterior wall, the in-flowing air velocity will depend primarily on the outdoor wind velocity. If the doorway separates two interior spaces, the velocity of in-flowing air depends on the small difference in air pressure across the wall—which is difficult to predict. When both sides are at the same pressure, common practice in dehumidification design in the past has been to assume a velocity of at least 50 fpm [0.25 m/s] through the open area.

Door load reduction through positive internal air pressure

Logic suggests that if the building has a positive average internal air pressure, air should move out of the building—not inwards—when the door opens. That may indeed be the case when the air outdoors is still. But usually, wind presses against the exterior wall with a force far greater than any slight counteracting pressure inside the building. The more probable benefit of positive internal pressure is to reduce outdoor air infiltration into building cavities, and also to reduce the total annual dehumidification load which in turn reduces the cost of operation. However, for peak load calculations, most designers do not assume that positive air pressure will measurably reduce the load from door openings. They assume that some humid outdoor air will be entering the building every time a door opens.

Step 8 - Estimating the minor loads

Quantifying the loads described in steps 3 through 7 is usually the key to successful humidity control. But in some buildings, there are some *internal* humidity loads which require careful attention.

In particular, in buildings with a brick veneer, the vapor permeation through the exterior wall can be significant, if there is no waterproof and vapor proof barrier behind the brick. And in multi-family residential buildings, sometimes the prolonged domestic loads from showers and cooking become more worthy of attention than they would be in hotels or offices.

Vapor permeation

Water molecules migrate slowly through solid materials by diffusion, encouraged by the difference between water vapor pressures on each side of the material. Humid air has a high water vapor pressure. Dry air has a lower vapor pressure. Water vapor moves slowly through material in response to that pressure difference.

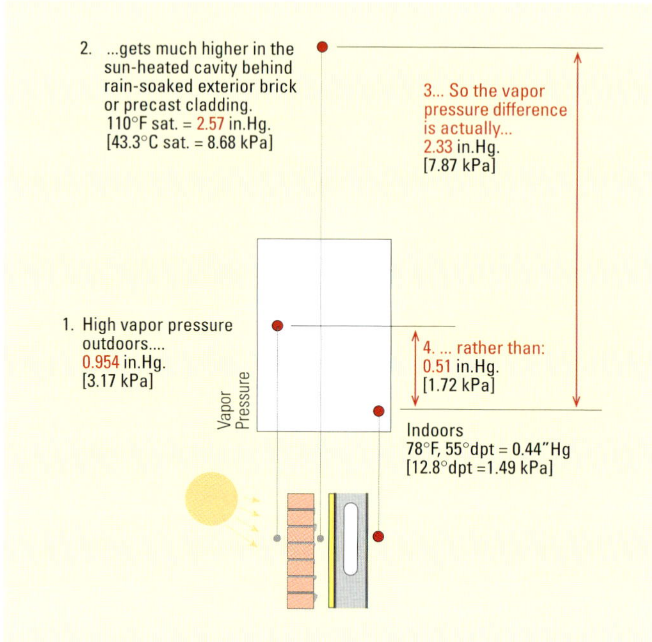

Fig. 11.16 Solar-driven vapor

Vapor Pressure at Saturation			
°F	**in.hg.**	**°C**	**kPa**
100	1.93	38	6.63
95	1.66	35	5.63
90	1.42	32	4.76
85	1.21	29	4.01
80	1.03	26	3.36
75	0.86	23	2.81
70	0.74	20	2.34
65	0.62	17	1.94
60	0.52	14	1.60
55	0.44	11	1.31
50	0.36	8	1.07
45	0.30	5	0.87
40	0.25	2	0.71
35	0.20		

Fig. 11.17
Vapor pressure of air at saturation

The molecules move slightly faster when the vapor pressure difference is high, and when the material is more porous. That's why fresh bread dries out more quickly when bagged in paper rather than plastic. Moisture passes more freely through porous paper than through dense plastic.

The unit that quantifies water vapor movement through a material with a defined thickness is called the "perm" (the *permeance*). In I-P units, one perm is the number of grains per hour that pass through one square foot of material when the vapor pressure difference is one inch of mercury. (1 perm = 1 gr/hr · ft² · in.hg.)

[In SI units, the permeance is the number of nanograms per second that pass through one square meter when the vapor pressure difference is one pascal.]

The designer will quickly perceive that the dehumidification load from permeation through solid materials will be a small fraction of the load from humid air infiltration.

Visualize the resistance encountered by individual water molecules as they slowly bump through all the other molecules in a solid material on their way towards the slightly lower vapor pressure. The process is slow, and the journey of each molecule is difficult! One should not waste too much time on it, given that the humidity infiltration load through even one narrow crack can outweigh the entire building's permeation load by a factor of 10. For dehumidification in commercial buildings, the load from permeance is nearly always so small that it's hardly worth bothering to calculate.

On the other hand, careful calculation of permeance is sometimes important in buildings which are clad with brick veneer, or hard-coat stucco or precast concrete panels. These act as a reservoir for the frequent rainwater which falls frequently in hot and humid climates.

The "*permeability*" describes how quickly water vapor moves through a given thickness of a material. In the I-P system, the permeability is the number of grains per hour, per square foot, per inch mercury of pressure difference, *per inch of material thickness*. [In the SI system, the permeability is the number of nanograms per second, per Pascal of pressure difference, per meter of material thickness.] Permeability can be used for calculating vapor movement through materials which vary in thickness, such as air gaps and insulation. On the other hand, most materials sold as vapor retarders are sold in a specific thickness. Such products carry a defined net *permeance* rating, based on their thickness.

The slowly diffusing moisture only becomes a load on the dehumidifier after it enters the inside face of the exterior wall, having traversed the layer with the lowest permeance (highest resistance). Therefore, to size the dehumidification system the designer only needs to calculate the transmission rate through the least-permeable layer—the material with the lowest perm rating.

Permeance or permeability ratings are given for a variety of building materials in figures 11.19 and 11.20. When the permeance and the total surface area are known, they can be used with the vapor

Vapor Permeance Load Per Hour (lb or kg)

$$W_{vp} = \frac{A \times Perm \times (p_h - p_d)}{\text{Divide by: 7000 to obtain lb/h; 1000 to obtain kg/h}}$$

- A: Material surface separating dry from humid air (ft^2 or m^2)
- Perm: Perm rating for material ($gr/(ft^2/in.hg.)$ or $ng/(s/m^2/Pa)$)
- p_h: Vapor pressure on the humid side (in.hg. or kPa)
- p_d: Vapor pressure on the dry side (in.hg. or kPa)

Fig. 11.18 Vapor permeance - Humidity load equation

pressure difference in equation 11.18 to estimate the moisture load that moves through the building's exterior wall by diffusion.

Figure 11.16 shows the potential problem. Rain soaks the cladding, which is then heated by the sun. The air behind that cladding becomes nearly saturated—at the temperature of the cladding, which is often over 100°F [38°C]. Then, the vapor pressure difference between indoors and the saturated air behind the cladding is very large—perhaps twice or three times larger than the vapor pressure difference between the outdoor ambient air and the indoor air. With such an unnaturally large vapor pressure difference, there will be a more significant amount of water vapor driven into the building, as can be seen by solving equation 11.18.

With masonry, or brick or stucco, it's useful for the HVAC designer to question the architectural designer quite closely about the water barrier which is usually installed behind such cladding to protect the

Material	Thickness in.	Thickness mm	Permeance $gr/h \cdot ft^2 \cdot in.hg.$	Permeance $ng/s \cdot m^2 \cdot Pa$
Concrete	8.000	203	0.40	23
Brick masonry	4.000	100	0.80	46
Concrete block, hollow	8.000	203	2.40	137
Gypsum wallboard	0.375	10	50.00	2860
Hardboard (Tempered)	0.125	3	5.00	286
Wood, sugar pine	1.000	25	0.4 to 5.4	23 to 309
Plywood (Douglas fir, exterior glue)	0.250	6	0.70	40
Plywood (Douglas fir, interior glue)	0.250	6	1.90	109
Sheet, glass fiber with acrylic resin	0.056	1	0.12	6.9
Sheet, glass fiber with polyester resin	0.480	12	0.05	2.9
Aluminum foil (Thick)	0.0010	0.03	0	0
Aluminum foil (Thin)	0.0004	0.01	0.05	2.9
Polyethelene (6 mil)	0.0060	0.15	0.06	3.4
Polyethelene (8 mil)	0.0080	0.20	0.04	2.3
Polyethelene (10 mil)	0.0100	0.25	0.03	1.7
Polyvinylchloride (PVC)	0.0040	0.10	0.8 to 01.4	46 to 80
Polyester	0.0076	0.19	0.08	4.6
Cellulose Acetate	0.0100	0.25	4.6	263
Vapor retarder paint	0.0031	0.08	0.45	26
Primer-sealer	0.0012	0.03	6.28	360
Vinyl-acrylic primer	0.0016	0.04	8.62	493
Semi-gloss, vinyl-acrylic enamel	0.0240	0.61	6.61	378
Exterior acrylic house and trim	0.0017	0.04	5.47	313
Flat paint (2 coats) on insulation board	-	-	4.00	229
Two coats of enamel on smooth plaster	-	-	0.5 to 1.5	29 to 86
Primers + sealer, on insulation board	-	-	0.9 to 2.1	51 to 120
Primers + flat oil paint on plaster	-	-	1.6 to 3.0	91 to 172
	Typical Density lb/ft²	**Typical Density kg/m²**		
Built-up roofing	-	-	0	0
Duplex sheet, asphalt, aluminum foil	8.6	0.42	0.176	10
Saturated and coated rool roofing	65.0	3.18	0.24	14
Kraft paper, asphalt-laminated, reinf.	6.8	0.33	1.8	103
Insulation backup paper, asphalt-coated	6.2	0.3	0.6 to 4.2	34 to 240
Asphalt-saturated, coated v. retarder ppr.	8.6	0.42	0.06	3.4
Asphalt-saturated, not coated ppr.	4.4	0.21	20.2	1155
15-lb asphalt felt	14.0	0.68	5.6	320
15-lb tar felt	14.0	0.68	18.2	1040
Single kraft paper, double layer	3.2	0.16	42	2400

Fig. 11.19
Vapor permeance of building materials

Material	Permeability Perm/in.	ng/s • m • Pa
Air (still)	120	174
Cellular glass	0	0
Mineral wool (unprotected)	116	245
Expanded polyurethane (board stock)	0.4 to 1.6	0.58 to 2.3
Expanded polystyrene (extruded board)	1.2	1.7
Expanded polystyrene (bead board)	2.5 to 5.8	2.9 to 8.4
Unicellular synthetic flexible rubber foam	0.02 to 0.15	0.029

Fig. 11.20 Vapor permeability of common insulation

wall's sheathing and the cooler inboard wall layers. If the waterproof layer is a true vapor barrier (below 1 perm), then the calculation will show that the vapor permeation DH load is still nearly insignificant, as is usually the case with non-reservoir claddings.

On the other hand, if the architectural designer has neglected to install a water-and-vapor barrier behind the reservoir cladding, the DH load calculation will show that the vapor permeation load is quite significant. Hopefully, the architectural designer will then realize the great importance of installing that barrier. If not, however, the HVAC designer can do an important service to both the architectural designer and the owner by asking them to read chapter 9 of this book (Keeping water out of the building).

If there is still no action to install a water-and-vapor barrier behind reservoir cladding, the HVAC designer can make the calculation, and point out that all that water vapor is likely to condense and help grow mold in the exterior wall, when the incoming vapor contacts the cool indoor surfaces of the interior wall.

From an HVAC perspective, if the interior surface is a vapor retarder such as vinyl wall covering, the water vapor will probably not become a load on the dehumidification equipment. The vapor will stay in the wall to help grow mold. But if the interior surface finish is permeable, then some of the vapor may become a dehumidification load in the conditioned space after it has traversed the permeable wall material.

But again, the vapor permeation load is not likely to be large unless the building has cladding which acts as a water reservoir which can be heated by the sun. Then, if that all-important water-and-vapor barrier is missing, the HVAC designer's time would be better spent by consulting his attorney and providing written warnings to the architectural designer and owner about the risks of mold, rather than worrying about whether the interior finish has a perm rating of 1.5 rather than 8.1.

Wet surfaces

As water evaporates from recently-cleaned carpets and floors or from pools and spas, it becomes a load on the dehumidification equipment. The evaporation rate determines the DH load, and in all cases, evaporation load is greater when the:

- Air flows quickly across the wet surface
- Water is hot (has a high surface water vapor pressure)
- Air has a low water vapor pressure (is dry)
- Surface area for evaporation is extensive

Swimming pools and spas

The moisture evaporating from swimming pools has been investigated frequently. The rate varies widely according to the type of pool, its use, and whether the area includes recreational equipment that increase the surface area available for evaporation.

The lowest load is an unoccupied pool in a private home, used a few minutes a day by a single occupant swimming slowly from end to end. Public indoor wave pools have the highest load, loaded to their capacity with frolicking teenagers, splashing each other and sliding down water slide tubes that induce evaporation air currents.

Swimming Pool Activity Factors

Type of Pool	Factor
Residential pool	0.5
Condominium	0.65
Therapy	0.65
Hotel	0.8
Public or school	1
Whirlpool, spas	1
Wavepools, water slides	1.5

Fig. 11.21 Swimming pool activity factors

Swimming Pool Load Per Hour (lb or kg)

$$W_{sp} = 0.1A \times (p_w - p_a) \times F_a$$

- A — Pool surface (ft² or m²)
- p_w — Saturation vapor pressure at water temperature (in.hg. or kPa)
- p_a — Saturation vapor pressure at air temperature (in.hg. or kPa)
- F_a — Activity factor

Fig. 11.22 Swimming pools - Humidity load equation

Equipment	Spray Rate gpm	Spray Rate l/s	Extraction Rate	Average Evaporation Rates lb/h [kg/h • m²]: Small Rooms w. Furnishings	Average Evaporation Rates lb/h [kg/h • m²]: Large, Open Spaces
Truck-mounted	1.5	0.095	50%	0.05 [0.25]	0.020 [0.10]
			85%	0.01 [0.05]	0.005 [0.02]
Portable	1.0	0.063	50%	0.06 [0.30]	0.020 [0.10]
			85%	0.02 [0.10]	0.006 [0.03]

Fig. 11.23 Evaporation rates from carpet after wet cleaning

For the same pool surface and other non-activity variables, a quiet private pool evaporates about 33% of the rate of evaporation in the busy water slide/wave pool. The activity within the space strongly influences the evaporation rate, because it determines how much surface is wet beyond the pool surface alone.

Recognizing this key variable, ASHRAE has added an "activity factor" to the classic wet surface evaporation equation developed by Willis Carrier in 1919. Like the Carrier equation, these activity factors are empirical, based on the experience of Engineers, owners and dehumidifier manufacturers. Factors based on a consensus of these groups are shown in Figure 11.21.[12] Given those values, the equation in figure 11.22 can be used to estimate moisture loads from different types of pools.

The designer might also consider the fact that moisture loads vary widely throughout 24 hours, depending on the actual use of the pool. The owner may not be especially concerned about maintaining humidity control in the pool area at peak load conditions. It may be sufficient to calculate the average load over 24 hours, and size the dehumidification equipment for that capacity, reasoning that over time, the equipment will "catch up" with the load. At peak load conditions, humidity will be above the set point, but the owner's purpose may not require continuous control.

Sizing the dehumidifiers for the average rather than the peak load provides substantial savings in both the construction budget and in annual operating costs. Assuming the owner agrees that continuous control is not needed, the designer can calculate the average load by estimating the number of full, partial and zero-use hours, making a calculation for each, summing the loads for all hours, then dividing the total by 24 to obtain the average hourly load. Nearly all manufacturers of swimming pool dehumidifiers provide software that helps a designer estimate these loads.

Still ponds & wet-mopped floors

Research has shown that evaporation rates from still ponds and wet floors are lower than from swimming pools and spas. Unlike swimming pools, floors and shallow ornamental ponds are seldom served with constantly flowing air, nor is the surface agitated. Also, the temperature at the water surface is likely to be near the wet bulb temperature of the air. So when the air velocity across the wet surface is below 25 fpm [below 0.13 m/s], the designer may choose to use an activity factor of 0.25 for better estimating the evaporation rate from such placid surfaces.[12]

Carpet cleaning

Commercial and institutional carpets are often heavily soiled, requiring cleaning by hot water extraction rather than by light shampoo or "dry" absorption techniques. Shampoo, foam and dry absorption leave very little moisture behind on the carpet, and what little remains should evaporate within 1 to 3 hours. In contrast, the hot water technique sprays water onto the carpet at a rate of either 8.3 or 12.5 lbs

Wet Carpet Load Per Hour (lb or kg)

Wet carpet surface (ft² or m²)

Evaporation rate (lb/ft²/h or kg/m²/h)

$$W_{cl} = A \times E_r$$

Fig. 11.24 Wet carpet - Humidity load equation

per minute [3.8 or 5.7 kg/min]. After application and extraction, the water left behind takes between 6 and 24 hours to evaporate.

In the cleaning process, hot water is sprayed onto the carpet and immediately pulled back off the surface by an "extractor head" attached to a vacuum blower. The blower may be located in a portable unit near the operator, or connected by long hoses to a unit located further away in a truck. The amount of water remaining on the carpet after "extraction", depends on the spray rate of hot water and on the amount of water recovered by the extractor head. Those variables depend on both the skill of the operator and the force of the vacuum at the extraction head. Skilled operators that follow standard guidelines will make a second, "dry pass" after each spray application to recover more water, leaving less to evaporate and become a dehumidification load. Also, truck-mounted vacuum blowers usually recover more water on each pass because they have greater suction than portable units.

Evaporation slows as the remaining moisture approaches the wet bulb temperature of the air, and as less water is available to evaporate. The values in Figure 11.23 are based on empirical observations and estimates by four carpet cleaning experts concerning spray rates, extraction rates and typical drying times. Experts all remark on the difficulty of obtaining accurate data on water extraction rates, which vary from 50 to 85% of the water under real world circumstances. They advise that the highest dehumidification load per hour occurs when the:

- Operator rushes to completion, and fails to make a second, "dry pass" after each "combined" application/extraction pass.
- Equipment has low suction capacity, as in portable units, or when truck units have long hoses or clogged filters.
- Air above the carpet is dry, so that evaporation takes place quickly, in spite of the usual lack of air movement at the surface.

The designer should note the informal character of the estimates in figure 11.23 and adjust his assumptions accordingly.

Intermittent domestic loads

In residential buildings like hotels, apartments, condominiums and dormitories, domestic operations can contribute a modest amount to the dehumidification load. Normally, however, the duration of these activities is seldom long. Over a 24-hour period the average hourly moisture load from showers, for example, is very small compared to loads from other sources.[13] Figure 11.25 shows the typical moisture loads from selected domestic activities.

As with other periodic loads, the designer might consider averaging such intermittent loads over the time the owner might allow humidity to be above the design set point, as opposed to designing the dehumidification equipment to maintain conditions throughout all peak load events. For example, does a hotel room *really* have to be kept below a 55°F [12.8°C] dew point while the guest is taking a morning shower? Or can the humidity be allowed to rise for a second hour so a smaller dehumidifier could slowly "catch up" with that one-time load? If the owner agrees to an hour above set point, the moisture load per hour is cut in half and the dehumidification capacity can be smaller and less costly to operate. An owner's preferences may be different in a health club, which could have heavy shower loads for long periods.

Load Source	Typical Moisture Load Per Event lb	kg	Hourly Load lb	kg	Comments & Assumptions
Bath (15 minutes)	0.12	0.06	0.03	0.01	One bath/hr
Shower (15 min)	1.40	0.66	5.60	2.55	Four showers/hr
Breakfast preparation (4 ppl)	0.35 to 0.58	0.16 to 0.27	0.61	0.28	Preparation lasts 20 minutes
Lunch preparation	0.53 to 0.68	0.25 to 0.32	0.80	0.36	Preparation lasts 20 minutes
Dinner preparation	1.22 to 1.58	0.55 to 0.72	0.92	0.42	Preparation lasts 40 minutes
Breakfast dishes (4 ppl, by hand)	0.21	0.10	0.04	0.02	Requires 10 minutes
Lunch dishes	0.16	0.07	0.03	0.01	Requires 10 minutes
Dinner dishes	0.68	0.31	0.23	0.10	Requires 20 minutes
Simmering pot (6 in. diam., per 10 min)	0.13	0.06	0.04	0.02	Simmers for 20 minutes
Boiling pot (6 in. diam., per 10 min)	0.57	0.26	0.19	0.09	Boils for 20 minutes
Floor mopping (dries over one hour)	0.03	0.014	0.03	0.013	Load per hour, per ft2 or m2
Unvented clothes dryer (per load)	4.7 to 6.2	2.13 to 2.90	3.60	1.64	Average load, dry in 40 minutes
Typical total daily load (4 ppl)	**3.5 to 10.9**	**1.6 to 5.0**	**0.3**	**0.14**	**Includes respiration & perspiration**
Green firewood (128 ft^3 over 6 months)	800	363.6	0.18	0.08	Assumes hardwood. Softwood is ±50% less
Crawl space (1000 ft^2, uncovered, per day)	95	43.2	0.004	0.002	Load per hour, per ft2 or m2
Crawl space (covered)	9.5	4.3	0.0004	0.0002	Load per hour, per ft2 or m2
1 ton A/C unit without drain (per day)	6	2.7	0.25	0.11	Moisture re-evaporated from pan
Construction moisture (1st year, per day)	21	9.5	0.88	0.40	Full concrete basement and wood framing.
Damp basement (1000 ft^2, per day)	21	9.5	0.001	0.0004	Load per hour, per ft2 or m2
10 Medium-size ferns (per day)	5.7	2.6	0.24	0.11	Typical of medium-size house plants
One 6ft tall beech tree (per day)	106	48.2	4.42	2.01	Typical of ornamental indoor trees

Fig. 11.25 Domestic humidity loads

Transient product loads

Humidity from wet building materials and from humid packaging is often a load in commercial buildings. These loads are periodic, rather than continuous. So the designer often uses a time-weighted average for these loads, based on their magnitude and frequency.

DH loads from humid packaging materials can be significant in retail stores and warehouses, where a great deal of cardboard packaging may flow through the building in a short amount of time. The outer packaging absorbs moisture during transit to the building, and releases it when the product enters a humidity-controlled storage space. The amount of water given off depends on the sorption characteristics of the material, and on the difference between the humid transit environment and the dry storage environment. Figure 11.26 shows the moisture equilibrium curves for a variety of materials.[14] If the designer knows the nominal weight of the product flowing through the building and the time it resides there, he can estimate the desorption load.

The time required to release the absorbed moisture depends largely on how fast air moves across the moist surface. Stacked cartons will dry quickly on the outside—perhaps in a few hours. But the cardboard between the packages may take days or weeks to dry out. Filing cabinets full of moist papers can take weeks or months to reach equilibrium with the dry air outside the cabinet. When the potential load is small, the designer can apply engineering judgment to estimate the time needed to dry specific materials, and in some cases the owner or the suppliers of the products may be able to assist in these estimates.

When the potential load is really large, the designer can test his assumptions by using an environmental simulation test chamber. These are accessible at reasonable cost through packaging engineering firms and through organizations that provide military-grade stress-testing or industrial material testing services.

DH loads from walls, furnishings, papers and books

The building and all its contents adsorb moisture when humidity rises, and release it when humidity falls. Some call this effect the "moisture capacitance" of the building. If the relative humidity indoors stays

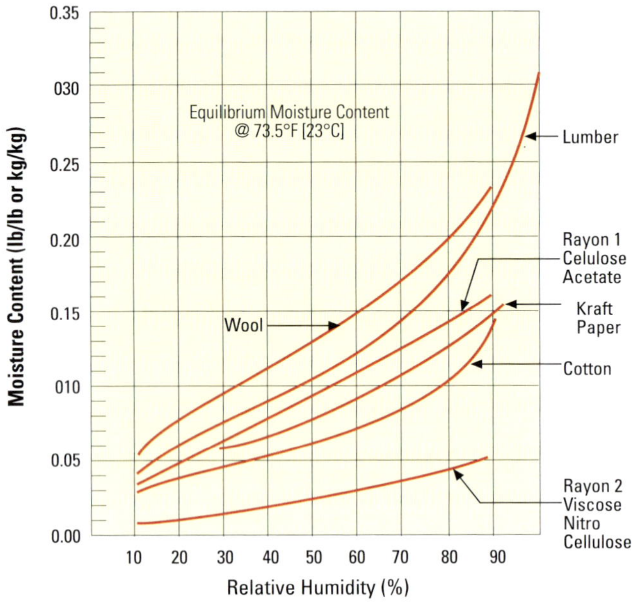

Fig. 11.26 Sorption curves of different materials

Humid Product Load Per Hour (lb or kg)

F_r — Product flow rate through the building (lb/day or kg/day)

MC_1 — Moisture content of the product when at equilibrium with high humidity (lb/lb or kg/kg)

MC_2 — Moisture content of the product when at equilibrium with low humidity (lb/lb or kg/kg)

$$W_{pr} = \frac{F_r}{24} \times (MC_1 - MC_2)$$

Fig. 11.27 Humid products - Humidity load equation

constant, there is little or no desorption, and therefore no dehumidification load from this source.

On the other hand, some designers in cool or mixed climates may have used cool outdoor air in evenings to precool the building for the following day, making it ready to absorb the high heat loads of the afternoon and reducing the net annual cost of sensible cooling. This technique is often called "free cooling", or "air-side economizer cooling."

This practice is very unwise in humid climates. Unfortunately for humidity control, cool outdoor air often carries massive amounts of excess moisture into the building. The relative humidity rises, so the building and its contents adsorb moisture which is then released to become a DH load as the system start up in the morning. There is no "free cooling" when the incoming outdoor air is humid.

When for some reason a designer needs to estimate such a load, reference material is available in the form of the final report from ASHRAE research project 455.[15] That report is available from ASHRAE headquarters.

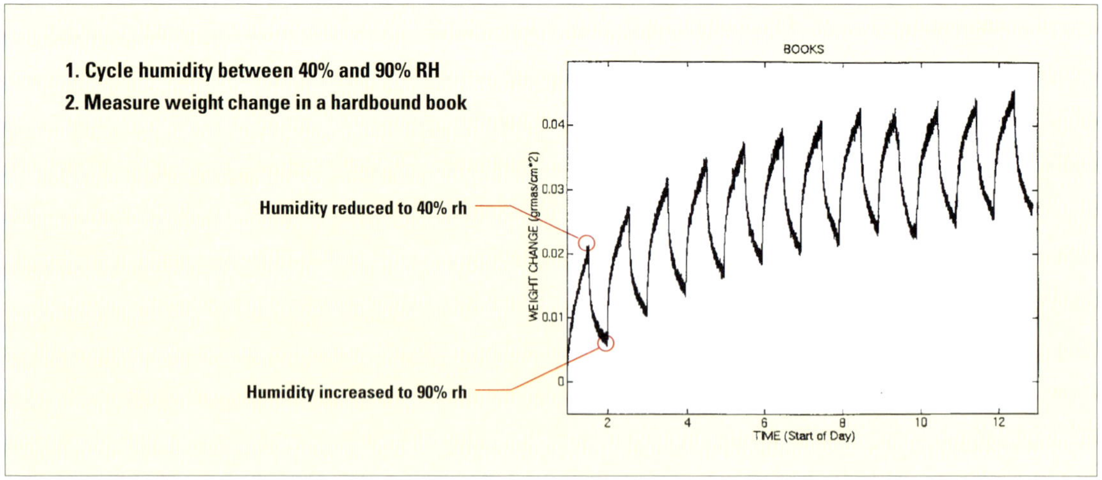

Fig. 11.28 Sorption and desorption from books—when relative humidity cycles between 40% and 90%

GENERAL NOTES

1. **Not identical to Table 6-1:** The rates in this table are based on table 6-1 of Standard 62.1-2007. HOWEVER, THE TABLE HEADINGS AND NOTES SHOWN HERE ARE NOT THE SAME. THESE WERE MODIFIED FOR CLARITY, IN COMPENSATION FOR THE ABSENCE OF THE COMPLETE TEXT OF THE STANDARD. FOR FULL GUIDANCE, CONSULT THE STANDARD ITSELF.

2. **Related requirements:** The rates in this table are based on all other applicable requirements of Standard 62.1-2007 being met.

3. **Smoking:** This table applies to non-smoking areas. Rates for smoking areas must be determined by other methods. See section 6.2.9 for ventilation requirements for smoking areas.

4. **Air density:** Volumetric airflow rates are based on an air density of 0.075 lb_{da}/ft^3 [1.2 kg_{da}/m^3], which corresponds to dry air ayt a barometric pressure od 1 atm [101.3 kPa] at an air temperature of 70°F [21°C]. Rates may be adjusted for actual density, but such adjustment is not required for compliance with this standard.

5. **Default occupant density:** The default occupant density shall be used when the actual occupnat density is not known.

6. **Default assumptions:** These rates are based on the assumed minimum occupant densities. ASHRAE Standard 62.1-2007 states that these assumed minimum densities *shall* be used whenever the actual occupancy is not known. The rates in these columns include the ventilation air required to dilute contaminants emitted by people (at that assumed density), plus the air needed to dilute contaminants emitted by the materials and contents of the building itself. For occupancy categories without an assumed minimum occupant density, refer to the columns labeled "...outdoor air per unit of floor area" to calculate the minimum amount of outdoor air required for the space in question.

7. **Unlisted occupancies:** If the occupancy category for the proposed space is not listed, the requirements for the occupancy category most similar to the proposed use in terms of occupant density, activities and building construction shall be used.

8. **Health-care facilities:** Rates shown here reflect the information provided in ASHRAE Std 6.1-2007, Appendix E. They have been chosen to dilute human bioeffluents and other contaminants with anadequate margin of safety and to account for health variations between different people and activity levels.

9. **Occupancy-specific requirements:** Notes A - K provide additional clarification of outdoor air requirements shown in this table.

MINIMUM VENTILATION RATES IN BREATHING ZONE (Based on—but not identical to—Table 6-1 of ASHRAE/ANSI Standard 62.1 - 2007)

Occupancy category	Outdoor air per occupant, plus...		...outdoor air per unit of floor area		Notes	Default Assumptions (For use when the actual occupancy is not known)			Air Class
						Occupants per 1000ft² or 100m²	Combined minimum outdoor air[5]		
	cfm/person	L/s • person	cfm/ft²	L/s • m²			cfm/person	L/s • person	
Correctional Facilities									
Cells	5	2.5	0.12	0.6		25	10	4.9	2
Dayrooms	5	2.5	0.06	0.3		30	7	3.5	1
Guard Stations	5	2.5	0.06	0.3		15	9	4.5	1
Booking/waiting rooms	7.5	3.8	0.06	0.3		50	9	4.4	2
Educational Facilities									
Daycare (though age 4)	10	5	0.18	0.9		25	17	8.6	2
Daycare sickroom	10	5	0.18	0.9		25	17	8.6	3
Classrooms (ages 5-8)	10	5	0.12	0.6		25	15	7.4	1
Classrooms (ages 9 & older)	10	5	0.12	0.6		35	13	6.7	1
Lecture classroom	7.5	3.8	0.06	0.3		65	8	4.3	1
Lecture hall (Fixed seats)	7.5	3.8	0.06	0.3		150	8	4.0	1
Art classroom	10	5	0.18	0.9		20	19	9.5	2
Science laboratories	10	5	0.18	0.9		25	17	8.6	2
University/College laboratories	10	5	0.18	0.9		25	17	8.6	2
Wood/metalworking shop	10	5	0.18	0.9		20	19	9.5	2
Computer lab	10	5	0.12	0.6		25	15	7.4	1
Media center	10	5	0.12	0.6	A	25	15	7.4	1
Music/theater/dance	10	5	0.06	0.3		35	12	5.9	1
Multi-use assembly	7.5	3.8	0.06	0.3		100	8	4.1	1
Food & Beverage Service									
Restaurant dining rooms	7.5	3.8	0.18	0.9		70	10	5.1	2
Cafeterias/quick-service dining	7.5	3.8	0.18	0.9		100	9	4.7	2
Bars/cocktail lounges	7.5	3.8	0.18	0.9		100	9	4.7	2

Important note: These data reflect the recommendations of ASHRAE Standard 62.1-2007 as of July, 2008. However, the standard is in continuous maintenance, which means it's recommendations are likely to change more frequently than this book can be updated. When compliance with ASHRAE Std 62.1 is important, the reader should consult the most current edition of that standard (along with any ammendments).

MINIMUM VENTILATION RATES IN BREATHING ZONE (Based on—but not identical to—Table 6-1 of ASHRAE/ANSI Standard 62.1 - 2007)

Occupancy category	Outdoor air per occupant, plus...		...outdoor air per unit of floor area		Notes	Default Assumptions (For use when the actual occupancy is not known)			Air Class
						Occupants per 1000ft² or 100m²	Combined minimum outdoor air[5]		
	cfm/person	L/s • person	cfm/ft²	L/s • m²			cfm/person	L/s • person	
General									
Break rooms	5	2.5	0.06	0.3		25	10	5.1	1
Coffee stations	5	2.5	0.06	0.3		20	11	5.5	1
Conference rooms/meeting rooms	5	2.5	0.06	0.3		50	6	3.1	1
Corridors	-	-	0.06	0.3		-	See note K	See note K	1
Storage rooms	-	-	0.12	0.6	B	-	See note K	See note K	1
Hotels, Motels, Resorts, Barracks & Dormitories									
Bedroom/sleeping area	5	2.5	0.06	0.3		10	11	5.5	1
Barracks sleeping areas	5	2.5	0.06	0.3		20	8	4.0	1
Laundry rooms (central)	5	2.5	0.12	0.6		10	17	8.5	2
Laundry rooms in dwelling units	5	2.5	0.12	0.6		10	17	8.5	1
Lobbies/prefunction areas	7.5	3.8	0.06	0.3		30	6	2.8	1
Multipurpose asembly areas	5	2.5	0.06	0.3		120	6	2.8	1
Office Buildings									
Office space	5	2.5	0.06	0.3		5	17	8.5	1
Reception areas	5	2.5	0.06	0.3		30	6	3.0	1
Call center/data entry clusters	5	2.5	0.06	0.3		60	6	3.0	1
Main entry lobbies	5	2.5	0.06	0.3		10	11	5.5	1
Miscellaneous Spaces									
Bank vaults/safe deposit vaults	5	2.5	0.06	0.3		5	17	8.5	2
Computer rooms (no printers)	5	2.5	0.06	0.3		4	20	10.0	1
Electrical equipment rooms	-	-	0.06	0.3	B	-	See note K	See note K	1
Elevator machine rooms	-	-	0.12	0.6	B	-	See note K	See note K	1
Pharmacy prep area	5	2.5	0.18	0.9		10	23	11.5	2
Photo studios	5	2.5	0.12	0.6		10	17	8.5	1
Shipping/receiving areas	-	-	0.12	0.6	B	-	See note K	See note K	1
Telecom closets	-	-	0.00	0.0		-	See note K	See note K	1
Transportation waiting areas	7.5	3.8	0.06	0.3		100	8	4.1	1
Warehouses	-	-	0.06	0.3	B	-	See note K	See note K	2

Important note: These data reflect the recommendations of ASHRAE Standard 62.1-2007 as of July, 2008. However, the standard is in continuous maintenance, which means it's recommendations are likely to change more frequently than this book can be updated. When compliance with ASHRAE Std 62.1 is important, the reader should consult the most current edition of that standard (along with any ammendments).

OCCUPANCY-SPECIFIC NOTES

A. For high school and college libraries, use the values shown for public assembly spaces-libraries.

B. Rates may not be sufficient when stored materials have potentially-harmful emissions.

C. Rate does not allow for humidity control. Additional dehumidification may be required to keep the indoor dew point low enough to prevent structural damage to the building enclosure.

D. Rate does not include dilution and exhaust of pollutants from special effects such as dry ice vapor (CO_2) or theatrical smoke.

E. When combustion equipment is used on the playing surface (such as ice-resurfacing vehicles) additional ventilation and/or source control shall be provided beyond the rates shown in this table.

F. Default occupancy for dwelling units shall be two people for studio and one-bedroom units, with one additional person for each additional bedroom.

G. Air from one residential dwelling unit shall not be recirculated or transferred to any other space outside of that dwelling unit.

H. Floor area for estimated maximum occupancy for health care facilities is based on the net occupiable area rather than the gross floor area.

I. Special requirements or codes or required air pressure relationships between adjacent spaces in health care facilities may determine ventilation rates and filter efficiencies which are different from the values shown in this table. Also, medical or other procedures which generate contaminants may require higher rates than those shown in this table.

J. Air shall not be recirculated from autopsy rooms into other spaces.

K. ASHRAE Standard 62.1-2007 has not provided a minimum assumed occupancy for this space. However, outdoor air remains a requirement, in order to dilute contaminants generated by the building itself and it's contents. Refer to the columns labeled "...outdoor air per unit of floor area" to calculate the minimum outdoor air requirement for this space.

OCCUPANCY-SPECIFIC NOTES

A. For high school and college libraries, use the values shown for public assembly spaces-libraries.

B. Rates may not be sufficient when stored materials have potentially-harmful emissions.

C. Rate does not allow for humidity control. Additional dehumidification may be required to keep the indoor dew point low enough to prevent structural damage to the building enclosure.

D. Rate does not include dilution and exhaust of pollutants from special effects such as dry ice vapor (CO_2) or theatrical smoke.

E. When combustion equipment is used on the playing surface (such as ice-resurfacing vehicles) additional ventilation and/or source control shall be provided beyond the rates shown in this table.

F. Default occupancy for dwelling units shall be two people for studio and one-bedroom units, with one additional person for each additional bedroom.

G. Air from one residential dwelling unit shall not be recirculated or transferred to any other space outside of that dwelling unit.

H. Floor area for estimated maximum occupancy for health care facilities is based on the net occupiable area rather than the gross floor area.

I. Special requirements or codes or required air pressure relationships between adjacent spaces in health care facilities may determine ventilation rates and filter efficiencies which are different from the values shown in this table. Also, medical or other procedures which generate contaminants may require higher rates than those shown in this table.

J. Air shall not be recirculated from autopsy rooms into other spaces.

K. ASHRAE Standard 62.1-2007 has not provided a minimum assumed occupancy for this space. However, outdoor air remains a requirement, in order to dilute contaminants generated by the building itself and it's contents. Refer to the columns labeled "... outdoor air per unit of floor area" to calculate the minimum outdoor air requirement for this space.

MINIMUM VENTILATION RATES IN BREATHING ZONE (Based on—but not identical to—Table 6-1 of ASHRAE/ANSI Standard 62.1 - 2007)

Occupancy category	Outdoor air per occupant, plus...		...outdoor air per unit of floor area		Notes	Default Assumptions (For use when the actual occupancy is not known)			Air Class
						Occupants per 1000ft² or 100m²	Combined minimum outdoor air[5]		
	cfm/person	L/s • person	cfm/ft²	L/s • m²			cfm/person	L/s • person	
Public Assembly Spaces									
Auditorium seating area	5	2.5	0.06	0.3		150	5	2.7	1
Places of religious worship	5	2.5	0.06	0.3		120	6	2.8	1
Courtrooms	5	2.5	0.06	0.3		70	6	2.9	1
Legislative chambers	5	2.5	0.06	0.3		50	6	3.1	1
Libraries	5	2.5	0.12	0.6		10	17	8.5	1
Museums (children's)	7.5	3.8	0.12	0.6		40	11	5.3	1
Museums/galleries	7.5	3.8	0.06	0.3		40	9	4.6	1
Residential									
Dwelling unit	5	2.5	0.06	0.3	F,G	See note F	See note F	See note F	1
Common corridors	-	-	0.06	0.3		-	See note K	See note K	1
Retail									
Sales (except as below)	7.5	3.8	0.12	0.6		15	16	7.8	2
Shopping mall common areas	7.5	3.8	0.06	0.3		40	9	4.6	1
Barbershop	7.5	3.8	0.06	0.3		25	10	5.0	2
Beauty & nail salons	20	10	0.12	0.6		25	25	12.4	2
Pet shops (animal areas)	7.5	3.8	0.18	0.9		10	26	12.8	2
Supermarket	7.5	3.8	0.06	0.3		8	15	7.6	1
Coin-operated laundries	7.5	3.8	0.06	0.3		20	11	5.3	2
Sports and Entertainment									
Sports arena (playing area)	-	-	0.3	1.5	E	-	See note K	See note K	1
Gymn/stadium (playing area)	-	-	0.3	1.5	K	30	See note K	See note K	2
Spectator areas	7.5	3.8	0.06	0.3		150	8	4.0	1
Swimming pool (pool and deck)	-	-	0.48	2.4	C	-	See note K	See note K	2
Dance area	20	10	0.06	0.3		100	21	10.3	1
Health club/aerobics room	20	10	0.06	0.3		40	22	10.8	2
Health club/weight room	20	10	0.06	0.3		10	26	13.0	2
Bowling alley (seating)	10	5	0.12	0.6		40	13	6.5	1
Gambling casinos	7.5	3.8	0.18	0.9		120	9	4.6	1

Important note: These data reflect the recommendations of ASHRAE Standard 62.1-2007 as of July, 2008. However, the standard is in continuous maintenance, which means it's recommendations are likely to change more frequently than this book can be updated. When compliance with ASHRAE Std 62.1 is important, the reader should consult the most current edition of that standard (along with any ammendments).

MINIMUM VENTILATION RATES IN BREATHING ZONE (Based on—but not identical to—Table 6-1 of ASHRAE/ANSI Standard 62.1 - 2007)

Occupancy category	Outdoor air per occupant, plus...		...outdoor air per unit of floor area		Notes	Default Assumptions (For use when the actual occupancy is not known)			Air Class
						Occupants per 1000ft² or 100m²	Combined minimum outdoor air[5]		
	cfm/person	L/s • person	cfm/ft²	L/s • m²			cfm/person	L/s • person	
Sports & Entertainment (Continued)									
Game arcades	7.5	3.8	0.18	0.9		20	17	8.3	1
Stages, studios	10	5	0.06	0.3	D	70	11	5.4	1

Health Care Facilities (Summarizing Appendix E - ASHRAE Standard 62.1-2007 - See general note 7 and occupancy-specific note H)									
Patient rooms	-	-	-	-	I	10	25	13	-
Medical procedure	-	-	-	-	I	20	30	15	-
Operating rooms	-	-	-	-	I	20	30	8	-
Recovery and ICU	-	-	-	-	I	20	15	8	-
Autopsy rooms	-	-	0.5	2.5	J	20	See note K	See note K	-
Physical therapy	-	-	-	-	I	20	15	8	-

Important note: These data reflect the recommendations of ASHRAE Standard 62.1-2007 as of July, 2008. However, the standard is in continuous maintenance, which means it's recommendations are likely to change more frequently than this book can be updated. When compliance with ASHRAE Std 62.1 is important, the reader should consult the most current edition of that standard (along with any ammendments).

OCCUPANCY-SPECIFIC NOTES

A. For high school and college libraries, use the values shown for public assembly spaces-libraries.

B. Rates may not be sufficient when stored materials have potentially-harmful emissions.

C. Rate does not allow for humidity control. Additional dehumidification may be required to keep the indoor dew point low enough to prevent structural damage to the building enclosure.

D. Rate does not include dilution and exhaust of pollutants from special effects such as dry ice vapor (CO_2) or theatrical smoke.

E. When combustion equipment is used on the playing surface (such as ice-resurfacing vehicles) additional ventilation and/or source control shall be provided beyond the rates shown in this table.

F. Default occupancy for dwelling units shall be two people for studio and one-bedroom units, with one additional person for each additional bedroom.

G. Air from one residential dwelling unit shall not be recirculated or transferred to any other space outside of that dwelling unit.

H. Floor area for estimated maximum occupancy for health care facilities is based on the net occupiable area rather than the gross floor area.

I. Special requirements or codes or required air pressure relationships between adjacent spaces in health care facilities may determine ventilation rates and filter efficiencies which are different from the values shown in this table. Also, medical or other procedures which generate contaminants may require higher rates than those shown in this table.

J. Air shall not be recirculated from autopsy rooms into other spaces.

K. ASHRAE Standard 62.1-2007 has not provided a minimum assumed occupancy for this space. However, outdoor air remains a requirement, in order to dilute contaminants generated by the building itself and it's contents. Refer to the columns labeled "... outdoor air per unit of floor area" to calculate the minimum outdoor air requirement for this space.

References

1. *Paragraph 1.3.2, page 17* of *Volume 2: Guidelines & Practice, Final Report of Annex 14: Condensation and energy*, International Energy Agency. 1991. Energy conservation in buildings and community systems programme. Hugo Hens, Ph.D; Operating Agent. Laboratory for Building Physics, Catholic University of Leuven, Leuven, Belgium

2. ASHRAE Standard 160p - *Criteria for analyzing moisture in buildings*. ASHRAE, Atlanta, GA

3. Harriman, Brundrett & Kittler, 2008. *ASHRAE Humidity Control Design Guide*, ISBN 1-883413-98-2 ASHRAE, Atlanta, GA

4. ASHRAE Standard 62.1-2007 - *Ventilation for acceptable indoor air quality*. ASHRAE, Atlanta, GA

5. Persily, A., "Myths About Building Envelopes". *ASHRAE Journal.* March, 1999 pp.39-47.

6. Cummings, J.B, Withers, C.R, Moyer, N, Fairey, P, McKendry, B. *Uncontrolled Air Flow In Non-Residential Buildings*. Final Report, FSEC-CR-878-96. April, 1996. Florida Solar Energy Center, 1679 Clear Lake Road, Cocoa, FL. 32922.7.

7. Treschel, H., Lagus, P., Editors. *Measured Air leakage of Buildings*. 1986. ASTM STP 904 American Society of Mechanical Engineers, 1916 Race St, Philadelphia, PA 19103 www.astm.org. ISBN 0-8031-0469-3)

8. SMACNA. *HVAC Duct Systems Inspection Guide* (15D, 1989 The Sheet Metal Manufacturers and Air Conditioning Contractors National Association 8224 Old Courthouse Rd., Tyson's Corner, Vienna, VA. 22182 (703) 790-9890 www.smacna.org.

9. ASHRAE Handbook—Fundamentals. 2005. Commercial Building Envelope Air Leakage Chapter 27 (Ventilation and Infiltration) page 23.

10. Jones, B., Sipes, J., Quinn, H., Mcollough, E., *The Transient Nature of Thermal Loads Generated by People*, Final Report of ASHRAE 619-RP, ASHRAE Transactions, 1994, V.100, Pt.2

11. ASHRAE Handbook—Fundamentals. 2005. Water Vapor Transmission Data for Building Components. Chapter 25, Tables 7a, 7b and 8, pages 15-17.

12. *ASHRAE Handbook—Applications*. 2007. Load Estimation. Chapter 4. Places of Assembly, (Natatoriums), p.6-7.

13. Christian, Jeffrey E. "A Search for Moisture Sources". 1993. Proceedings of the Bugs, Mold & Rot II Conference pp.71-81. National Institute of Building Sciences. Washington, DC (202) 289-7800

14. Harriman, L.G. III, Ed. *The Dehumidification Handbook*, Second Edition. 1990. Munters Corporation. 79 Monroe St., Amesbury, MA 01913.

15. Bailey, D.W., Bauer, F.C., Slama, C., Barringer, C., Flack, J.R., *Investigation of Dynamic Latent Heat Storage Effects of Building Construction and Furnishings.* Final report of ASHRAE 455-RP. June, 1994.

Chapter 12
Estimating Cooling Loads

By Lew Harriman

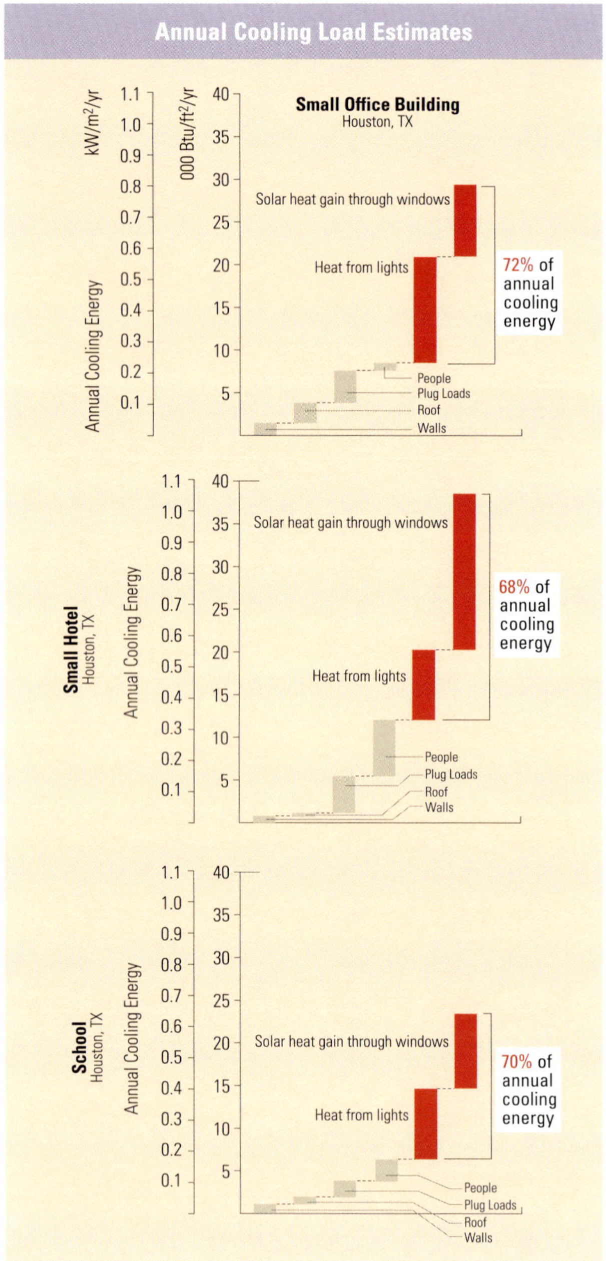

Huang & Franconi - Lawrence Berkeley National Laboratory - U.S. DOE - LBL-37208 - 1999

Fig 12.1 Estimating sensible cooling loads - It's all about the glass

In hot and humid climates, the glazing decisions dictate the annual sensible cooling loads. The photo above shows a common and very bad alternative—a high percentage of glass, unshaded, facing west, which is not configured for daylighting and which transmits a great deal of solar heat. HVAC designers can help prevent such high-cost, high-energy, low-comfort buildings by quantifying, for the owner and architect, the cooling loads for enclosure alternatives ***before*** *the design of that enclosure is "set in stone."*

Key Points

In current HVAC design practice, a computer will be used to help estimate cooling loads. This chapter is not a substitute for computer-powered calculations. Instead, we focus on a few of the designers' basic assumptions about sensible loads and about the cooling systems which remove them. The designers' assumptions, sometimes unstated and unrecognized, will always strongly influence the results from computer-assisted calculations, for better or for worse.

Suggestions and cautions for calculating sensible cooling loads in hot and humid climates include:

- First—Glass decisions rule the sensible cooling loads. Calculate glass loads early, to benefit the owner by minimizing sensible cooling loads.
- Next—separate and calculate the dehumidification loads independently from sensible cooling loads, to avoid common errors in HVAC design in humid climates.
- The peak outdoor temperature is not the worst case for the ventilation load. The highest total ventilation load (highest enthalpy) comes at part-load sensible cooling conditions, when the outdoor *dew point* reaches its peak.
- Peak ventilation loads can be greatly reduced with exhaust air heat exchangers.
- Use ASHRAE research to avoid the common mistake of overestimating "plug loads."

Quantify glass-related loads early and often, to improve the architectural design

The solar load through glass—combined with its consequences for lighting energy and the heat generation from those lights—dominates the sensible cooling loads. The HVAC designer's most influential (and very brief) opportunity for improving the building is when he or she informs the owner and architect of the dimension of those loads, along with their implications for energy consumption, thermal comfort and construction cost.

If the owner and architect want a low-energy building, the HVAC designer is in the best position to help them accomplish that goal *before the building enclosure design is settled*. The owner and architect seldom understand the fatal energy consequences of their glazing decisions. The HVAC designer can help them make more informed design decisions. An early cooling load calculation tells them the size of the cooling equipment for different glazing choices.

After the architectural decisions have been made concerning the size, shape, location, exterior shading and number of windows, the look of the building and the space-planning for each floor are basically fixed. These will be very difficult for the owner to change. The marketing for the building strongly depends on its look-and-feel, and so do its regulatory approvals. Also, the architect will not gratefully embrace changes after the look-and-feel of the building has been established and agreed to by the owner and the regulatory authorities. The architect begins to lose money quickly when the building's entire exterior and its interior space planning must be redesigned and re-approved because of different locations, aspect ratios, shading or percentage of glazing on exterior walls.

So the magic moment for improvement is at schematic design—before design development and long before construction documents. Early in schematic design of the building's exterior, the HVAC designer's load calculations can make a real improvement. After the glazing decisions are finished (if those glazing decisions were poor ones) all the HVAC designer can do are the usual things—moan and groan, enhancing his reputation for being negative and uncreative, and then try to make the best of a bad situation, using a budget which won't be adequate for all the equipment and controls needed to provide comfort, much less the mechanical space needed to service them.

That all-too-common situation can be avoided when the HVAC designer makes the decision to participate (positively, creatively and firmly) in the architectural design decisions at an early stage. At that point, quantification of the glazing loads can make the difference between a low energy building and a wasteful one.

Rough out the loads quickly to start the conversation along productive lines, informing and guiding the glazing decisions. Then take more time later, during the design development stage, to make the detailed load calculations based on actual details and specifications. It's difficult and uncomfortable for technical professionals to do this. No engineer is comfortable guessing about the thermal characteristics of walls and windows before the architect has decided what they are, exactly, and defined them in detail. But that's the point. At this stage, the architectural design is more flexible.

Early glazing decisions that make big differences in cooling loads

Some glazing decisions make a bigger difference than others. Here are several which come early in the schematic design of the building's exterior. These usually have the greatest influence on the sensible cooling loads, for better or for worse.

A lot of glass is bad

Thermally, the best building is one with little glass instead of a lot. And the reduced amount of glass must exclude most of the solar heat gain.

There have been major improvements in glass technology in recent years. Architectural designers have been especially impressed by these improvements. Many have designed buildings which are basically large glass boxes, under the apparent misimpression that lots of glass has somehow become a low-energy technology. It isn't. Glass is an exceptionally bad insulator[1], and also, even modern glass still lets in far too much radiant heat.

Excellent modern glass might now have a solar heat gain coefficient as low as 0.35. But that must be compared to an insulated wall at 0.0 solar heat gain. Zero percent of the solar heat is a lot less than 35% of the solar heat. Therefore, the more glass on the building, the more money and the more energy it will take to cool it. In other words, lots of glass surface on the building is bad, from a cooling load perspective.

The HVAC designer can make this fact clear to the owner and architectural designer through early load estimates using different percentages of glazing on each exterior wall.

After the amount of glass has been reduced to its minimum, the next issue is the thermal quality of the glass, and the amount of shading which should be above that glass.

A first step is glass which has a low solar heat gain coefficient (SHGC). Then add shading to reduce the SHGC still further, as illustrated in figure 12.2. The problem with specifying an extremely low SHGC for the glass (and leaving off the shading) is that such glass will be so dark that it may not be pleasant for occupants to look though. Or it may be so reflective that nearby drivers can occasionally be blinded by the reflected glare outdoors. So selecting glass with a moderately-low SHGC combined with a horizontal window overhang of more than one meter often provides a more congenial way to reach a low net SHGC.

Reduce the glazing towards 20% of the wall surface

The ASHRAE Advanced Energy Guides[3,4,5,6] provide benchmark target values for glazing as a percent of the wall, and for the net solar heat gain coefficient of those windows, including shading. In hot and humid climates, the guides call for a target of 20 to 30% glazing on any wall. Also, the solar heat gain coefficient of those windows should be less than 0.31.

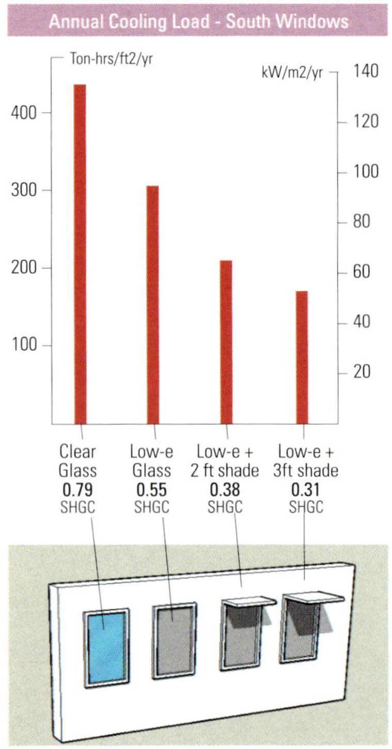

Fig. 12.2 SHGC and its effect on annual cooling load[2]

The lower the solar heat gain coefficient, the lower the annual cooling load. The HVAC designer can help improve the architectural design by quantifying this difference at an early stage, before the exterior fenestration has been "set in stone."

Fig. 12.3
ASHRAE Advanced Energy Guides
Target values for all components of the exterior enclosure are clearly outlined in the ASHRAE Advanced Energy Guides, which are available—at no cost—for downloading from the ASHRAE website.

As of the publication date of this book, Advanced Energy Guides are available for schools, offices, warehouses and small business hotels.[3,4,5,6]

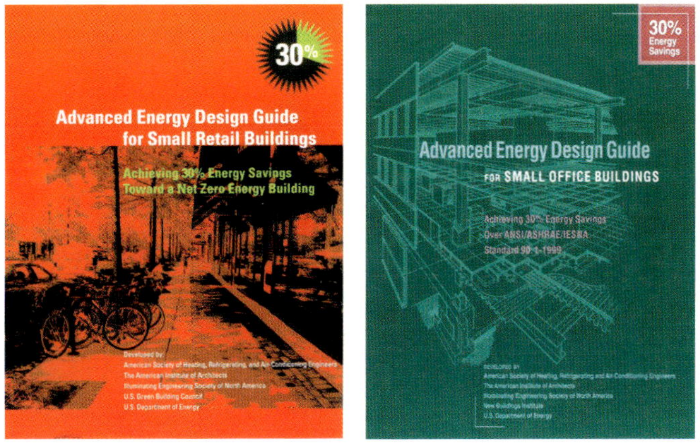

In cool or mixed climates, guidance to architectural designers is a bit different. In climates far from the equator, glass that lets in some solar heat is sometimes not all bad, because it can, depending on how it is arranged, reduce the need for winter heating. But in hot and humid climates, the basic guideline is simple: less glass is better. Then make sure the glass itself has a low SHGC and shade it if necessary, for a combined total SHGC of less than 0.31. The lower the better.

Using calculations to show the hidden energy benefit of better glass

Another important-but-seldom-recognized benefit of glass with a low solar heat gain is better thermal comfort near windows without the need to drop the thermostat setting. This is a big benefit that seldom shows up in typical single-point load calculations, because it involves human response, over time, to radiant heat from solar-heated window surfaces.

A single-point load calculation usually assumes that the building thermostat is set somewhere near 78°F [26°C]. But those loads would increase substantially if the thermostat setting were pushed down to 71°F [21.7°C]. Such thermostat twiddling often happens in the real world, especially when the glazing has a high solar heat gain coefficient. Here's why.

With poor glazing (glass with a high SHGC) the radiant heat from hot window surfaces overheats the nearby occupants. In response, they turn down the thermostat. At lower temperatures, the AC system uses much more energy to cool the building. At the same time, the temperature further away from those windows (in the core of the building) often becomes far too cold for comfort, sometimes even triggering the need for supplemental heat. The people near the windows are being slowly broiled while the people in the core are being flash-frozen. Nobody is comfortable, and they all blame the HVAC system, even though the high solar heat gain coefficient of large, low-budget windows is often the cause of the problem.

To capture and quantify the equipment and energy cost reduction that comes from installing fewer and better windows, the HVAC designer can use the computer to quickly run the loads at 71°F vs. 78°F thermostat set points. [At 21.7°C vs. 25.5°C] The reduced energy and smaller equipment for the warmer set point will add to the arguments in favor of better windows, and smaller ones.

Understanding modern glass

If the HVAC designer wants to make a big improvement in the sustainability of the building, he or she would be wise to become intimately familiar with modern glass and window technology. With that understanding, the HVAC designer can become a more useful resource (and a more persuasive advocate) during those critically important early conversations with architects and owners.

Reference 7 provides an excellent starting point for this education. It is a brief, engagingly-written and well-illustrated description of current window terminology, technology and issues. After reading that introduction, the recommendations in the ASHRAE Advanced Energy Guides will be easier to understand for those who may not have extensive experience with recent advances in window technology.

Then, references 8 and 9 go beyond basics to more detailed information, to help the design team implement the specifics of the low-energy glazing recommendations of ASHRAE Std. 90.1-2007.

Western glass is the worst

When calculating the loads from windows, look very carefully at the load from windows on the western walls. The west face of the building is the worst place to locate windows, from a thermal perspective. There, the cooling loads come at the worst-possible time—after the building has been heated up all day long by the sun and by the occupants' activities. Figure 12.4 shows the load from both south and west-facing windows. Note that during the hottest months of the year, the load through the west-facing windows is 2.7 times larger than the load through the south-facing windows.

This fact will be most apparent when the designer looks at the *peak hour loads*, rather than the total load for the entire day or year.

Over 24 hours, the cooling load from western windows is the same as the load from eastern windows. But the eastern windows generate that load during the early morning, when the internal loads are very small or non-existent. So the cooling system can handle those loads using very little of its capacity. Later in the day, when all the sensible loads are peaking at the same time, the cooling system will struggle. (The usual occupant complaints about temperatures are: too cold in the morning, too hot in the afternoon—never just right.)

So, the computerized calculations can be very helpful in understanding how important the western windows are to the peak load. Calculating different percentages of glazing, and calculating the effect of different shading geometry will be especially helpful to the owner and architect as they plan the look and feel of the building, and the uses of the spaces which would be affected by western windows.

Thermally, the western side of the building would be a good location for storage rooms, mechanical rooms and similar uses which don't benefit from windows. When windows are necessary on the western wall, it's best to make them high on the wall, small and horizontal, so they will be effective in daylighting. But when view windows are also necessary on the western wall these should:

- Cover the least-possible area of the wall
- Have glass with a very low SHGC
- Be shaded with vertical louvers

During peak cooling load hours, (afternoon and early evening) the sun is closer to the horizon as it beats on the western wall. So a shading device made with vertical blades, set at an angle and extending down the full face of the window will be helpful in limiting the solar heat gain during peak cooling load hours.

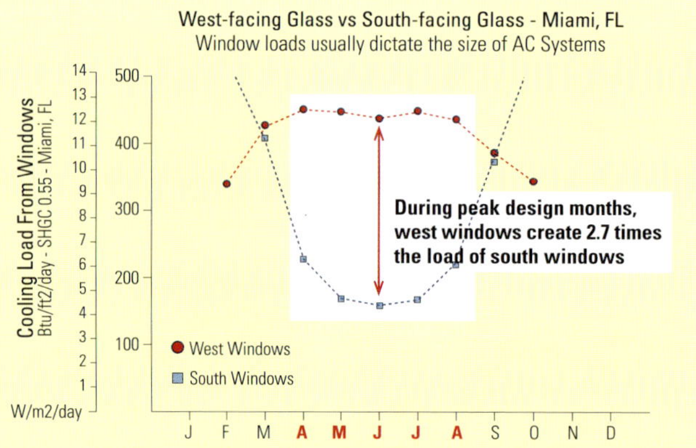

		Jan	Feb	Mar	Apr	May	Jun	Jul	Aug	Sep	Oct	Nov	Dec
South	Btu/ft2/day	609	537	404	236	173	159	174	221	380	543	616	604
	W/m2/day	17	15	11	6	5	4	5	6	10	15	17	16
West	Btu/ft2/day	271	339	424	465	458	433	453	432	392	345	276	237
	W/m2/day	7	9	12	13	12	12	12	12	11	9	8	6

Fig. 12.4 Western walls are the worst place to locate windows

That's because during the hottest months of the year, and at the hottest time of the day (the afternoon), they allow more than 2.7 times more heat into the building than windows located on the south wall.[2]

Daylighting can reduce peak cooling loads

When the building is equipped with well-designed daylighting, its cooling loads can be significantly reduced. Daylighting windows are small, narrow, horizontal and set near the ceilings. (See figure 12.5 for an example.) These can have a higher solar heat gain coefficient, in order to transmit the maximum amount of visible light. But the view windows, usually much larger and set below the daylighting windows, will have a very low SHGC to limit heat gain.

Even more significantly, with effective daylighting the electric power needed for lights in the perimeter zones is greatly reduced during peak cooling load hours. And when that lighting power is reduced, so too is the cooling load generated by lighting.

It is still true that full lighting power will be installed in the building, for use at night. But that cooling load comes long after the peak cooling load hours, at a time when the cooling system has a great deal of unused capacity.

Since daylighting is becoming more popular with owners and architects, it's useful for the HVAC designer to clearly understand its thermal importance, and to design the HVAC equipment accordingly. It would not be wise to assume, as many have in the past, that the daylighting will be ineffective at reducing the cooling loads. Done right, daylighting greatly reduces the loads in the perimeter zones. The cooling equipment should be reduced accordingly, to avoid the overchilled occupants, high indoor humidity and needless reheating of the supply air that comes from an oversized cooling system.

Fig. 12.5 Daylighting design

Daylighting reduces both the solar heat gain and the heat from the interior lights it displaces. These reductions are especially welcome during the afternoon, when sensible heat loads are peaking.

Glazing substitutions can ruin HVAC designs and occupant comfort

Running multiple cooling load calculations can help the HVAC designer head-off last-minute substitutions. Glass, especially good glass, is expensive. And it's often easy for the owner to substitute lower-cost glass without major architectural design changes when the construction bids come in over budget.

However, for buildings with a great deal of window area, glass substitutions could be catastrophic for the HVAC budget and system design. The sensible cooling loads could be far in excess of what the HVAC designer expected when the original calculations were run. With multiple computer load calculation runs at an early stage, the HVAC designer will have a faster and more informed response (with more useful suggestions) if the client needs to substitute lower-cost glazing. Figure 12.2 showed a small example of the difference that good glass makes in reducing the cooling load.

Separate and then calculate the dehumidification loads

In hot and humid climates, the dehumidification (D/H) loads are large and nearly continuous. D/H loads also peak at different times than the sensible cooling loads. That's why, especially in hot and humid climates, it is wise to calculate dehumidification loads separately from sensible loads. Also, it's best to calculate the lbs or kg which must be removed per hour—rather than thinking about Btu's per hour or kWh or the sensible heat ratio. Tracking the D/H loads in pounds or kg per hour allows the designer to deal with the key variable directly.

Estimating the dehumidification loads can be done quickly, using hand calculations guided by Chapter 11 of this book. Then, after the three major dehumidification loads of *people*, *ventilation air* and *infiltration* are clearly identified and quantified, the designer can return to the computer to calculate the sensible cooling loads.

Costly experiences with mold, wasted mechanical system budgets, occupant complaints of swampy, cold buildings and outlandish costs for energy all underline the importance of quantifying the D/H and sensible loads separately.

Dehumidification loads will peak at moderate outdoor temperatures, when cooling coils are rarely operating continuously. Without the dehumidification provided by active cooling coils, the indoor dew point can rise too high. That high indoor dew point leads to the problems described in Chapter 2 (Improving Thermal Comfort), Chapter 3 (Reducing Energy Consumption), and Chapter 5 (Avoiding Bugs, Mold & Rot).

After making a separate D/H load calculation, the designer will quickly see and understand the astonishingly important dimension of the ventilation dehumidification load, and the unexpectedly large D/H loads from air infiltrating into the building through joints. These will not be as apparent when the computer is calculating the sensible load at the peak outdoor dry bulb temperature.

D/H load calculations allow better design for humidity control

As suggested by the comparison in figure 12.6, in most hot and humid climates, the primary dehumidification loads (ventilation) will be 30 to 40% higher at the peak outdoor dew point than it will be at the peak outdoor dry bulb condition. If the cooling system design is based only on the peak dry bulb temperature and its average coincident wet bulb, the design will be missing 30 to 40% of the dehumidification load. Without the capability to remove that load, the system is not likely to make the building either comfortable or energy-efficient.

Another helpful benefit of a separate D/H load calculation is that one can more easily *visualize* the hourly humidity load. Pounds or kilograms of water which must be removed from the air *every hour* are easier to visualize than are Btu's and kW of "latent cooling" load. With that weight of water in the mind's eye, the water's volume every hour also becomes part of the designer's thinking. (See figure 12.8) More attention will likely be given to components like drain pans which really drain, and condensate piping which has functioning traps rather than ineffective bends in undersized flexible tubing. The weight and volume of the hourly D/H load also reminds the designer that *all that water has to go someplace*. So it's less likely that the designer will forget to specify how the condensate drain should get to the storm or sanitary sewer instead of spilling out of the drain pan and onto the floor.

Using the D/H load from computerized load calculations

All computerized load calculations will calculate the latent load (the dehumidification load) along with the sensible cooling loads. In some programs, that latent load is not reported separately from the total load. Other programs will quantify and report a separate estimate of the latent load. Unfortunately, these values are often wrong.

As of the publication date of this book, computerized load programs do a reasonably good job of quantifying a few parts of the D/H load, such as vapor from occupants or the load from vapor

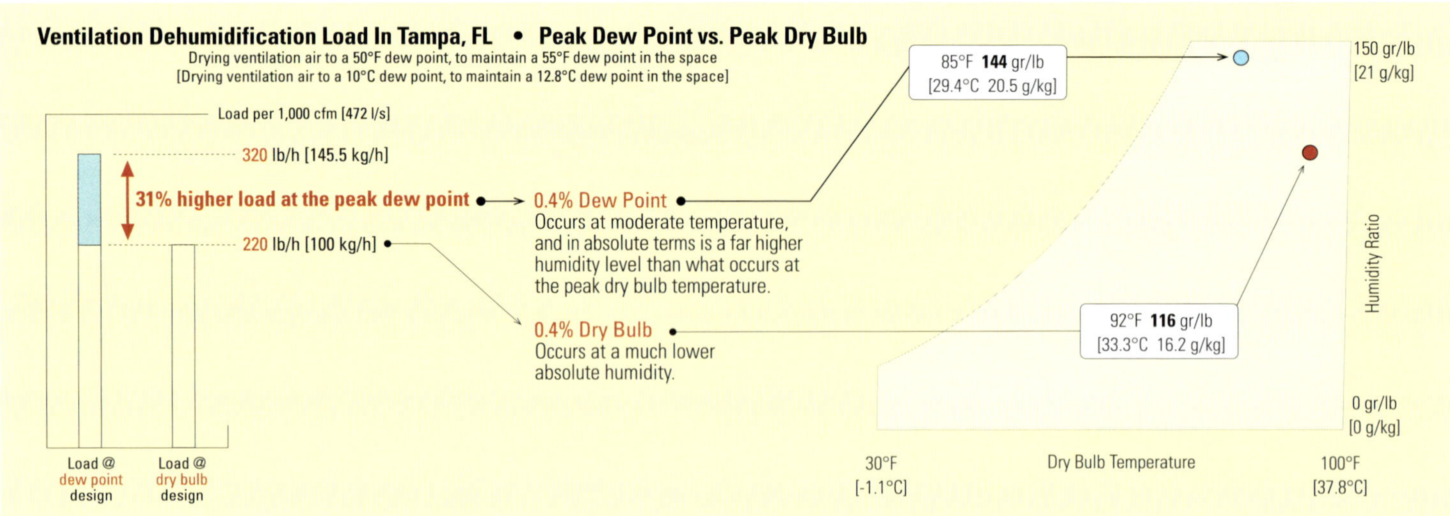

Fig. 12.6 Use the peak dew point to calculate the peak ventilation load

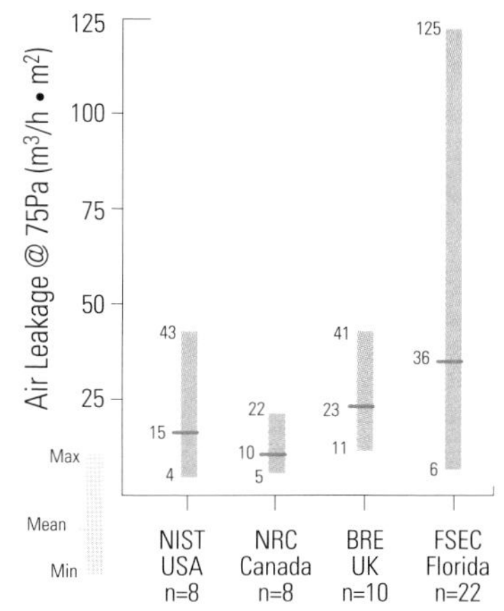

Fig. 12.7 Buildings leak air[10]

Typical low-rise commercial and institutional buildings leak air, especially when it is windy. In hot and humid climates, that infiltration air brings in a large dehumidification load, along with a somewhat smaller sensible cooling load.

permeation through solid materials. But those are the small parts of the total D/H load. They usually account for less than 10% of the total D/H load. The much larger and more important components—75 to 90% of the total D/H load—are the humid air infiltration into the building and the humid ventilation air.

In most computerized load programs, the ventilation load is calculated only at the peak dry bulb temperature instead of at the peak dew point, when the humidity load is 30% higher. Nor do most programs provide any estimate of the infiltration load if the designer's input says the building will be under net positive air pressure. Unfortunately, one thing we know from extensive field measurements is that buildings under average positive air pressure still have a high rate of air exchange with the outdoors.[10,11,12] Buildings measured in the field consistently show air leakage rates from 0.25 to 2 complete air changes per hour—vastly more than HVAC designers expect.

As just one set of examples, consider the buildings described by Figure 12.7. Note the wide range of variation between the tightest and leakiest buildings. Note also that *all buildings* had a significant amount of leakage.[10] It is unwise for the designer to assume there will no air infiltration when the building is designed for positive air pressure. Positive pressure reduces infiltration, to be sure. But some leakage still occurs. More information about air leakage is contained in Chapter 11 (Estimating Dehumidification Loads). In hot and humid climates, outdoor air infiltration makes a much bigger contribution to the D/H load than it does to the sensible cooling load.

So while the *sensible* load calculations in currently-popular programs are becoming very sophisticated and presumably more accurate, those same programs still miss, or often grossly underestimate the two largest D/H loads on typical buildings: the amount of humidity brought into the building by the ventilation air and by infiltration air.

Still, the latent loads calculated by such programs can be useful as a reality check for hand calculations of the D/H load.

If the computer program says the latent loads are higher than the total from the hand calculations, the designer should re-check the hand calculations. Something important was probably missed. Hand calculations, because they can account for the ventilation D/H load and for air infiltration at the peak dew point, generally yield higher D/H loads than what are calculated by most computer programs.

That said, one can improve the utility of computerized load calculations with respect to ventilation loads by simply running the program at two different peak design conditions, as described next.

Calculate ventilation cooling loads twice—at peak dew point and at peak dry bulb design

Ventilation air carries both excess heat and excess humidity. It's a large part of both loads. But it's important to keep in mind that the annual ventilation load is mostly dehumidification rather than sensible cooling.[13,14] Figure 12.8 provides a visual reminder of the size of the ventilation dehumidification load.

Ventilation D/H load peaks at relatively moderate outdoor temperatures—either first thing in the morning as the overnight dew evaporates, or in the afternoon after a heavy rain results in evaporation from lawns, trees and hot pavement. At the same time, the air temperature is being cooled by that evaporation. Figure 12.6 illustrated this effect by comparing the peak dry bulb to the peak dew point design conditions for Tampa, FL. When the temperature is high, the humidity is lower; at 92°F, the average humidity is only 116 gr/lb. But when the humidity is high, the temperature is lower; *at 144 gr/lb., the average temperature is 85°F instead of 92°F.*

[When the temperature is high, the humidity is low (at 33.3°C, the average humidity is 16.2 g/kg). But when the humidity is high,

Fig. 12.8 Visualizing the D/H load of ventilation air

Imagine pouring 72 gallons of water, every hour, into a school... or 330 gallons every hour into a gambling casino or small hospital.[14] Those are typical of D/H loads from ventilation air, and they occur during ***part-load conditions*** *for the cooling system. Making a separate D/H load calculation helps the designer remain conscious of what must be done to remove humidity* ***when the cooling system is only partly loaded.***

the temperature is lower; *at 20.5 g/kg., the average temperature is 29.4°C, not 33.3°.*]

Consequently, when calculating cooling loads with a computer program, it's wise to run the calculation at least twice—once at the peak outdoor dry bulb to size the cooling equipment, and then *again* at the peak outdoor dew point to understand how that equipment (and the surrounding system) must behave at part-load conditions.

The difference in loads will be quite striking. At the peak dew point (the peak D/H load condition), the sensible heat loads from outside the building will be comparatively low. With heavy cloud cover, the solar heat gain through windows will be lower, and the exterior surfaces of the building walls and roof will not be as hot as they are at the peak dry bulb temperature. That's one reason that the peak dew point condition is an excellent point at which to check the "part-load" performance of the cooling system.

When the designer quantifies the loads from both peak sensible and peak dew point hours, it becomes much more obvious how the cooling system must be arranged if is to keep the indoor temperature comfortable at part-load conditions.

Part-load hours are, by definition, much more typical of the running load than are the peak design conditions. Peak outdoor temperatures occur for less than 0.4% of the hours in the year (35 hours per year). Part-load conditions last for the rest of the year (99.6% = 8,725 hours).

Most readers will probably recall, from personal experience, buildings which felt too cold for comfort during the early mornings in the summer. Those HVAC systems were probably not designed to gracefully unload sensible cooling capacity during such "off-peak" hours.

By making at least two complete computer runs of load calculations—one using the peak dry bulb and the other using the peak dew point—the designer will be aware of how far the cooling equipment must *unload* to avoid overchilling the occupants.

Enthalpy heat recovery reduces peak cooling loads

Installing an enthalpy heat exchanger is an excellent way to reduce the peak sensible cooling load. Enthalpy exchangers use the cooling effect of the exhaust air to precool the incoming air. Calculating the load with and without that device can show the strong benefit of the technology, which also helps meet the requirements of ASHRAE Std 90.1-2007 for energy efficient buildings.[15]

The time to consider this device is during the load calculations, when its beneficial effect of reducing the peak cooling loads is most obvious. Later, after the load estimates are done and the equipment and systems laid out, it may be difficult to seriously consider adding the equipment, no matter how beneficial it might be. That's because an enthalpy heat exchanger must be physically installed between the exhaust and supply air streams, as shown in Figure 12.9. So those two ducts must come together at the heat exchanger location. The load calculation stage comes early. So the physical constraints of the building do not yet preclude the addition of an enthalpy heat exchanger.

In hot and humid climates, adding a rotary enthalpy heat exchanger may cut the size of the cooling system by more than 30%, because the hot and humid ventilation air accounts for such a large percentage of the peak cooling load.

Like any equipment, there are many types of enthalpy heat exchangers, and there are technical and commercial considerations both for and against these devices. But the point is, evaluating the potential load reduction is best done at the load calculation stage.

Fig. 12.9 Rotary enthalpy heat exchanger to reduce ventilation loads

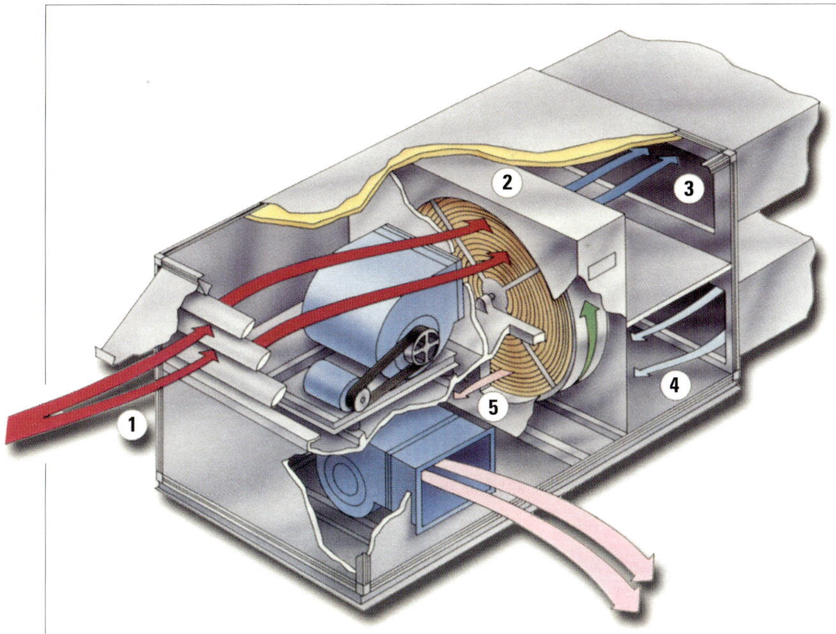

1. Hot, humid outdoor air enters the unit.
2. The air gives up part of its heat and moisture as it flows through the passive desiccant wheel.
3. Leaving the wheel, the ventilation air carries a much smaller heat and humidity load.
4. Relatively cool and dry exhaust air leaving the building strips heat and moisture from the passive desiccant wheel.
5. The exhaust air, now warm and moist, leaves the system, exhausted to the outdoors.

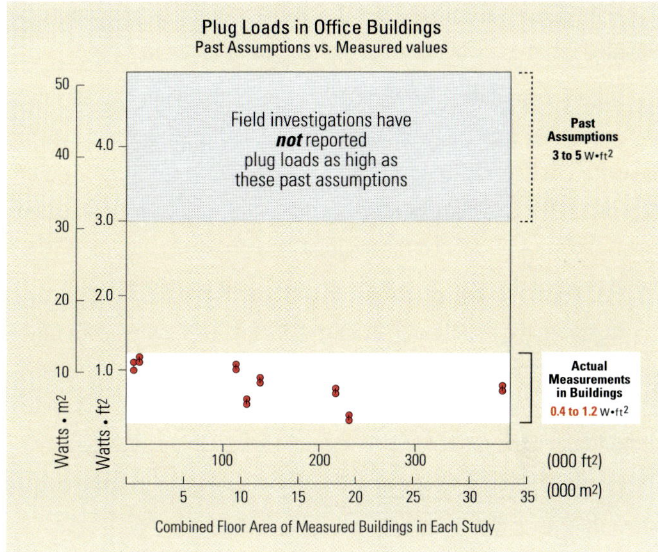

Fig. 12.10 Assumptions vs. measured plug loads

Past assumptions have consistently been three to five times larger than actual plug loads. ASHRAE research shows measured values, to help avoid the many problems of oversized cooling equipment.[16,17,18,19]

Don't overestimate office plug loads

Sensible heat loads from computers, copiers, task lighting and other electrical equipment have proven difficult to estimate accurately. Over the last 15 years, ASHRAE and others have funded extensive field research to provide measured values of such equipment. These results help the designer avoid assuming plug loads which are far larger than what actually occur. Using these field-researched values can help avoid oversizing the cooling system, saving construction cost and improving thermal comfort.

For example, the common assumption before the mid-1990's was that plug loads in offices added between 3 to 5 Watts per square foot to the cooling load. In fact, the measured loads in offices ranged between 0.44 W/ft2 and 1.1 W/ft2.[16]

[Plug loads assumed to be 21.5 to 53.8 W/m2 turned out to be actually 4.3 to 11.8 W/m2 when measured in the field. [16]]

Taking computers, printers and copiers as being typical office appliances, the heat they contribute to the cooling load, when actually measured, is between 25% and 50% of their nameplate power rating. The 50% reflects the largest measurement on specific pieces of equipment, and the 25% reflects the net load for a group of equipment when the measured diversity of equipment use is taken into account.

In summary, traditional assumptions for heat gains from office plug loads are far larger than what actually occurs in real buildings. To avoid the problems of oversized cooling systems, the HVAC designer should adjust the input to the computer program accordingly.

References 16,17,18 and 19 will be very helpful to designers when they need full details about real-world plug loads.

Notes and References

1. Architectural designers are often misled in thinking that high-quality glazing, which is usually insulated glass, is not as useful in hot and humid climates because the average annual air temperature difference is very small compared to the temperature difference in cooler climates. That's true.

   But focusing on the insulating value of the glass misses the *far* more important issue for hot and humid climates: the Solar Heat Gain Coefficient (SHGC). It's almost always the high-quality double-glazed windows which also have the lowest (best) solar heat gain coefficients—usually 0.4 or lower, compared to 0.86 for clear single glazing. You'll want glass with a low SHGC. But it's expensive and it *still* lets in more than 30 times the solar heat gain of a solid wall. So whenever possible, encourage the architectural designer to exercise creativity using walls instead of extra glass.

2. Gronbeck, Christopher *Window Heat Gain Calculator* 2007. http://www.susdesign.com/windowheatgain/
3. McBride, Merle, Project Chair. *Advanced Energy Design Guide for Small Retail Buildings* 2006. ASHRAE. 1791 Tullie Circle, NE. Atlanta, GA 30329 ISBN 1-933742-06-2
4. Jarnigan, Ron, Project Chair. *Advanced Energy Design Guide for Small Office Buildings* 2004. ASHRAE. 1791 Tullie Circle, NE. Atlanta, GA 30329 ISBN 1-931862-55-9 (Available for downloading as a PDF file, without cost at: http://ashrae.org/publications/page/1604)
5. Torcellini, Paul, Project Chair. *Advanced Energy Design Guide for K-12 School Buildings*. 2007. ASHRAE. 1791 Tullie Circle, NE. Atlanta, GA 30329 ISBN 1-978-933742-21-2 (Available for downloading as a PDF file, without cost at: http://ashrae.org/publications/page/1604)
6. Torcellini, Paul, Project Chair. *Advanced Energy Design Guide for Warehouses and Self-storage Buildings*. 2008. ASHRAE. 1791 Tullie Circle, NE. Atlanta, GA 30329 ISBN 978-1-933742-22-9 (Available for downloading as a PDF file, without cost at: http://ashrae.org/publications/page/1604)
7. McGowan, Alex. 2008. Introduction to green window design and performance. Journal of Green Building, Volume 3, Number 2, Spring 2008 pp.3-12 www.collegepublishingus/journal.htm
8. O'Connor, Jennifer, Lee, E. Rubenstein, F. & Selkowitz, Stephen, *Tips for Daylighting with Windows - The Integrated Approach* Report no. LBNL-39945 1997. Building Technologies Program. E.O. Lawrence Berkeley National Laboratory, Berkeley, CA
9. Carmody, John; Selkowitz, Steven; Lee, Eleanor; Arasteh, Dariush and Wilmert, Todd. *Window Systems for High Performance Buildings* 2004. Norton & Company, 500 5th Avenue, New York, NY. 10110 ISBN 0-393-73121-9
10. Persily, Andrew. "Myths About Building Envelopes." 1999. *ASHRAE Journal*, pp. 39-45. March, 1999 ASHRAE 1791 Tullie Circle, NE Atlanta, GA 30329

11. Cummings, James; Withers, Charles; Moyer, Neil; Fairey, Philip; McKendry, Bruce. *Uncontrolled Air Flow in Non-Residential Buildings*. 1996. Final Report, FSCEC-CR-878-96, 1996. Florida Solar Energy Center, 1679 Clearlake Rd., Cocoa, FL. 32922

12. Henderson, Hugh; Cummings, James; Zhang, Jian Sun; Brennan, Terry. *Mitigating the Impacts of Uncontrolled Air Flow on Indoor Environmental Quality and Energy Demand in Non-Residential Buildings*. 2007. Final Report - NYSERDA Project # 6770. New York State Energy Research & Development Authority, 17 Columbia Circle, Albany, NY 12203-6399

13. Harriman, Lewis G. Kosar, Douglas and Plager, Dean. 1997. "Dehumidification and Cooling Loads from Ventilation Air." ASHRAE Journal, November, 1997 pp.37-45. ASHRAE, Atlanta, GA. www.ashrae.org

14. See figure 11.1 in Chapter 11 of this book: "Dehumidification Load Estimates for Tampa, FL" It shows the relative size of the different D/H load components, for ten common types of commercial and institutional buildings.

15. ASHRAE Standard 90.1-2004 *Energy Standard for Buildings Except Low-Rise Residential Buildings*. 2004. ASHRAE. 1791 Tullie Circle, NE. Atlanta, GA 30329 ISSN 1041-2336

16. Komor, Paul. 1997. Space cooling demands from office plug loads. ASHRAE Journal, December 1997. pp.41-44. www.ashrae.org

17. Wilkins, Christopher and Hosni, M.H., 2000. "Heat gain from office equipment." ASHRAE Journal, June 2000 pp.33-39 www.ASHRAE.org

18. Pratt, Robert G, "Errors in audit predictions of commercial lighting and equipment loads and their impacts on heating and cooling load estimates." ASHRAE Transactions, 1990. AT-90-11-2 pp.994-1003 www.ashrae.org

19. Chapter 30 - Nonresidential heating and cooling loads. *ASHRAE Handbook—Fundamentals 2005*. www.ashrae.org

Image Credits

Fig. 12.5 - Tips for Daylighting. O'Connor, Lee, Rubenstein & Selkowitz. U.S. Department of Energy - Lawrence Berkeley National Lab, Berkeley, CA

Fig. 12.9 - © 1999 Semco/Flakt Inc.

Chapter 13

Designing Dehumidification Systems

By Lew Harriman

Fig. 13.1 Humidity control

When year-round humidity ***control*** *is the goal rather than just humidity moderation when it's especially hot outdoors, the system will need dedicated components and controls which measure humidity and take excess moisture out of the air on demand.*

Measuring humidity and controlling the dehumidification components based on the dew point makes the whole system simpler and more stable than if it were controlled based on the relative humidity.

Key Points

HVAC designers quickly learn how to keep a building cool. That job is part of everyday practice. But dehumidification (DH) is a different matter. It's not always given much attention in engineering courses.

ASHRAE recognized this problem, and has published a 500-page book on the subject.[1] This chapter will not repeat the contents of that book. Instead, we will focus on the issues that many designers find confusing about DH design. The information will be especially useful when designing for hot and humid climates, where dehumidification is a year-round concern.

To avoid the most common problems with high humidity, keep these key points in mind:

- The supply air must be delivered below the humidity control set point. When air is delivered at the same dew point as the control level, humidity in the space will rise out of control, because internal loads will push up the humidity above the desired limit.
- Actual *control* of humidity (as opposed to intermittent *moderation* of humidity) requires dedicated DH components, along with controls which measure humidity. If the only control is a thermostat and the only equipment is a cooling coil, one can't expect the system to keep humidity below a defined limit, except intermittently.
- DH components must be sized to remove the loads which occur at the *peak dew point.* That will always be when the outdoor temperature is cooler than at peak sensible design. So make sure the DH components will remove the humidity loads at the ASHRAE design dew point, when the outdoor temperature is moderate.
- Calculate the DH equipment's performance based on lbs or kg removed per hour—rather than using the equipment's sensible heat ratio. Absolute, weight-based moisture removal values avoid confusion, and they avoid common misunderstandings about the dehumidification performance of cooling equipment.
- Humidity cannot be controlled until the ventilation air has been dried. If ventilation air is allowed to enter the space carrying its full humidity load, humidity in the space will vary widely because of humidity variations in the incoming ventilation air. To ensure humidity control below a defined limit, dry the ventilation air *below* the desired control condition *before* it enters the controlled space.
- It's simpler to control humidity based on dew point than on relative humidity. The dew point in a controlled space stays fairly constant, unlike the relative humidity, which varies with air temperature. Controlling humidity based on dew point will generally provide the most stable result and a simpler control system—and it can reduce the risk of problems in the building enclosure.

Essential Elements of DH Design

Dehumidification can be confusing, even to experienced HVAC designers. As one becomes expert in cooling a building, one tends to use the same "mental toolkit" for other aspects of HVAC design. But in the case of dehumidification, not all tools in the cooling toolkit are helpful. Some are, but others get in the way of an efficient thought process and can often create confusion.

In cooling design, one thinks in terms of degrees of temperature and Btu's or kW of cooling load and cooling capacity. But in dehumidification, the most efficient thought pattern uses grains per pound or grams per kilogram of humidity and lbs or kg of both DH load and DH capacity.

To reduce confusion and to arrive quickly at an efficient DH design, begin the design process by estimating the dehumidification loads in lbs or kg per hour, as shown in Chapter 11. Then continue by selecting equipment which will remove that weight of water vapor from the air, every hour, no matter what the air's temperature might be, and no matter what the cooling loads might be at the same time. Focus on the weight of the water vapor loads, and on the weight of water vapor removed by the equipment. Don't let the cooling issues (or the units used when designing cooling systems) confuse the dehumidification design process.

Deliver air drier than the control condition

Moving on to specifics, we'll begin with the most basic and critical concept in controlling humidity, namely: you can't deliver air at the same condition you expect to maintain.

This fact translates perfectly from one's knowledge of cooling design. No designer would ever deliver air at 77°F [24°C] and expect that air to keep the room at 77°F [24°C] . The supply air must be cooler than the desired control condition so that the air can absorb the cooling loads in the space itself.

The same thinking must be applied when designing a dehumidification system. One must not expect that air delivered at a condition of 65 gr/lb [9.3 g/kg] will keep the space at 65 gr/lb [9.3 g/kg]. The supply air must be drier than the control condition in the space so it will absorb the internal DH loads.

To those who have been involved with DH design, this caution is so basic, so obvious, that it sounds insulting even to mention it. But really, it's a common mistake and one which is understandable. For designers who are experts in cooling and heating, it's not obvious which concepts from those areas automatically apply to dehumidification and which do not. Also, it's not obvious that the units must be different in order to do the work of design efficiently. In the confusion of dealing with an unfamiliar topic, unfamiliar units and new equations, even very experienced designers may lose track of this essential point.

Don't let that happen to you. Keep a clear focus on the fact that to maintain a space below a given humidity level, you must deliver the supply air below that level. And the degree of dryness required of that supply air will depend on how high or low the dehumidification loads are in the controlled space. Higher loads require drier supply air, and lower loads will allow supply air which is not quite so dry. In concept, it's just like when you're designing for cooling—except that you will use water vapor units (gr/lb or g/kg) instead of degrees.

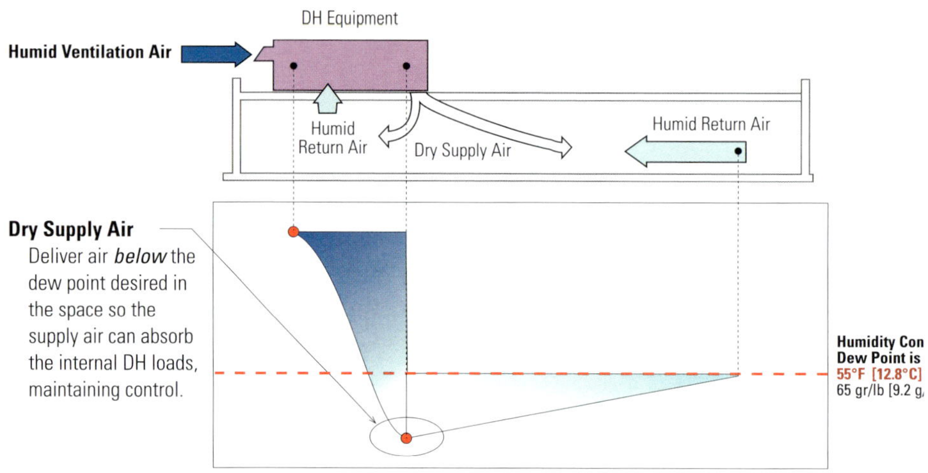

Fig. 13.2 Supply air must be dry

To keep humidity below a defined limit, the dew point of the supply air must be lower than the dew point you want to maintain in the room. That way, the supply air will absorb the internal humidity loads.

This is exactly like designing for cooling, when the supply air temperature must be cooler than the thermostat set point.

Control requires dedicated DH components

Controlling humidity below a set maximum requires humidity sensors, humidity controllers and some type of HVAC component which removes humidity *on demand*—not as an occasional artifact of the cooling process.

This point must be prefaced with the distinction between "humidity control" and "humidity moderation." One can expect that any cooling equipment will, by itself, *moderate* some of the extremes of humidity caused by weather variations. But that moderation will occur only when cooling loads are also high—when the cooling equipment is running for long periods (long enough to remove moisture). With cooling equipment alone, one can*not* expect that humidity will be held below a defined maximum *at all times* unless the designer adds a sensor which measures humidity, and ensures that some component will take moisture out of the air when humidity rises above the building's control set point.

Fig. 13.3

Use the peak dew point for DH design

When designing for humidity control, make sure to use the peak dew point for your load calculations—NOT the peak dry bulb temperature with it's average wet bulb temperature. Those values are fine for cooling design, but they typically underestimate the most important DH loads by more than 30%.

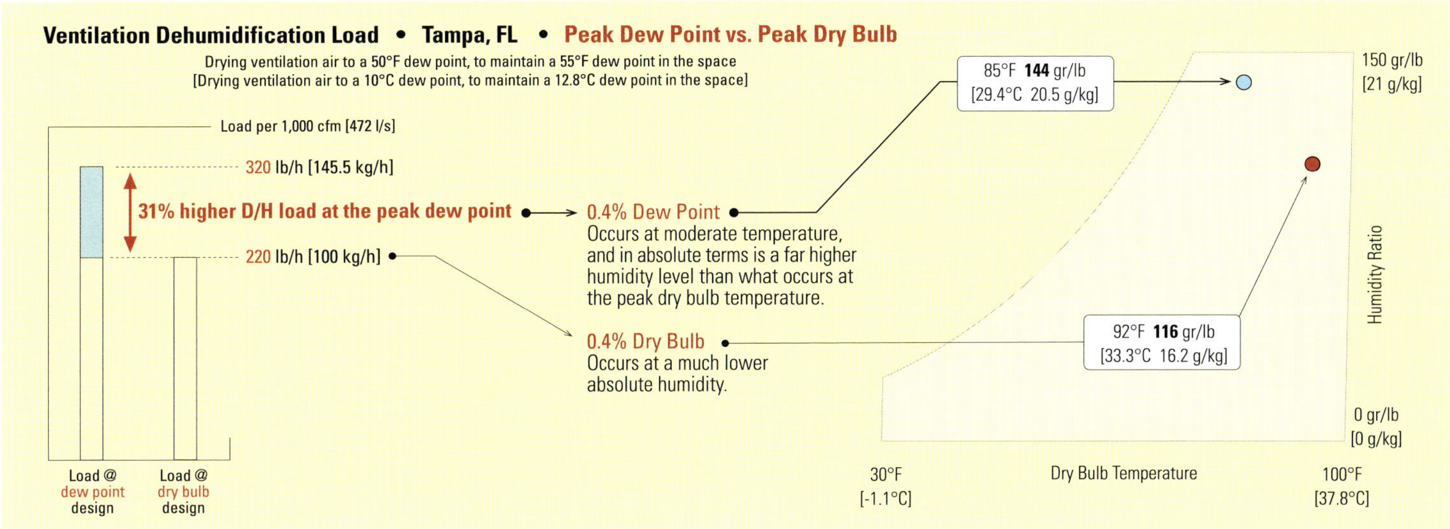

This point is useful to keep in mind not only when designing for humidity control, but also when designing the cooling system. For example, if a variable air volume system is designed so that the coil *always stays cold enough* to dry the air, there may be no need to add a separate dehumidifier. When the supply air is always cold and therefore always dry, adding reheat (ideally using waste heat from condensers) may be enough to keep the spaces from getting too cold. On the other hand, if the VAV coil temperature is allowed to rise (reset higher, to provide energy savings or to reduce reheat), then the VAV cooling system's coils may no longer be cold enough to condense moisture and dry the air.

So in hot and humid climates, where humidity moderation is really always a baseline requirement, the designer has two choices:

- Add a dedicated dehumidification component, humidity sensor and controls to achieve control or...
- As a minimum, add a sensor, controller and reheat coil to the cooling systems so they can dry the air independently of the need for cooling, without overcooling the space.

Note that to comply with energy codes, the reheat energy will need to come from a waste heat source such as the cooling system's condensers, unless the system qualifies for an exemption under the applicable code. For example, ASHRAE Standard 90.1 is the basis for many codes in North America, and that standard does allow exemption from the waste heat requirement for certain types of systems and occupancies.[2] For example, Std 90.1 allows new energy for reheat when the system is very small, or when the system will reduce its air flow to less than 50% of full flow before applying reheat, or where there are "process" requirements for humidity control, such as in data centers. But as a general rule, waste heat is preferred for reheat in all cases and is required for most.

Size DH equipment based on the peak outdoor dew point

Another very important fact—seldom obvious to most designers—is that the peak dehumidification load does *not* occur at the same time as the peak sensible cooling load.

Outdoors, the dry bulb temperature is usually highest during the afternoon, after the sun has taken all day to heat the building.

However, the highest outdoor *humidity* levels happen at other times. The true peak humidity is 25 to 35% higher than the humidity which happens to occur at the peak dry bulb temperature.[3,4]

Often, the outdoor *dew point* reaches its peak just after a late afternoon or early evening rain, which evaporates into the warm air. The sensible heat in the air powers most of the evaporation. So as the dew point *rises*, the dry bulb temperature *falls* to more moderate levels. The falling rain acts as an evaporative cooler for the air.

The outdoor dew point can also reach its peak early in the morning, as the sun evaporates overnight condensation (the morning dew). Again, sensible heat from the air is consumed as it powers all that moisture evaporation.

Here's the key point of both examples: the dew point peaks when the outdoor air temperature is moderate—*not* when it's really hot outdoors. This fact has many important implications for DH system design.

First and most important, *don't* design the DH system based on the ASHRAE design *dry bulb* temperature and its mean coincident wet bulb temperature. (The values used for sizing cooling equipment.) If you do that, you'll be grossly underestimating the peak dehumidification loads. Figure 13.3 shows one example of the potential magnitude of the deficit. For sizing the DH components, estimate the loads using the ASHRAE *peak dew point design value*, with *its* coincident dry bulb temperature.

Next, note that the worst case for dehumidification loads will occur when the cooling system *won't be* running, or if it's running, it won't be running at full cooling power. That means the cooling equipment won't be *drying* that highly-humid air unless you force it to. The thermostat won't often be calling for cooling during part-sensible-load hours—which are actually the hours when the *peak* dehumidification loads occur.

This part-load-high-humidity problem is most severe with constant-volume DX cooling equipment. But constant-volume chilled water systems suffer from a similar problem. When the chilled water cooling coil's temperature rises as its capacity is being reduced, its leaving air temperature rises, which means the supply air dew point also rises. In other words, the supply air is no longer as dry as when the coil-leaving air temperature was low (when the sensible cooling loads were so high that the coil needed to supply the coldest air.

Basically, the best DH performance of any cooling system is during those few hours each year when the sensible loads are at their peak. At all other times (during the 99.6% of the year when the sensible cooling loads are not at their peak), the cooling system is not as effective as a dehumidifier. And in fact, the cooling system is probably *least* effective as a dehumidifier at precisely the time when the DH loads are at their *highest* point - the peak outdoor dew point conditions.

Next, understand that because a DX cooling system's compressors won't be running for long periods when the DH loads are high, it's a very *bad* idea to oversize the cooling equipment in the hope that the excess capacity will dry the air. It won't. When DX cooling equipment is oversized, its compressors run for even shorter periods. Oversized DX cooling equipment is *even less* effective as a dehumidifier than

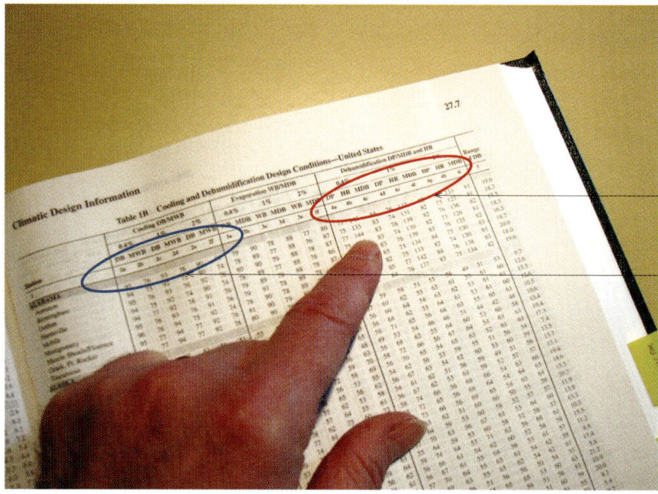

Fig. 13.4

Locating peak dew point design data

In the ASHRAE Handbook—Fundamentals, the peak dew point design data is in the chapter titled "Climatic Design Information." The chapter number is different in each edition, and the data format has changed in each volume since the peak dew point first appeared in 1997.

The key point is to use the peak dew point data, rather than the sensible cooling design data or the evaporative cooling design data. All three groups of data usually appear on the same page, as in this 2001 edition of the Handbook.

normally-sized cooling equipment, because it runs for even shorter periods when the outdoor temperature is moderate.[5]

To summarize this point, size and select the DH components based on the worst case for dehumidification loads, not the worst case for sensible heat loads. In most applications, that worst case occurs at the *peak dew point* condition, which is neatly displayed for you in the *ASHRAE Handbook—Fundamentals*. Figure 13.4 shows where to locate that data in Chapter 27 of the 2001 Fundamentals volume. In the 2004 edition, all the climatic design data is on the CD which accompanied the printed volume. In the 2009 volume of Fundamentals, the climatic design data appears in Chapter 14, and also on the CD which accompanies the printed volume.

Unfortunately, there is *no peak dew point data in earlier Handbook volumes*—only peak sensible design data. So don't use ASHRAE climatic data from before 1997 for DH design. Using the peak dry bulb conditions will grossly underestimate the DH loads, as explained above.

One final note on this subject is that in high-density occupancies the "worst case" DH loads can occur at even lower outdoor temperatures than at the ASHRAE peak outdoor dew point. For example, in a classroom, the student density may be so high that the internal DH load (the water vapor from students) may combine with the ventilation load to create a *total* DH load which is highest when outdoor temperatures are even lower than the temperature which coincides with the peak outdoor dew point.[6]

But in most cases, unless the need for humidity control precision is very high (as in the case of industrial applications), sizing the DH equipment based on loads calculated at the ASHRAE peak outdoor dew point and its coincident dry bulb temperature will result in ample DH capacity to remove the DH loads. In the real world, all the individual load elements rarely *all* peak during the *exact* same hour.

Determine DH performance based on weight of water removed per hour... not the sensible heat ratio

The sensible heat ratio is another variable which is useful for cooling system design, but which leads to destructive confusion and a false sense of security when used in dehumidification design.[7]

Don't rely on the sensible heat ratio of a cooling coil to quantify the equipment's DH performance. Instead, for any type of equipment used as a dehumidifier, use the entering and leaving air conditions and air volumes to calculate the number of *lbs* or *kilograms* of water vapor the equipment actually removes from the air every hour. The diagrams and equations shown in figure 13.5 explain how this is done. It's the simplest way to calculate DH performance with certainty. And using weight units lets you easily compare the equipment's humidity removal to the DH loads you calculate with the equations in Chapter 11.

There are several problems with using the sensible heat ratio for quantifying DH performance. First, it expresses the latent portion (the dehumidification portion) of the coil's performance as that portion of the total cooling performance which is not used for sensible cooling. That's confusing. It's sort of like rating light bulbs by how much dark they don't produce.

Even more confusing, the sensible heat ratio quantifies DH performance in the same units used for describing sensible cooling capacity—Btu's or kW. These units become very misleading when applied to a DH process in actual hardware. Btu's and kW are very familiar units to cooling system designers. Their very familiarity is the problem. When the units in question are Btu's or kW, it seems like, if you have enough cooling capacity, the cooling system should be able to take water vapor out of the air. But it won't, unless the air is cooled down low enough to actually condense moisture.

Not just any Btu's of capacity will do the job. You have to take those latent Btu's (the water vapor) out of the air *after* the air has been chilled down to its dew point. Btu's removed above the dew point don't remove water—they only remove sensible heat. In theory, the rated sensible heat ratio of the coil should quantify its DH

Equation 13.1
Dehumidification Capacity
What is the dehumidification capacity of the supply air? Said another way; how much moisture can a known amount of air with a known humidity ratio remove from the conditioned space every hour?

Q (supply air flow rate - scfm or l/s)
w_1 (humidity ratio of air in the controlled space - gr/lb or g/kg)
w_2 (humidity ratio of the supply air - gr/lb or g/kg)

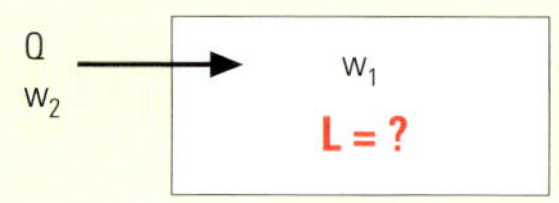

$$\textbf{lbs/h} = \frac{Q \times (w_1 - w_2) \times 4.5}{7000} \qquad \textbf{kg/h} = \frac{Q \times (w_1 - w_2) \times 4.32}{1000}$$

Equation 13.2
Required Supply Air Humidity Ratio
If the supply air flow rate and the indoor humidity ratio are fixed, how dry does the supply air have to be in order to remove a known internal moisture load?

Q (supply air flow rate - scfm or l/s)
w_1 (humidity ratio of air in the controlled space - gr/lb or g/kg)
L (moisture load in the controlled space - lb/h or kg/h)

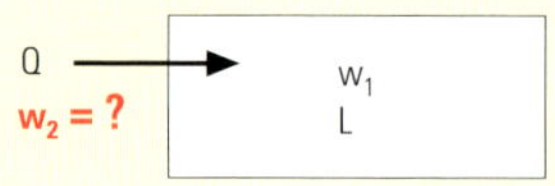

$$\textbf{gr/lb} = w_1 - \left(\frac{7000 \times L}{Q \times 4.5} \right) \qquad \textbf{g/kg} = w_1 - \left(\frac{1000 \times L}{Q \times 4.32} \right)$$

Equation 13.3
Required Supply Air Flow Rate
If the supply air humidity ratio and the indoor humidity ratio are both fixed, how much supply air is needed to remove a known internal moisture load?

w_1 (humidity ratio of air in the controlled space - gr/lb or g/kg)
w_2 (humidity ratio of the supply air - gr/lb or g/kg)
L (moisture load in the controlled space - lb/h or kg/h)

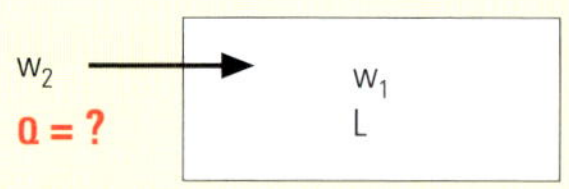

$$\textbf{scfm} = \frac{7000 \times L}{(w_1 - w_2) \times 4.5} \qquad \textbf{l/s} = \frac{1000 \times L}{(w_1 - w_2) \times 4.32}$$

Equation 13.4
Mixed-air Humidity Ratio
When two air flows with different humidity ratios are blended, what will be the humidity ratio of the resulting mixture?

Q_1 (Air flow rate of the first air stream - scfm or l/s)
Q_2 (Air flow rate of the second air stream - scfm or l/s)
w_1 (humidity ratio of air in the controlled space - gr/lb or g/kg)
w_2 (humidity ratio of the supply air - gr/lb or g/kg)

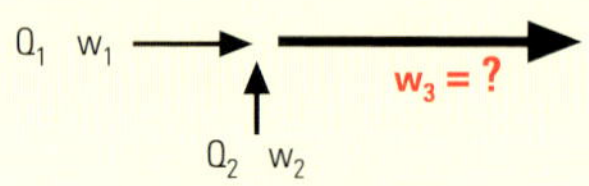

$$\textbf{gr/lb} = \frac{(Q_1 \times w_1) + (Q_2 \times w_2)}{Q_1 + Q_2} \qquad \textbf{g/kg} = \frac{(Q_1 \times w_1) + (Q_2 \times w_2)}{Q_1 + Q_2}$$

Equation 13.5
Dehumidified Air vs. Bypass Air
Usually, a dehumidifier only needs to dry ***part*** *of the supply air rather than its entire volume. How much of the supply air should be dried by the dehumidifier in order to bring the entire supply air flow to the required humidity ratio?*

Q_3 (Supply air flow rate - scfm or l/s)
w_1 (Humidity ratio of air leaving the dehumidifier - gr/lb or g/kg)
w_3 (Required humidity ratio of the supply air - gr/lb or g/kg)

$$\textbf{scfm} = \frac{(Q_3 \times w_2) - (Q_3 \times w_3)}{w_2 - w_1}$$

$$\textbf{l/s} = \frac{(Q_3 \times w_2) - (Q_3 \times w_3)}{w_2 - w_1}$$

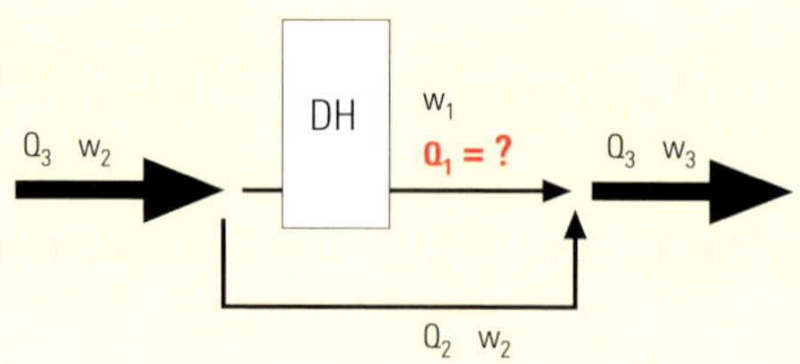

Fig. 13.5
Equations for DH design
Since the DH load calculations are best done with units of water vapor weight rather than their energy equivalent, these equations will be useful in designing systems which are effective for dehumidification. Use the table in figure 13.6 to find the weights of water to use in these equations, after the control dew point has been established.

Fig. 13.6

Dew points and humidity ratios

The control level is best specified in terms of dew point, but the calculations are done in units of water vapor weight. This table shows the weight of water vapor in air at each dew point—the humidity ratios at each dew point. (The humidity ratio is the weight of the water vapor in each lb [or in each kg] of air if it were perfectly dry.)

Precision of this table

At a constant dew point, the weight of water vapor at saturation (at the dew point) does change slightly at higher or lower pressures. This fact becomes important in compressed air design, or when designing HVAC systems at either high altitude locations or down in deep mines.

But most hot and humid climates are at near sea level pressures. So this table will be accurate enough for designing most commercial, institutional and residential HVAC systems. When greater precision is needed (usually for very low dew point industrial applications), the ASHRAE Handbook—Fundamentals can provide the appropriate psychrometric precision.

°F	gr/lb	in.hg.	°C	g/kg	kPa
84	179	1.176	28.89	25.58	3.98
83	173	1.139	28.33	24.74	3.85
82	167	1.103	27.78	23.92	3.73
81	162	1.067	27.22	23.13	3.61
80	156	1.033	26.67	22.36	3.49
79	151	1.000	26.11	21.61	3.38
78	146	0.967	25.56	20.89	3.27
77	141	0.936	25.00	20.19	3.16
76	136	0.905	24.44	19.51	3.06
75	132	0.876	23.89	18.85	2.96
74	127	0.847	23.33	18.21	2.86
73	123	0.819	22.78	17.59	2.77
72	119	0.792	22.22	16.99	2.68
71	115	0.765	21.67	16.41	2.59
70	111	0.740	21.11	15.85	2.50
69	107	0.715	20.56	15.30	2.42
68	103	0.691	20.00	14.77	2.33
67	100	0.667	19.44	14.26	2.26
66	96	0.645	18.89	13.76	2.18
65	93	0.622	18.33	13.28	2.10
64	90	0.601	17.78	12.82	2.03
63	86	0.580	17.22	12.37	1.96
62	83	0.560	16.67	11.93	1.89
61	80	0.541	16.11	11.51	1.83
60	78	0.522	15.56	11.10	1.76
59	75	0.504	15.00	10.70	1.70
58	72	0.486	14.44	10.32	1.64
57	70	0.469	13.89	9.95	1.58
56	67	0.452	13.33	9.59	1.53
55	65	0.436	12.78	9.24	1.47
54	62	0.420	12.22	8.90	1.42
53	60	0.405	11.67	8.58	1.37
52	58	0.391	11.11	8.27	1.32
51	56	0.376	10.56	7.96	1.27
50	54	0.363	10.00	7.67	1.23
49	52	0.349	9.44	7.38	1.18
48	50	0.337	8.89	7.11	1.14
47	48	0.324	8.33	6.84	1.10

°F	gr/lb	in.hg.	°C	g/kg	kPa
46	46	0.312	7.78	6.59	1.05
45	44	0.300	7.22	6.34	1.02
44	43	0.289	6.67	6.10	0.98
43	41	0.278	6.11	5.87	0.94
42	39	0.268	5.56	5.64	0.91
41	38	0.258	5.00	5.43	0.87
40	37	0.248	4.44	5.22	0.84
39	35	0.238	3.89	5.02	0.81
38	34	0.229	3.33	4.82	0.77
37	32	0.220	2.78	4.64	0.74
36	31	0.212	2.22	4.46	0.72
35	30	0.204	1.67	4.28	0.69
34	29	0.196	1.11	4.11	0.66
33	28	0.188	0.56	3.95	0.64
32	26.5	0.1805	0.00	3.79	0.610
31	25.3	0.1724	-0.56	3.62	0.583
30	24.2	0.1646	-1.11	3.46	0.556
29	23.1	0.1572	-1.67	3.30	0.531
28	22.0	0.1501	-2.22	3.15	0.507
27	21.0	0.1432	-2.78	3.01	0.484
26	20.1	0.1367	-3.33	2.87	0.462
25	19.1	0.1304	-3.89	2.74	0.441
24	18.3	0.1244	-4.44	2.61	0.420
23	17.4	0.1186	-5.00	2.49	0.401
22	16.6	0.1131	-5.56	2.37	0.382
21	15.8	0.1078	-6.11	2.26	0.365
20	15.1	0.1028	-6.67	2.16	0.347
19	14.4	0.0980	-7.22	2.05	0.331
18	13.7	0.0933	-7.78	1.96	0.315
17	13.0	0.0889	-8.33	1.86	0.300
16	12.4	0.0847	-8.89	1.77	0.286
15	11.8	0.0806	-9.44	1.69	0.272
14	11.2	0.0767	-10.00	1.61	0.259
13	10.7	0.0730	-10.56	1.53	0.247
12	10.2	0.0695	-11.11	1.46	0.235
11	9.7	0.0661	-11.67	1.38	0.224
10	9.2	0.0629	-12.22	1.32	0.213
9	8.8	0.0598	-12.78	1.25	0.202

Fig. 13.7

Dry out the ventilation air *before* it mixes into the room air

*This visual analogy helps explain why drying ventilation air **early** is best. After all those leaves mix into the lake, it will take more people and more work to chase them all down and fish them out.*

In the same way, drying the humid ventilation air before its huge water vapor load mixes into the room air makes for simpler and less costly DH systems.

performance. But it doesn't in reality, because the SHR of the coil only predicts performance at steady-state conditions. But in the real world, the cooling coil's DH performance will change with entering conditions, it will change as the coil is warming up and cooling down, and it changes whenever cooling capacity is modulated.

So for DH system design, stay way from the sensible heat ratio. Keep all the DH calculations and the required DH equipment performance expressed in terms of the weight of the water vapor added or removed per hour. When you specify a certain number of lbs or kg per hour removal (at a defined inlet temperature, humidity and air flow rate), your system will behave predictably. The equipment vendor will be obliged to provide you with a submittal which shows *predictable and dependable* DH performance for your defined inlet conditions which is *independent of* any sensible cooling effect that the equipment might also provide.

No humidity control until the ventilation air is dry

In hot and humid climates, the ventilation air can impose a dehumidification load during some part of *every* day, all year long. To gain control of humidity in the building, the DH load brought inside by the ventilation air must be removed. And removing that load *sooner* is better than removing it later. Here's why.

When humid ventilation air enters the controlled space, its effect is immediate. That load is so large compared to any other source of humidity that the dew point will quickly rise above the set point. But it will take time for the DH equipment to respond and remove that load from the space. In the meantime, humidity continues to rise towards the level in the outdoor air. When the DH equipment responds, the DH effect won't always be immediate. With some types of DH equipment, it typically takes *longer* to remove moisture from the air (and from the building itself) than it takes for cooling equipment to cool the air. So

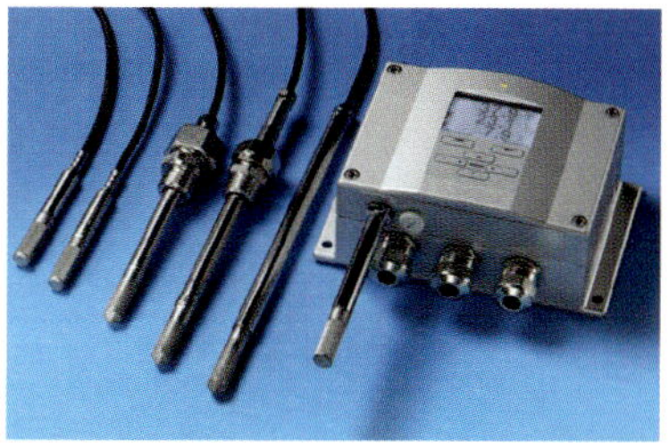

Fig. 13.8 Dew point transmitters
In earlier times, it was not so easy to use the dew point to control the DH components. Now, however, there is a rich variety of instruments and controls which measure, display and control based on dew point. In fact, any building automation system can easily convert the signals of air temperature and relative humidity to a dew point value, which can then be used to control the DH equipment.

the DH system response time is probably longer when overly-humid ventilation air is injected directly into the space.

The much better approach is to dry the ventilation air *before* it gets to the conditioned space. This will provide more stable humidity levels, because the internal DH loads are quite small compared to the load in the ventilation air. With low internal loads and a constant stream of *dry* air from the system instead of a constant injection of *humid* air, the indoor humidity level stays quite even.

Also, in terms of equipment it's much easier to remove the ventilation DH load while it is still "highly concentrated"—before it is diluted by the dry indoor air. With pre-drying, the DH equipment can be smaller and more economical, and it can run at higher loadings, which makes it more efficient and effective.

A useful visual analogy is a small stream which carries fallen leaves into a lake. It's much easier to filter all the leaves out of that small stream—rather than working much harder and longer to remove those same leaves after they have spread out into the lake. (Figure 13.7)

Design for dew point control instead of rh control

Most humidistats sense the relative humidity of the air. But these days, that no longer means you must control the DH system based on rh. It's a small matter to electronically convert the rh and temperature to the *dew point* value and then use that value as the control signal. In most buildings, there's really no need to control the system based on relative humidity. Especially since there are so many benefits to controlling the system based on the air's dew point.

First among those benefits is that the dew point is an indicator of current risk to the building. When you know the air's dew point, you also know what surface temperature will condense moisture from that air. So you will quickly realize, for example, that when the dew point is 65°F and the supply air is 55°F, the outside of a steel duct which carries that cold air will sweat (condense large amounts of moisture). Condensation and moisture absorption lead to the mold growth and corrosion which cause so much damage to buildings, as explained in Chapter 5 - Avoiding Bugs, Mold & Rot.

The relative humidity of the air does not give you that same useful information. The relative humidity at the *sensor* is not the same as the relative humidity at the cool *surfaces* of ducts, pipes and walls washed by cold air. The dew point tends to be nearly the same throughout the building.

Another good reason for controlling on dew point is that it governs thermal comfort more than does the rh. The body's ability to release heat is a function of the difference in dew point between the saturated air near the surface of the skin and the surrounding air in the building. All other variables being equal, the greater that dew point difference, the more readily the body will release metabolic heat and remain thermally comfortable.

The relative humidity also plays a role, but the dew point is the principle humidity variable which governs comfort, as described in more detail in Chapter 2 - Improving Thermal Comfort.

Next is the benefit of the dew point as an aid to managing the systems after hours, when the temperature can be allowed to rise to save energy. If you know the dew point, you know how much water vapor is building up indoors—water vapor that will have to be removed when the building is cooled back down for occupancy. If instead you monitor the relative humidity (which goes down as that air temperature rises), you might falsely assume, as many have in the past, that the building is safe from mold and condensation. A relative humidity reading of 65% seems quite reassuring, until one realizes that at 85°F, a 60%rh reading means the dew point has risen to 74°F. When the cooling systems turn on and start chilling the building, all sorts of surfaces will begin to condense moisture and can eventually grow mold, as explained in Chapter 5.

[A relative humidity reading of 65% seems quite reassuring, until one realizes that at 30°C, a 60%rh reading means the dew point has risen to 23°C. When the cooling systems turn on and start chilling the building, all sorts of surfaces will begin to condense moisture and grow mold, as explained in Chapter 5.]

Finally, by controlling on dew point rather than on relative humidity the DH equipment will provide a more stable absolute humidity level. The system is no longer confused by small changes in the air's dry bulb temperature. This makes the placement of the humidity sensors much less critical.

For example, when the system is controlled on rh, if the humidistat is washed by cool supply air, the rh in that location will be quite high. That means the DH components may switch on needlessly, wasting energy. Conversely, if sunlight or another heat source warms the rh sensor, the relative humidity will seem too low, and the DH components may not switch on at all, resulting in a humidity build-up in cooler parts of the building.

Controlling the system based on the dew point in the space—which does not change with air temperature—will:

- Provide the most relevant signal for controlling dehumidification components,
- Indicate clearly the risk of condensation, and...
- Provide more stable humidity conditions.

Avoiding common problems in DH design

Our discussion now proceeds to avoiding common problems in maintaining a given humidity control level. The subject begins with four unexpected problems which can ruin the control level.

Measure and control ventilation air flow, continuously

As noted many times throughout this book and as shown again in figure 13.9, the largest dehumidification load is carried by the ventilation air. Small changes in the ventilation air volume will make big changes in the DH load. Therefore, excess ventilation air can make it difficult or impossible for the DH equipment to maintain a given dew point in the space.

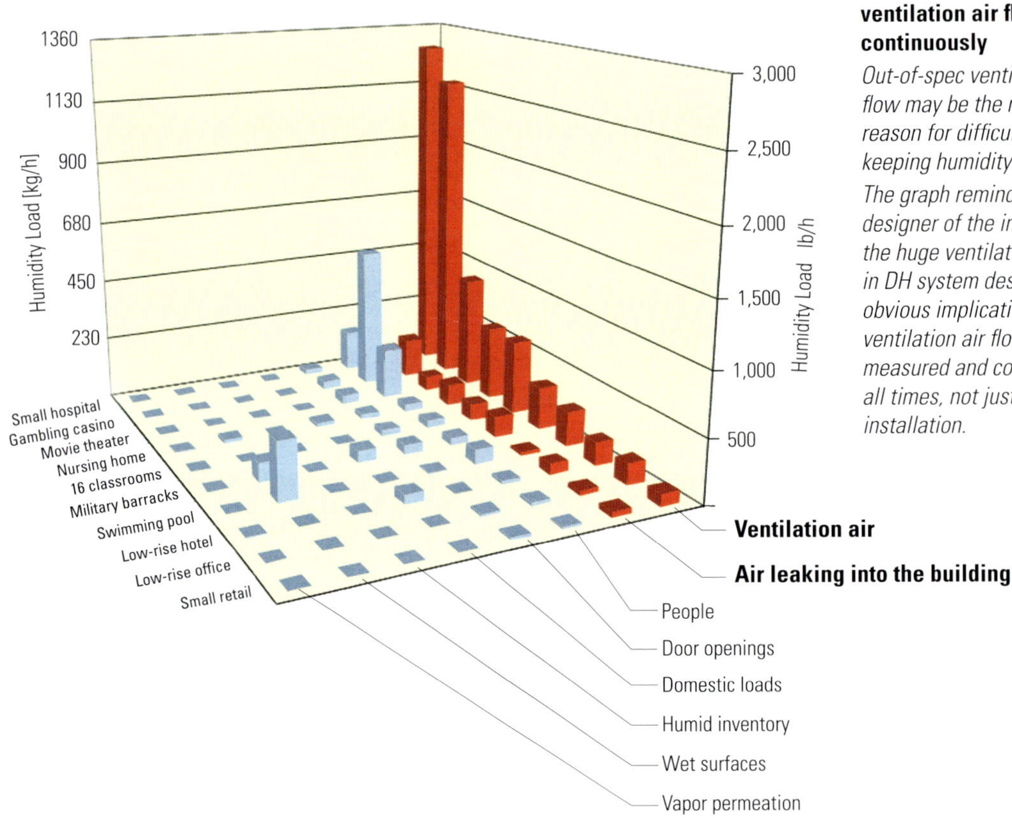

Fig. 13.9
Measure and control the ventilation air flow—continuously
Out-of-spec ventilation air flow may be the most common reason for difficulties in keeping humidity under control.
The graph reminds the designer of the importance of the huge ventilation air load in DH system design. The obvious implication is that the ventilation air flow should be measured and controlled—at all times, not just after installation.

This is intuitive when one thinks about it. Nobody would expect the *cooling* equipment to be effective if the calculated cooling load were somehow doubled. In the same way, one cannot rely on DH components to keep a constant dew point indoors if the ventilation air flow is twice the value used to calculate the loads.

What most designers miss is the fact that unless the ventilation air is explicitly measured and controlled, the air flow volume may be quite random. Field measurements in hundreds of buildings show that buildings are consistently over-ventilated and under-ventilated. [8,9,10,11]

Fig. 13.10
Watch out for unexpected exhaust
It's obvious that the combined ventilation and make up air flows must be larger than the combined exhaust air flows. But sometimes, people other than the HVAC designer will add extra exhaust fans to the building. For humidity control, it's very important to be aware of all exhaust flows, and to measure and control them, so the building air pressure stays positive.

Most HVAC designers do not specify air flow measurement devices and modulating air flow control dampers on ventilation air inlets. Usually, they assume that a fixed-position damper will be accurately set, for all time, by the balancing contractor, if there is one. This is generally a false hope on two levels. First, air balancing is very difficult under the best of circumstances, and setting the outdoor air flow accurately is especially difficult because of all the turbulence and pressure changes in an air handler. Second, with a fixed-position damper, the air flow changes greatly with changes in system pressures and wind pressures.[12] More details of this problem are described in Chapter 16 - Air-Tight HVAC Systems.

So if the designer needs to keep humidity under a fixed maximum, he or she must (really must) *measure* and *control* the volume of ventilation air explicitly. An excess of ventilation air is the most common reason for problems in maintaining control of humidity.

Extra exhaust vent fans will ruin control

The next most-common problem is an exhaust fan (or two or three) which somehow escape the attention of the HVAC designer. With unexpected exhaust fans, the makeup air system may not have enough capacity. That means the building will—literally—suck. The exhaust fans will pull in untreated outdoor air through cracks, penetrations, construction joints and around doors and windows. That humid infiltration air will play havoc inside the cool cavities of the building enclosure. And of course this infiltration air adds to the internal dehumidification load, making it very difficult for the DH equipment to keep the indoor dew point below reasonable limits.

When the HVAC designer has full control of all air-side components, exhaust fans probably won't be a problem. The designer can simply count up the fan capacities and then ensure there is enough dry makeup air to more than balance the total exhaust. That's the first and fairly obvious task the HVAC designer must accomplish, but it is not always easy. (See figure 13.10)

Three problems often occur with exhaust system. First is the matter of *others* adding exhausts and not informing the HVAC designer. Second is the fact that few HVAC designers remember to *control* the air flow leaving through the exhaust systems so they don't pull too much air out of the building. Third is the fact that many HVAC designers don't understand how important it is to *seal up* all the connections in the exhaust air systems. Any one of these issues will create an obstacle to humidity control by increasing infiltration of humid outdoor air.

There are two frequent suspects when it comes to adding extra exhaust fans after the fact: kitchen designers and indoor swimming pool designers. Neither of these spaces is cooled to the same level as the rest of the building, and both have large exhaust fans. Also, both kitchens and pool enclosures *need* to operate at a slight negative pressure to keep odors and humidity out of the rest of the building. So their designers don't always have a habit of providing enough dry makeup air to balance their exhaust flows. The kitchen and the swimming pool enclosure will always need to "steal" dry makeup air from the rest of the building.

The HVAC designer who wants to control humidity must take time to coordinate with the other professionals who might be specifying exhaust air fans. Make sure that overall, the building has enough dry makeup air to balance the sum of the exhaust air flows and keep the bulk of the building under a net positive pressure, using dry air.

Another problem comes up with large central exhaust fans, such as the fans used to exhaust odors from toilets in hotels, hospitals, nursing homes and other large buildings. It's important to *seal up* all the duct connections of those central systems. Otherwise, they will pull in air from the building cavities they pass through on their way out of the building. When that happens, the air in the cavities is often replaced by untreated outdoor air, pulled into the building by the suction of the exhaust fan. Sealing up the duct connections ensures that the air will be drawn from the bathrooms, and won't lead to humid air infiltration into building cavities.

One caution for central exhaust systems will be obvious to most designers, but perhaps not to all. It's important to *measure* and *set* any exhaust air flows, so that the total exhaust does not accidentally overwhelm the makeup air system and lead to building suction. When it comes to exhaust air flows, bigger is *not* better, from the perspective of humidity control in hot and humid climates. This caution applies especially to hotels, high-rise residential buildings and nursing homes, where continuous toilet exhaust using central systems is a common design practice. Central exhaust systems with rooftop fans keep the fan noise far away from sleeping room occupants. But these exhaust systems are not usually given enough attention by the designer or the installer or the air balance contractor. The designer needs to specify balancing dampers to make sure the air flows leaving each toilet can be adjusted individually by the air balance contractor.

One last exhaust fan caution applies to buildings in which the positive building air pressure is managed partly by a modulating, pressure-controlled exhaust fan. In some buildings, the systems must supply *so much* ventilation air during peak occupancy that unless the designer adds an exhaust fan to relieve excess positive pressure, the building will be so over-pressurized that the exterior doors won't close. Here's the caution. Such pressure-relief exhaust fans must match the reductions in incoming ventilation air flow. For example, they should not be set up to maintain a minimum exhaust flow if the incoming ventilation air flow will sometimes be reduced to zero. If any exhaust fan operates when the ventilation system does not, that exhaust fan can pull hot and humid outdoor air into the building through cracks and joints instead of through a system which could dry that air.

Fig. 13.11
Wassat? Reseal the air plenums?... Not my job, man. I got cable to run.
It's very difficult to make sure building cavities like return air plenums really stay airtight. Even if all the plumbers, electricians, communications technicians and security installers are told of its importance, it probably won't happen. So for best humidity control, use tightly-sealed ducts instead of leaky building cavities for air distribution and air flow control.

Avoid return air plenums and make sure all air-side connections are sealed up, air tight

This subject is covered in detail in Chapter 16 - Air-Tight HVAC Systems. So it's not necessary to repeat all the reasons why sealed air systems are so important. It's enough to note that the suction created by any leaking return air connections can lead to infiltration of humid outdoor air through the building enclosure—just like any leaking connections in exhaust duct systems.

So it's important to seal up *all* duct connections to *all* components so they are air tight. This will save energy, and will also avoid creating unexpected dehumidification loads from infiltrating outdoor air.

For similar reasons, it's useful to avoid return air plenums. Instead, use ducted returns which are well-sealed. It's very, very difficult to seal up return air plenums so they are airtight. And if they are air-tight at system start-up, they may not remain airtight after a technician cuts holes through the plenum's walls to run electrical cable, or security wiring, or an extra drain pipe. (Figure 13.11)

When one must use return air plenums, recognize there is a risk to humidity control from outdoor air infiltration. One can reduce this risk by first specifying that return air plenums should be sealed using spray-applied fire sealant. Next, keep the building under a net positive air pressure. Finally, add some extra DH capacity to remove any unexpected loads which might come from air infiltration. Usually, these measures cost more in both construction costs and operating costs than simply ducting the returns.

Don't assume cooling systems are effective dehumidifiers

Not all cooling systems will dry the air effectively when you need them to do so. The reasons for this fact are described in detail in Chapter 14 - Designing Cooling Systems. But in this chapter where we discuss dehumidification, it's worth repeating that if a cooling system *is* effective at drying on demand, its manufacturer should be able to quantify, in lbs or kg per hour, how much moisture the equipment will remove even when there is no need for cooling.

If the manufacturer cannot provide a credible estimate of the equipment's moisture removal performance at peak dew point design conditions, one could assume the equipment is optimized for sensible cooling, not dehumidification. In the absence of credible, quantified DH performance from the cooling system, it would be wise to add a dedicated DH component which *does* have quantified performance at peak dew point design conditions.

This may not be a problem when the designer's goal is simply to moderate indoor humidity during the few hours when the outdoor temperatures reach their peak. But if the designer needs humidity control during the entire year, then it's important to select equipment which has a defined DH capacity at all part-load conditions.

Note that many types of cooling systems *can* be quite effective at drying the air on demand. Variable air volume systems, for example, usually operate with a constantly cold coil which can be made cold enough to provide effective dehumidification. And other types of cooling equipment can be made effective by the addition of supplemental components such as heat pipes or plate exchangers which precool and reheat the supply air before and after their cooling coils. Also, some types of cooling equipment can provide dehumidification without excessive cooling by reducing the air flowing though the coil, rather than by reducing the coil temperature. Variable refrigerant flow DX systems (sometimes called "VRF" or "mini-split or "multi-split" systems) usually use this strategy to improve DH performance at part-sensible-load conditions.

It is fairly easy for the designer to be more certain which cooling systems will also be effective as dehumidifiers. They share *all* of the following characteristics:

- A constantly-cold cooling coil, which provides a leaving air dew point which is *always below* the control dew point in the space.
- The ability to respond to a humidity control signal and provide dehumidification on demand, without respect to the temperature requirements in the space.

- Their manufacturers will be willing to provide moisture removal performance—expressed in lbs or kg removed per hour—for any defined set of inlet conditions.

Load reducers cannot dry air by themselves

Heat exchangers are also used to reduce the dehumidification and cooling loads of ventilation air. These recover the energy used for cooling and dehumidification by *transferring* the heat and humidity of the incoming air to an outgoing exhaust air stream.

The most popular and widely-applied form of these heat exchangers are also known as "enthalpy wheels" or "total heat wheels" and sometimes as "energy recovery ventilators." They save a tremendous amount of energy in both cooling and dehumidification, and they can reduce the size of the installed cooling equipment by more than enough to pay for their installation. In effect, they more than pay off their cost before the system even starts up.

However, there is no free lunch in engineering. For these devices to pre-dry the air as intended, they need to have *dry* exhaust air. If the exhaust air is not dry, then it can't absorb the water vapor that the heat exchanger is trying to transfer out of the incoming air.

That means that an enthalpy wheel *can't do the job by itself*. There must be a dehumidification component operating someplace else in the system so that the exhaust air remains dry enough to be useful whenever ventilation air is brought into the building—including after hours and during vacations.

Vented attics and soffit vents can ruin humidity control

Vented attics are a common feature of residential and light commercial construction. (Attic venting is often done because some building codes still require it. In the past, the theory was that attic venting will keep asphalt roof singles cool enough to help them last longer.) But vented attics create problems for humidity control.

With a vented attic the architectural designer locates the insulation between the attic and the occupied floors. The hope is that insulation will isolate the *unconditioned* attic from the *conditioned* space below. But humid air from the attic still gets down into the occupied spaces, creating unexpected dehumidification loads. And these loads from the attic can be very large.

For example, in one school in Florida the measured passive air leakage of the building was reduced from 7300 cfm to 3300 cfm [3445 l/s to 1560 l/s] by sealing up the open soffit vents which provided the attic ventilation.[13] That architectural designer certainly did *not intend* for humid air to get down into occupied space, but it did. And it often does, as documented by that case and others.[10, 14,15]

Attic air flows downward for many reasons. First, the HVAC systems push and pull air around the building. The systems create many pressures, some of which pull or push air out of the attic, especially when ducts located there are not airtight. Also, some air flows downward because there is rarely if ever a tight air barrier between the attic and the occupied floors. Often, the insulation separating the unconditioned attic from the floors below consists of loose glass fiber batts, laid on top of dropped-ceiling acoustic tiles. Such tiles are supported by a grid which has no gaskets around the edges.

In other cases, fire codes require a continuous ceiling of fire-rated sheet material, such as gypsum wall board. But there are still so many penetrations for electrical wiring, plumbing and AC ducts that the "continuous" ceiling still allows a great deal of leakage from the attic to the floors below. Attic air also flows down through cracks around "can light" fixtures set into the ceilings. Light fixtures are seldom airtight. And even when they are, the penetrations they make are seldom sealed up airtight.

Even without HVAC-induced pressures, humid air from the attic flows downward, pulled by falling cool air below. This is sometimes called the "reverse stack effect." It's one reason for the mold which sometimes grows around ceiling light fixtures. As the humid air falls down through the gap around the light fixture, moisture is absorbed by the cool gypsum wall board near the gap. Eventually, the paper face of that wall board absorbs enough moisture to support mold growth. It usually grows in a ring pattern around the light fixture.

To avoid this problem, the architectural designer must decide to seal up the attic rather than venting it. That takes money. It's more expensive to seal the attic and insulate the bottom of the roof sheathing rather than laying loose batts on top of ceiling tile. But that's the best way to design the building so that it uses less energy and provides better comfort with reduced mold risk, through lower dehumidification loads.

At the same time, although building codes are changing rapidly, some may still require vented attics. And not all architectural designers and owners are convinced that sealed, semiconditioned attics are affordable. So when a vented attic *is going to be* the plan, the HVAC designer can take these steps to reduce the damage that the vented attic will inflict on the building's humidity control:

- Make absolutely certain that all duct connections located in the vented attic are sealed up, airtight, to minimize the forces which pull and push humid air down into the occupied spaces below.
- Make certain that any fire-code-mandated-airtightness standards for separation of the attic from the occupied floors are *enforced* by the owner and architectural designer.
- Add a large safety factor to your assumptions about air infiltration when estimating the building's dehumidification loads.

The cases described by references 10,13,14 and 15 suggest that *at least* doubling the passive infiltration rate would be a reasonable assumption for single-story buildings with vented attics. In the single-story school described in reference 13, the initial infiltration rate into the building was measured at 0.087 cfm • ft^2 of floor area. When the soffit vents and other attic penetrations were sealed up, the infiltration rate fell to 0.04 cfm • ft^2. [Air leakage of 0.44 l/s • m^2 of floor plan area declined to 0.2 l/s • m^2 after the attic soffit vents were sealed].

Ways to reduce DH-related energy

Removing water vapor from air requires energy. Quite a lot if it, in fact. So dehumidification is not free. But at the same time, the cost of humidity control must be weighed against the cost of increased cooling costs when humidity is not controlled. When humidity builds up inside a building, occupants typically force down the thermostat setting to regain comfort (as described in Chapter 2 - Improving Thermal Comfort). With colder indoor temperatures, the cooling energy will go up by more than the amount of energy that would have been used to keep the air dry.

When humidity control will be part of the system, the HVAC designer can take advantage of these next suggestions to minimize the *total* amount of energy consumed by the building.

Keep the dew point below 55°F... then let the dry bulb rise

In the absence of a specific client requirement, or when you need a rule of thumb which optimizes the balance between comfort and energy, set the maximum humidity at a 55°F dew point, and set the thermostat at 78°F. [Set the maximum humidity at a dew point of 12.8°C, and set the thermostat at 25.5°C.] Field measurements have suggested these levels may save 15 to 20% of the net annual air conditioning energy, while providing comfort for a variety of different occupants with differing clothing and activity levels.[16,17]

Also, limiting the indoor dew point to 55°F [12.8°C] will minimize the risk of condensation and mold growth in the building, as described in more detail in Chapter 5.

Reduce ventilation when people leave the building

Energy consumption depends on the dehumidification load, and the largest DH load is carried by the ventilation air. So, if you reduce the ventilation air flow as people leave the building, you will reduce the energy needed to keep the building dry.

Ventilation air flow can be controlled with time clocks for more predictable and regular occupancies, such as classrooms in

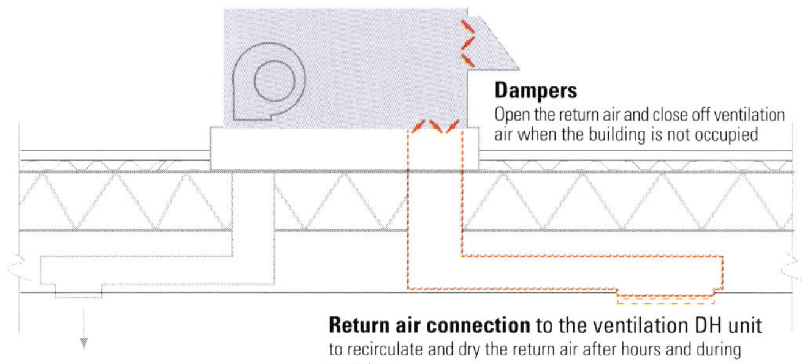

Fig. 13.12 Add a return air connection to ventilation DH equipment
That way, when the building is unoccupied the unit can dry some return air, keeping the building dry at very low energy cost.

elementary schools and high schools. Then, for less-predictable occupancies such as hotel conference rooms, occupancy sensors are a good choice. Finally, for highly variable occupancies in large spaces such as gymnasiums, auditoriums, health clubs and sports facilities, CO_2 sensors may be the optimal choice to help avoid over-ventilation when the occupancy is low.

Recirculate dry air when the building is unoccupied

Keeping the dew point low at *all* times minimizes the risk of condensation and mold growth, as mentioned often throughout this book. But also, when the building is unoccupied, as long as the dew point is still held below 55°F [12.8°C], the dry bulb temperature can be allowed to rise quite high without mold risk. There's seldom any need to keep the building cool when it's not occupied, as long as it stays dry.

To take advantage of the energy savings, shut off the ventilation air, and arrange the dehumidification components so they can dry a steady stream of recirculating air which flows throughout the building. If the DH component is located on the ventilation air, as is often the case, make sure to provide a return air connection to that system, as shown in figure 13.12.

Use condenser heat for reheat and desiccant reactivation

Dehumidification components use a considerable amount of energy. Cooling-type dehumidifiers use energy to cool the air before the coil and to reheat it after the coil has dried the air. Desiccant-type dehumidifiers use heat to reactivate the desiccant (dry it out) so it can dry the supply air continuously.

Condenser heat is an excellent source of the heat for both types of dehumidifiers, as shown in figure 13.13. Using this source of waste heat avoids the need to use new (expensive) energy for either reheat or desiccant reactivation. And using waste heat to provide dehumidification meets the requirements for low-energy buildings as outlined by ASHRAE Standard 90.1

Use air-to-air heat exchangers for precooling and reheat

Cooling-type dehumidifiers chill the air down below its dew point. Often, this cool air can help remove the sensible cooling loads from the space, so it can be used directly. At other times (when the space does not need cooling) these units will reheat the air after it has been dried, so that the supply air will not be too cold to send into the controlled space. Precooling and reheating requires energy. But both functions can be performed by a single, passive device: an air-to-air heat exchanger. Several types of heat exchangers are used for this purpose, including plate exchangers, heat pipes and runaround coil loops—basically any type of air-to-air heat exchanger which does not leak humidity into the dry air stream. Figure 13.15 shows how this works with a plate-type heat exchanger. The warm humid air entering the system is pre-cooled by the heat exchanger before it flows through the cooling coil. Then the re-heat following the coil is provided by that same heat exchanger.

The only energy used for pre-cooling and reheating is the fan power needed to overcome the air flow resistance of the heat exchanger. In moderate climates, the annual cost of that fan power sometimes exceeds the energy saved by precooling and reheating. But in hot and humid climates, the DH load is *high for nearly all*

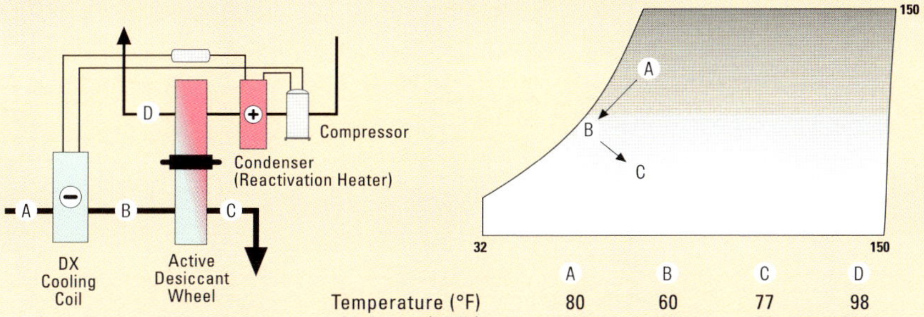

Desiccant Dehumidifier Using Condenser Heat Reactivation

	A	B	C	D
Temperature (°F)	80	60	77	98
Moisture (gr/lb)	124	77	54	144
Air Flow (scfm)	2,000	2,000	2,000	2,000
(°C)	26.7	16.0	25.0	37.0
(g/kg)	17.7	11.0	7.7	20.6
(l/s)	942	942	942	942

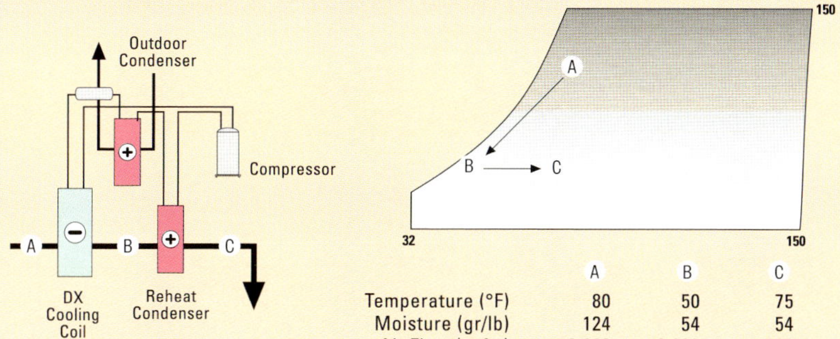

Mechanical Dehumidifier Using Condenser Reheat

	A	B	C
Temperature (°F)	80	50	75
Moisture (gr/lb)	124	54	54
Air Flow (scfm)	2,000	2,000	2,000
(°C)	26.7	10.0	23.9
(g/kg)	17.7	7.7	7.7
(l/s)	942	942	942

Fig. 13.13 Using condenser heat to reduce the cost of keeping the dew point low

Fig. 13.14
Condenser-heat-reactivated desiccant dehumidifier used in large-scale retail buildings

of the operating hours. So the overall energy economics usually favor the use of a heat exchanger for precooling and reheating in a cooling-based dehumidifier.

Install exhaust air energy recovery to minimize DH energy

When clean exhaust air leaves the building near where the ventilation air enters the building, installing a rotary enthalpy heat exchanger is an excellent way to reduce the dehumidification load. These devices, variously known as "enthalpy wheels", "heat wheels", "total heat wheels" or "passive desiccant wheels" operate as shown by the diagram in figure 13.16. The wheel rotates between the incoming humid air and the exhaust air leaving the building. The incoming air gives up some of its moisture to the surface of the wheel, which is coated or impregnated with a desiccant. The desiccant-coated wheel rotates into the exhaust air stream, which is dry enough to strip moisture off the desiccant and vent it back outdoors to the weather. The supply

Cooling-based DH with Air-to-Air Heat Exchanger for Precool-Reheat

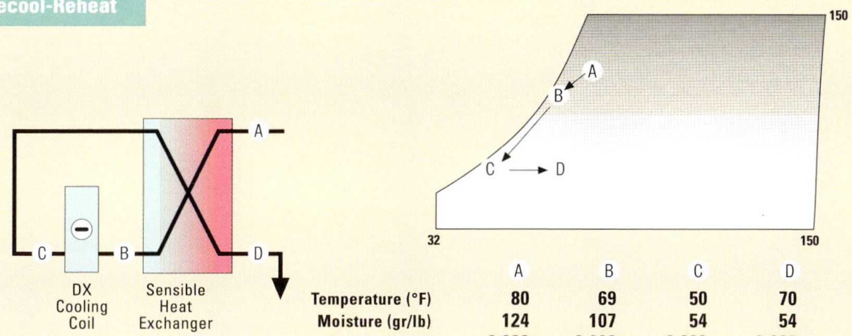

	A	B	C	D
Temperature (°F)	80	69	50	70
Moisture (gr/lb)	124	107	54	54
Air Flow (scfm)	2,000	2,000	2,000	2,000
(°C)	26.7	20.6	10.0	21.1
(g/kg)	17.7	15.9	7.7	7.7
(l/s)	944	944	944	944

Fig. 13.15
Cooling-based DH unit—with energy reduction via an air-to-air heat exchanger to precool and reheat dry supply air

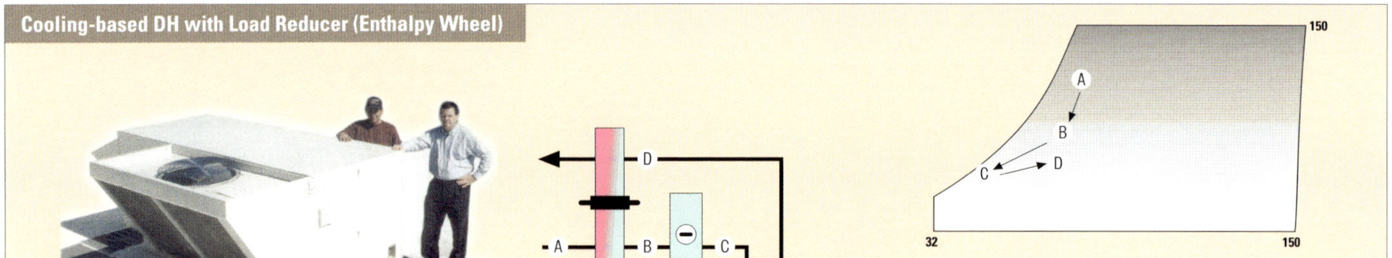

	A	B	C	D
Temperature (°F)	80	77	52	75
Moisture (gr/lb)	124	91	54	65
Air Flow (scfm)	2,000	2,000	2,000	2,000
(°C)	26.7	25.0	11.1	23.9
(g/kg)	17.7	13.0	7.7	9.2
(l/s)	944	944	944	944

Fig. 13.16
Cooling-based DH unit—with energy reduction via a rotary enthalpy heat exchanger which reduces the DH load.

air is predried and precooled by the rotary heat exchanger, greatly reducing the load on the cooling and dehumidification components downstream of the wheel.

By installing an enthalpy wheel as part of a new system, the peak sensible cooling load from ventilation air will probably be cut in half. And at peak sensible design conditions, the wheel also reduces the ventilation dehumidification load by about 50%. These benefits come at the very modest expense of the fractional horsepower motor used to spin the wheel, plus the cost of overcoming the pressure drop created by the wheel.

However, there is one important limitation of enthalpy wheel technology. As discussed earlier, the designer must not make the mistake of assuming the wheel can dry the incoming ventilation air if the exhaust air is not dry. In other words, there must be a DH component operating someplace in the building, so that the exhaust air leaving the building is dry enough to remove moisture from the desiccant wheel.

This is an important caution, because at the peak dew point design conditions when drying is most needed, the enthalpy wheel may not be operating. At those moderate temperatures the wheel is usually turned off, or air is bypassed around the wheel—so that it does not over-recover sensible heat.

For example, when the system's supply air temperature is set at 65°F [18.3°C], the designer will probably not want to heat the incoming air above that temperature. So in that system, the wheel probably will not operate during the thousands of hours each year when the outdoor air temperature is between 65°F [18.3°C] and the temperature of the exhaust air. If it *were* to operate, the wheel would add unwanted sensible heat to the incoming air. But when the wheel is not operating, it's not predrying the air, either.

This means that enthalpy exchangers won't provide *all* the dehumidification needed for the building. The designer will always need a dedicated dehumidification component which can remove the peak DH loads. But installing an enthalpy wheel *will* reduce annual energy consumption of that DH component. And the wheel will reduce the installed cost of the *sensible* cooling equipment as well as its annual operating costs. An enthalpy wheel is not the magic solution to all humidity problems. But it can be a very good investment for many types of heavily-ventilated buildings in hot and humid climates—especially those which are ventilated *continuously*, such as hospitals and nursing homes.

Fig. 13.17

Cooling-based DH unit—with a humidity concentrator-reheater

This configuration uses a desiccant-impregnated wheel to remove humidity after the DX coil, moving that moisture to the incoming air stream while slightly reheating the supply air after it leaves the coil. In effect, the wheel provides slightly drier and warmer supply air. It also concentrates the humidity of the incoming air, allowing the DX cooling coil to remove more moisture than it would without that desiccant wheel.

Cooling-based DH with Humidity Concentrator-Reheater

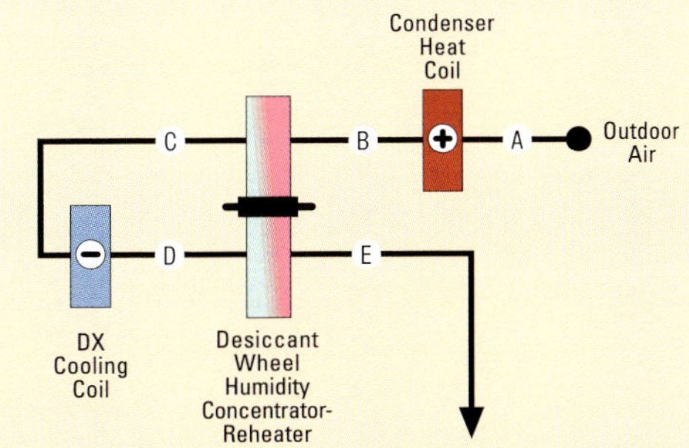

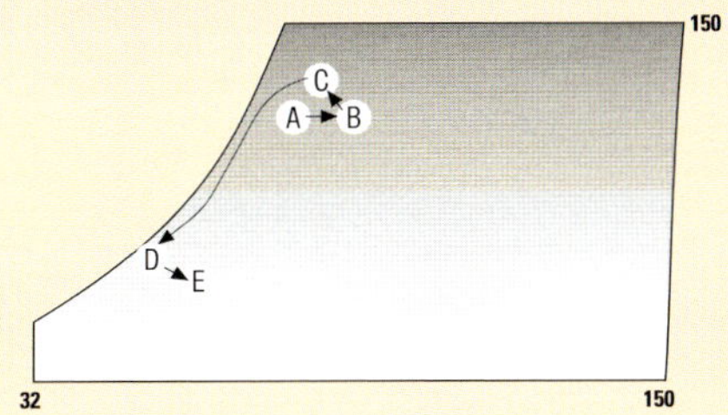

	A	B	C	D	E
Temperature (°F)	**80**	**90**	**85**	**55**	**60**
Moisture (gr/lb)	**124**	**124**	**133**	**63**	**54**
Air Flow (scfm)	**2,000**	**2,000**	**2,000**	**2,000**	**2,000**
(°C)	**26.7**	**32.2**	**29.4**	**12.2**	**15.5**
(g/kg)	**17.7**	**17.7**	**19.0**	**9.0**	**7.7**
(l/s)	**944**	**944**	**944**	**944**	**944**

Summary

Dehumidification is not especially difficult, as long as the designer *understands and acts* on the fact that DH loads are separate from—but just as important as—the building's sensible cooling loads. For effective and consistent control of humidity, make sure the overall system can provide both drying *and/or* cooling on demand, as each load changes.

References

1. Harriman, Brundrett & Kittler, 2008. *ASHRAE Humidity Control Design Guide*, ISBN 1-883413-98-2 ASHRAE, Atlanta, GA
2. ASHRAE Standard 90.1-2004 *Energy Standard for Buildings Except Low-Rise Residential Buildings*. 2004. ASHRAE. 1791 Tullie Circle, NE. Atlanta, GA 30329 ISSN 1041-2336
3. ASHRAE Handbook—Fundamentals 2009 - Chapter 14, Climatic Design Information. ASHRAE, Atlanta, GA. www.ashrae.org
4. Harriman, Lewis G, Colliver, Donald, Hart, K. Quinn. 1999. "New Weather Data for Design and Energy Calculations" ASHRAE Journal, March 1999 ASHRAE, Atlanta, GA. www.ashrae.org
5. Shirey, Don B. III, Henderson, Hugh and Raustad, Richard. 2004. "Understanding the Dehumidification Performance of Air Conditioning Equipment at Part-Load Conditions. Report # FSEC-CR-1537-05, (Department of Energy/National Energy Technology Laboratory Project Number DE-FC26-01NT41253). Florida Solar Energy Center, Cocoa, FL. www.fsec.org

6. Murphy, John and Bradley, Brenda. 2004. "Better part-load dehumidification." Trane Engineers Newsletter 33(2). Trane Commercial Systems, Ingersol-Rand Company, LaCrosse, WI. www.trane.com

7. The sensible heat ratio of a cooling coil is the result of dividing its sensible cooling effect by the sum of its sensible and latent cooling effects. In other words: SHR = (sensible cooling) ÷ (Latent cooling + sensible cooling). For example, if the coil removes a total of 12,000 Btu/h, and of that total, 10,000 Btu/h is sensible cooling, then the SHR of the coil will be: 10,000 ÷ 12,000 = 0.83. Then, you just need to keep in mind that the remaining heat removed by the coil (that which was *not* sensible heat) is latent heat. In other words, it's moisture.

   Usually, the manufacturer will provide the sensible heat ratio for a coil, based on how that coil performed at the steady-state ARI test conditions. For example, in a given system, the manufacturer might say the coil has a capacity of 12,000 Btu/h with a SHR of 0.83. To find the coil's latent performance (its dehumidification performance at the ARI test condition), first multiply the total capacity by the SHR, and then subtract the result from the total coil capacity. In this example, 12,000 x 0.83 = 9960, and 12,000 - 9960 = 2040 But/h. But that value is still not much help to the DH designer until it's divided by the latent heat of vaporization per pound of water, which is 970 Btu/h. Using that division, you can finally obtain the number of pounds of water the coil will remove, in this case: 2,040 ÷ 970 = 2.1 lbs of water per hour. (Unless you prefer to use the value for the latent heat of vaporization of water for the coil's leaving air temperature of 50°F. In that case you would divide by 1083 Btu/h instead of by 970 Btu/h, which is the latent heat of vaporization of water when it's boiling at 212°F. Then you would conclude that the coil will remove only 1.9 lbs/hr, not 2.1 lb/hr.)

   Everything clear now? Probably not, for most people. And actually, it gets even worse for packaged DX cooling units. The SHR of a DX cooling unit is not usually cataloged by the manufacturer for the typical air entering conditions the unit will see in service—it's only defined for the ARI test conditions. With respect to humidity, SHR is indirect, complicated and not usually even known for the conditions the cooling system will see during actual operation. SHR is just not an efficient way to think about DH performance, much less to reliably quantify it. That's why DH design will be much easier if you first *think about humidity levels in terms of dew point*—and then *make your calculations using the corresponding water weight units*, as shown in figure 13.5.

8. Persily, Andrew; Gorfain, Josh; Brinner, Gregory. "Ventilation Design and Performance in U.S. Office Buildings." *ASHRAE Journal*, April 2005, pp.30-35 ASHRAE, Atlanta, GA www.ashrae.org

9. Emmerich, Steven; McDowell, Timothy; Anis, Wagdy. "Simulation of the Impact of Commercial Building Envelope Airtightness on Building Energy Utilization." *ASHRAE Transactions, Volume 113, Part 2. 2007*. ASHRAE 1791 Tullie Circle, NE Atlanta, GA 30329

10. Cummings, James B, Withers, C.B, Moyer, N, Fairey, P., McKendry, B, "Uncontrolled air flow in non-residential buildings." April 15th, 1996, Final Report of FSEC project number FSEC-CR-878-96. Florida Solar Energy Center, 1679 Clearlake Rd, Cocoa, FL 32922.

11. Jacobs, Peter. 2003. "Small HVAC Problems and Savings Reports." California Energy Commission Technical Report No. P500-03-082-A-25, New Buildings Institute, White Salmon, WA www.newbuildings.org

12. ASHRAE Handbook—Fundamentals. 2005. Chapter 15, Fundamentals of control - Figures 13 and 14.

13. Äsk, Andrew. "Ventilation and Air Leakage" *ASHRAE Journal*, November 2003, pp.29-34 ASHRAE, Atlanta, GA

14. Bailey, Ronald, 2008 - Input from review committee member, recounting problems identified and solved in 1995 at a high school in Port Charlotte, a city which is located on the Gulf Coast of South Florida.

15. Toburen, Timothy, 2008. Cold Attic Syndrome; A Case Study of Unintended Consequences. Indoor Environment Connections, February, 2008. Indoor Air Quality Association, Rockville, MD. ieconnections.com

16. Fischer, John; Bayer, Charlene. "Failing Grade for Many Schools - Report Card on Humidity Control" *ASHRAE Journal*, May 2003. pp.30-37.

17. Spears, John; Judge, James. "Gas-Fired Desiccant System for Retail Super Center" 1997. ASHRAE Journal, October 1997 pp.65-69.

Image Credits

Figure 13.1 - Comet, Czech Republic, www.cometsystem.cz

Figure 13.7 - Mike Dater, Portsmouth, NH.

Figure 13.8 - Vaisala Inc. Woburn, MA www.vaisala.com

Figure 13.14 - Munters DryCool Products, Selma, TX www.munters.com

Figure 13.15 - (Inset photo) DesChamps Products, Buena Vista, VA, www.deschamps.com

Figure 13.16 - (Inset photo) AAON Inc, Tulsa, OK www.aaon.com

Chapter 14

Designing Cooling Systems

By Lew Harriman

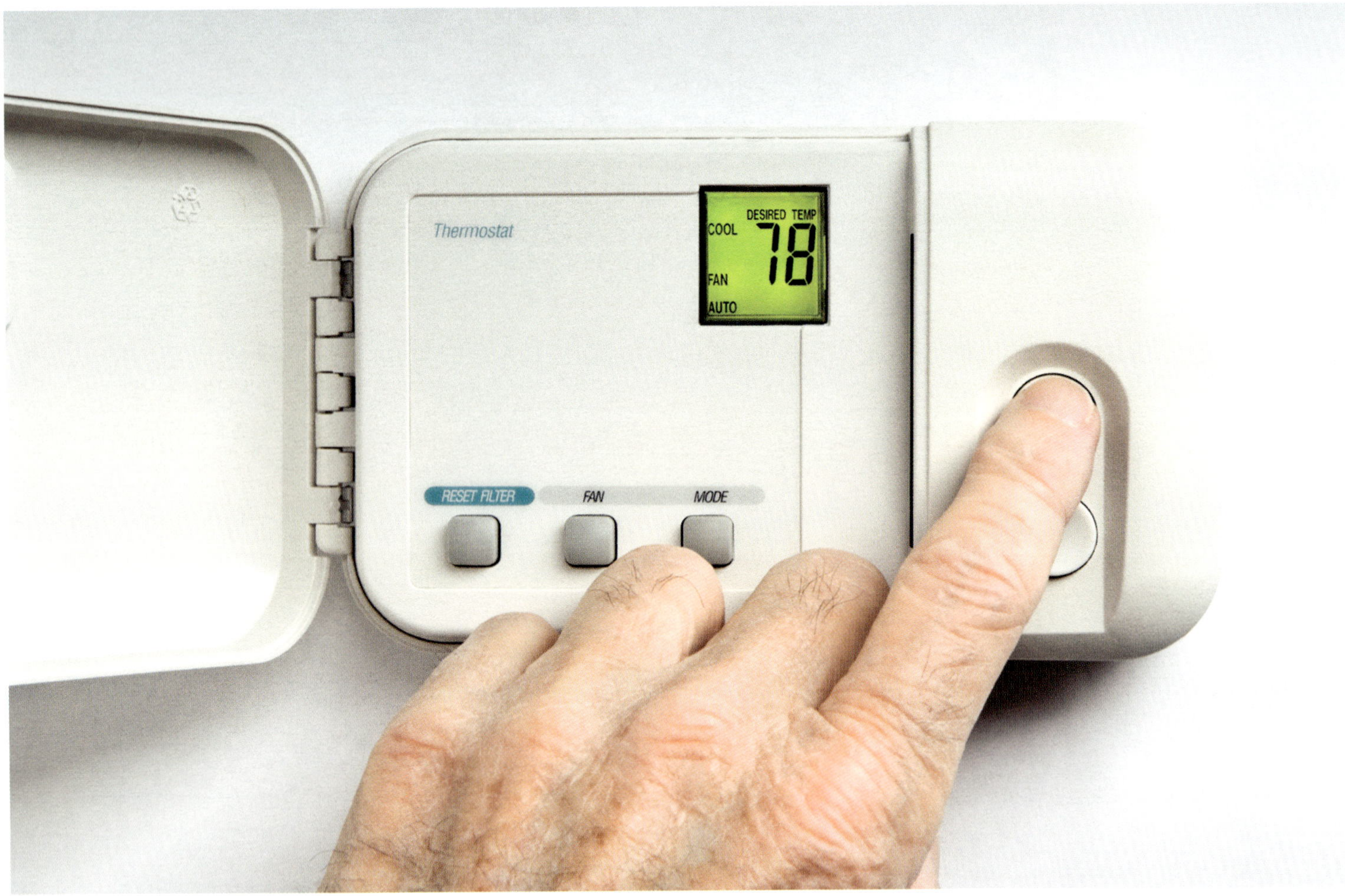

Fig. 14.1 Designing for comfort at energy-efficient temperatures

The key to low-energy cooling systems in hot and humid climates is to make sure the overall system can also dry and ventilate effectively even when there is no thermostat call for cooling. With dry air and adequate (but not excessive) ventilation, occupants will be comfortable without expensive overcooling.

Key Points

In hot and humid climates, the cooling and dehumidification loads are high during nearly every month of the year. Also, many architectural designs indulge in huge sheets of exterior glass, which create even higher cooling loads from solar heat gain. HVAC designers know these facts, of course. So a very common response is to "make sure the systems have plenty of cooling capacity."

Ironically, that well-intended, extra-powerful cooling capacity sometimes leads to the opposite result: cold, damp rooms which are seldom comfortable. Also, systems with "plenty of capacity" often use excessive amounts of energy to remove the normal cooling loads, which (by definition) are far below the building's peak loads for 99.6% of its life. For better cooling systems in hot and humid climates, the HVAC designer may wish to consider these suggestions and observations:

- Design so that dehumidification and ventilation can be accomplished any time they are needed—not just when a thermostat calls for cooling.
- Selecting cooling equipment "with lots of capacity" does not achieve dehumidification. In fact, it's usually the reason that humidity goes out of control.
- Adding "safety factors" when sizing equipment usually *creates* problems with comfort, energy and system maintenance instead of avoiding them.
- For better comfort and less energy consumption—measure, control and dry the ventilation air at all times.
- Focus carefully on the exterior glass—it controls both the sensible cooling loads and the comfort of nearby occupants. The glass controls the real-world thermostat set point and therefore the HVAC systems' energy consumption.
- Design air systems which are really air-tight.

Designing for independent dehumidification and ventilation

To understand why this advice is important, it's necessary to understand the limitations of the low-budget, more widely-applied solution: constant-volume, packaged cooling equipment for rooftops and for individual room air conditioning. These systems have many virtues, but unfortunately they often do a poor job of controlling humidity and providing economical ventilation in a hot and humid climate.

Such modern "packaged" cooling equipment is marvelously versatile and energy-efficient. In a constant-volume rooftop packaged unit, the designer can get components for cooling, dehumidification, ventilation, filtration and heating. The same features are sometimes combined in the packaged terminal air conditioning units (PTAC's) frequently seen in the rooms of hotels, dormitories, eldercare facilities and schools. Packaged equipment is both economical to install and energy-efficient with respect to sensible cooling. Sensible cooling is in fact its primary function, and is the only one of its functions which is controlled by the regulatory standards for manufacturers. That's the good news.

The bad news is that most such cooling-optimized and thermostatically-controlled cooling equipment does not provide any dehumidification until the thermostat calls for cooling—and it may provide very little dehumidification even during those short periods.

There are exceptions, of course. The range of packaged cooling equipment alternatives is very broad. These are the most widely-installed cooling systems in North America. But in many cases, packaged, constant-volume cooling equipment has been associated with mold and moisture problems in buildings in hot and humid climates.[1,2,3]

Here's the main problem; the ventilation humidity load is very high, *nearly all the time*. But the sensible cooling load varies widely over the day. In packaged equipment, cooling coils operate only for "short-cycles", which are separated by *long periods of non-cooling* during morning hours. Cooling cycles are even shorter and the period

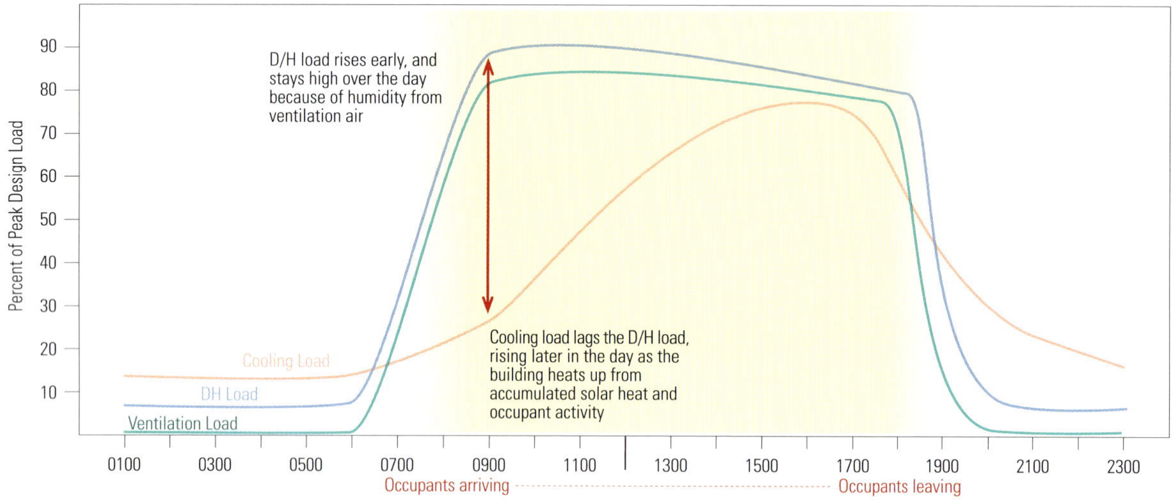

Fig. 14.2 Cooling, ventilation and DH loads peak at different times of the day

This graphic is an abstract representation of load profiles which are typical in an office building. Over the 24 hours of a typical day, the cooling, ventilation and dehumidification loads are seldom equal. That's why is wise to include equipment and controls which can measure and control each load ***independently*** *of the other two.*

between those cycles is even longer after the building is unoccupied and the thermostat is set back. Whenever the cooling coils are not cooling—such as when they are shut off for long periods between cooling cycles—the coils are not drying the air. Whenever the thermostat is satisfied, dehumidification stops and humidity rises.

Similar problems with high humidity can occur with constant-air-volume chilled water cooling systems.[4] Unless the cooling coils stay constantly cold (as would be typical of a variable air volume system), dehumidification performance can fall off during the morning, overnight and during weekends when sensible cooling loads are low. That's because when sensible loads are low, a constant-air-volume system slows down the flow of chilled water through the cooling coils, or to resets the chilled water temperature upward to save energy. Both of these strategies reduce or eliminate the system's ability to dry the air. The cooling coil is no longer cool enough to condense out enough humidity to keep the dew point low inside the building, especially when the ventilation air is adding so much humidity.

But these problems need not happen. They can be avoided when HVAC designers keep in mind that dehumidification (DH) and ventilation loads are *different* from the sensible cooling loads, and that they peak at *different* times of the day. Because these other two loads are so large and so persistent in hot and humid climates, *the HVAC system needs to be arranged to meet all three loads when they occur, not just when the thermostat senses a need for cooling.*

Figure 14.2 illustrates, for a generic office building, the typical pattern of the different times of the day when each of these three loads are likely to rise and fall. Overnight, when the building is not occupied the ventilation load is (or should be) nearly zero. Over the course of the day, the ventilation load should rise and fall with its occupancy. More people should mean more ventilation. And then when people leave, the system should provide less ventilation air.

The dehumidification load follows the ventilation load. The ventilation air is the largest component of the dehumidification load in most commercial and institutional buildings. So if the system is well-designed (so that it brings in only the amount of ventilation air required for number of current occupants, and to make up for any exhaust) the dehumidification load will rise and fall with the ventilation air load.

But note how the peak of the sensible cooling load comes at a very different time of the day compared to the peak ventilation and dehumidification loads. Early in the morning—before the sun has completely heated the building's roof and walls, and before the lights and occupants have generated heat inside the office, the sensible cooling loads are quite low. It's only later in the day that the sensible cooling loads reach their peak—after the building has heated up and when the sun shines through the western glass.

So the best HVAC system is one which can *separate* these three functions, allowing the air to be dried whenever it is needed, and allowing only enough ventilation air into the building to match the actual occupancy and the exhaust air loads. If the system's simpler equipment and controls only operate in response to a thermostat, then the system will be ventilating when it's not necessary, and not removing the peak dehumidification loads when they actually occur.

Figure 14.3 shows what happens when a cooling system—responding only to the temperature in the space—is also expected to provide dehumidification and ventilation. The picture shows the mold which grew on the walls of a classroom in Houston, TX during the summer vacation.[5] The system was configured to provide air circulation along with ventilation air continuously, rather than shut it off when the building was unoccupied. Also, the thermostat set point and the chilled water temperature were raised over the summer, to reduce energy consumption. Dehumidification only occurred during the very few (and very short) periods when the cooling coils called for chilled water. So the building filled up with humid ventilation air. Then when the cooling system operated, it chilled the walls. Their chilled surfaces absorbed enough moisture from the humid indoor air to grow mold.

Fig. 14.3 The results of a system which did not control ventilation and humidity independently of temperature

Mold grew over a summer vacation in this school near Houston, TX. The system could neither shut off ventilation nor control humidity independently of the operation of the cooling equipment. Dedicated components to vary the ventilation air flow and to dry the air would have prevented this problem.

If that system had been designed to *separate* the cooling, dehumidification and ventilation functions, and if it had *dedicated components and sensors to control those loads separately*, the mold problem could have been avoided. The overall control system could have *shut off* the ventilation as soon as the building was unoccupied, and it would have *dried out the air* instead of just chilling the walls occasionally.

There are as many ways to control and remove these three separate loads as there are creative manufacturers and creative HVAC designers. No one method will outperform the others in all circumstances. But the key point is that the overall system, no matter what types of equipment it eventually includes, should be capable of separately sensing the needs for dehumidification, ventilation and sensible cooling, and it should contain equipment which controls those three variables separately.

If additional direct evidence of the need is necessary, think about this issue as you travel in a hot and humid climate. The next time you try to sleep in a hotel room on damp sheets, with cold and clammy air which smells less than fresh and clean, you'll be reminded of the desirability of an HVAC system optimized for hot and humid climates—one which *separately* controls temperature, humidity and ventilation.

Extra cooling capacity does not provide dehumidification

One of the most difficult bad practices to dislodge from the HVAC industry is the habit of thinking that more cooling capacity will provide greater dehumidification effect. Usually, it does not.

In fact, the reverse is usually true. In most constant-volume DX cooling equipment, too much sensible cooling capacity (oversized equipment) is often the main reason for the system's *failure* to control humidity. Here's why.

When cooling equipment is oversized for the peak load, it has tremendous cooling power at normal loads, which means it can cool the space very quickly—so quickly that the cooling must be reduced or switched off entirely—before the air has been chilled low enough to remove any significant amount of humidity. Oversized DX cooling equipment cycles on and off quickly for very short periods to avoid overcooling the space. But it switches off *so soon*—and it *stays off so long* when the loads are lower than its capacity—that its cooling coils don't have enough time to actually remove much moisture. So humidity builds up *because* of that overly-powerful sensible cooling capacity.

This problem is counter-intuitive to most HVAC designers, who have been taught to think about dehumidification in terms of the equipment's "sensible heat ratio." That figure of merit often creates the misimpression that the cooling coil will always, instantly and continuously, remove a fixed amount of latent heat (water vapor) in addition to a fixed amount of sensible heat. But that only happens in the rating lab, and then only after steady-state load conditions have been achieved.

In an actual building, nothing is ever at steady state. The loads are never constant and certainly never as high as the equipment's capacity. So in constant-volume DX equipment, the cooling coil cycles on and off. And along with that cycling, the latent capacity—its dehumidification effectiveness—goes up and down. Here's the key point: any cooling equipment's actual sensible heat ratio over any given 60 minutes is not constant. The cataloged sensible heat ratio of a cooling coil does not reflect the real amount of water that coil will remove from the air during an hour when its cooling capacity is cycling on and off. Since oversized equipment is *always* cycling off quickly, its dehumidification effectiveness is *always much lower* than its sensible heat ratio would suggest. In fact, the dehumidification effect of oversized cooling equipment is often zero—the exact reverse of what the designer intended when he oversized it.

To illustrate how this problem happens, consider the graph shown in Figure 14.4. It shows laboratory tests of the latent capacity (dehumidification capacity) of a conventional constant-volume, 3-ton residential DX cooling system.[6] Note that as soon as the compressor turns on and begins to cool the coil, the latent capacity immediately rises—water vapor is indeed being removed from the air along with sensible heat. But then, after 10 minutes the compressor turns off because the thermostat is satisfied. Now, the water that had condensed onto the coil surfaces immediately begins to *re-evaporate*. The coil did not operate long enough for the water to flow down the fins and drip into the condensate pan. In other words, the wet (and no-longer-cooled) coil surfaces are now *adding* humidity to the air—not removing it. In those first ten minutes, the unit was a *de*humidifier. But over the next 26 minutes, the unit became a *humidifier*. And by the end of the hour, there was no net dehumidification effect. During that hour, the net dehumidification effect of that oversized system was *zero* pounds per hour.

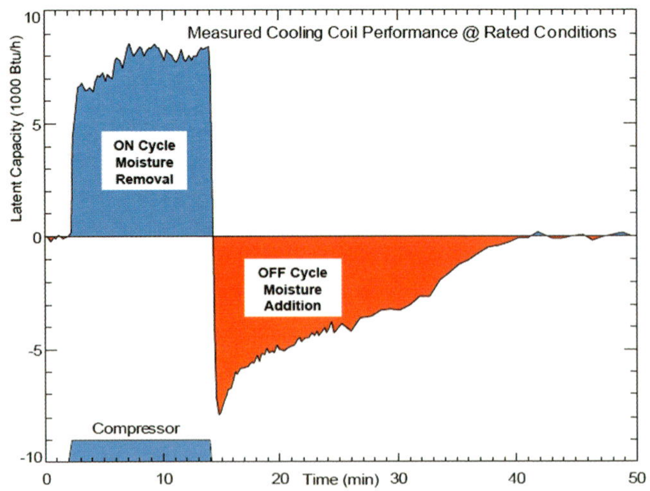

Fig. 14.4 Oversized cooling systems don't remove humidity continuously

Unless a cooling coil is continuously cold, it will not dry the air continuously. When DX equipment is oversized, there are long periods between cooling cycles. During those periods, no dehumidification occurs, no matter what the sensible heat ratio of the equipment might be. Oversized constant-volume DX cooling equipment is a very common cause of high indoor humidity in hot and humid climates.

So that's why the more the cooling equipment is oversized, the less effective it will probably be in keeping the dew point under control.

To be fair to cooling-based technology, there are *many* types of cooling-based equipment and systems that do an excellent job of controlling humidity. And there are many desiccant-based and hybrid equipment alternatives as well. But the *effective* alternatives will remove a defined weight of water vapor per hour, regardless of the sensible loads in the space. Manufacturers of such effective equipment will generally be able to provide defined DH performance, in lbs or kg removed per hour, based on the inlet air temperature, humidity ratio and flow volume specified by the designer.

As the old saying goes—you get what you measure. If a system is not designed to *measure* the indoor humidity, and if the equipment is not rated by its manufacturer to remove a *defined* number of lbs or kg per hour regardless of the sensible cooling load, it's not likely that the system will really be effective in controlling humidity year round.

Therefore, for dehumidification, avoid being distracted by the sensible heat ratio. Focus instead on the lbs or kg of water vapor that the manufacturer says the equipment will remove when the humidity loads are high and the sensible loads are low. In other words, at the peak outdoor *dew point* design conditions as opposed to when the outdoor temperature is at its peak. This is the approach described by ASHRAE Standard 62.1.[8] It's a much more effective design strategy than oversizing equipment which is optimized for sensible cooling.

Don't double-up the safety factors

"Safety factors" are a real problem in cooling system design. Adding extra loads to cover uncertainties usually results in poor cooling systems rather than good ones. Oversized systems make a building less comfortable and increase the risk of humidity problems, as described in the previous sections of this chapter, and as described in Chapter 2 (Improving Comfort) and Chapter 5 (Avoiding Bugs, Mold and Rot).

In spite of these facts, adding "safety" factors to cooling loads is such a powerful, primal urge among building owners and some HVAC designers that one might suspect it may be embedded in the human genetic code. It seems rather like the urge that sends thousands of lemmings over a cliff. Maybe they all really believe it's the best choice given the circumstances. And perhaps they take comfort from the fact (as they fall to the rocks below) that most of their peers have made the same decision.

Our suggestion is to avoid safety factors in cooling load calculations entirely. But if one must succumb to that urge, it's best to cover any uncertainties *after* the *known* loads have been totalled.

That way, it's easier to keep in mind how oversized the equipment is. The designer can then consciously design the system as a whole for a *great deal of capacity modulation*. Smooth modulation of capacity, or many discreet stages of capacity, can mitigate the comfort, humidity and energy problems of installing the oversized equipment—even if it won't mitigate such a system's oversized installed cost, high operating cost and its maintenance complexity.

To avoid doubling the safety factors, focus on some of the key decisions which must be made during load calculations. For example; what will be the maximum occupancy of an office building? There are all sorts of answers to that question, and the different estimates have very different effects on cooling, dehumidification and ventilation loads. The number of people governs the amount of ventilation air required by codes. So if you estimate that the occupancy will be the maximum allowed for safe egress by the fire safety code, the ventilation requirement will be huge, and so too will be the sensible cooling loads theoretically generated by those imaginary crowds.

Assuming that the maximum egress occupancy is the actual occupancy effectively adds a huge false load to the load calculation itself. Then after the total load comes to, for example, 195 tons, the designer might think "I better install a 220-ton chiller, to give the system 10% spare capacity." That means the system has been oversized twice; first by the estimate of the number of people in the building, which inflates its ventilation and air circulation requirements, and then again when 10% more excess capacity is piled on top of those inflated loads.

Another classic question is the sensible heat from "plug loads"—the lights, computers, fans, cell phone chargers and other appliances plugged into the electrical sockets of a typical office. In the past, designers have assumed that office plug loads total about 3 to 5 Watts/ft² of net floor space. In fact, ASHRAE research[9,10,11] has firmly established that office plug loads, when actually measured, are less than 1.2 W/ft², and often do not exceed 0.5 W/ft². So assuming 5.0 W/ft² instead of the typical value of 0.8 W/ft² grossly oversizes the cooling load during the calculation.

[In the past, designers have assumed that office plug loads total about 32 to 55 Watts/m² of net floor space. In fact, ASHRAE research[9,10,11] has firmly established that office plug loads, when actually measured, are less than 12 W/m², and often do not exceed 6 W/ft². So assuming 55 W/m² instead of the typical value of under 10 W/m² grossly oversizes the cooling load during the calculation.]

The result is even worse if the designer takes the nameplate power rating from office equipment. Again, ASHRAE Research has shown that the actual draw of office equipment is less than 50% of the nameplate power. And that's before the reality of use-diversity is applied. Not all of the appliances in the building will draw their full running power at the same moment. After use-diversity is applied, the maximum power consumption of typical office equipment is less than 25% of the combined nameplate load.[10]

Lastly there's the even more important issue of the total *running* load compared to the maximum *design* load.

The peak load calculation is based on those few hours when the internal loads, combined with the outdoor temperature and solar load is higher than it will be for the other 8,725 hours in a typical year.

Consider that time period. The equipment is selected at peak load, which will only occur for about 35 hours each year—if the load is not overinflated. If the load is inflated, then the calculated peak load will never occur, at all, ever. Then the system designed for that imaginary peak load will be very poorly matched to the loads that really do occur during normal operation.

The building becomes like the situation shown in Figure 14.5. Its cooling capacity is like a high-powered engine bolted onto a normal car. It's true that the car will never lack for engine capacity, even if the passengers should invite *all* their friends, neighbors and their favorite football team for a drive. But until that day, the engine will overpower the car. It will be tough to control for the short distances between stoplights, the passengers are likely to have a bumpy ride and the *net energy consumed* will be very wasteful for the actual distances travelled.

In buildings with oversized cooling systems, the typical running load may well be less than half of the peak load estimated when sizing the system. If the system has to modulate down to 50% capacity just to avoid overcooling during *normal* operating hours, it's not likely to be able to modulate smoothly (if at all) when the loads go still lower, as during the early morning and overnight.

Fig. 14.5 The result of "safety factors" applied to load calculations
Adding safety factors during cooling load calculations results in oversized equipment. The common result is a system which is indeed powerful. But that excessive power is difficult to control during normal operation. Also, an oversized system is difficult to fit into the available space. And its cost is so high that the other functions of dehumdification, ventilation control and filtration may not fit into the remaining budget. Wise cooling system designers avoid wasting the HVAC budget by adding safety factors during cooling load calculations.

So for all these reasons, it's best to avoid adding "safety factors" to the system's cooling capacity. But when the owner feels an irresistible compulsion to have excess capacity, at least don't double up the safety factors. Add any excess capacity at the *end* of a calculation which uses reasonable, real-world assumptions about the known loads, rather than overestimating those loads before the calculation is complete.

Measure, control and dry the ventilation air–at all times

Usually, cooling systems have no difficulty keeping a building cool. As discussed above, they often have so much extra sensible cooling capacity that any complaints are likely to be about feeling too cold rather than feeling too hot.

But humidity and ventilation are different matters. Keeping the dew point low enough for comfort and making sure the ventilation air is adequate (but not excessive) are more of a challenge. That's why this book has separate chapters focused on those challenges.

But it's also useful to keep in mind, when designing cooling systems, that ventilation and its dehumidification load will affect temperature control and thermal comfort if those loads are *not* removed. Here's how.

What usually happens is that the high ventilation rates required to maintain indoor air quality in densely-occupied spaces adds a large dehumidification load to the room's cooling system. If the system is not designed to remove that dehumidification load, the occupants' only choice to maintain thermal comfort is to drive the thermostat set point down. That's when such rooms get overcooled, or are overcooled by the building managers in anticipation of a comfort problem later, as the room fills up with people.

Figure 14.6 shows a recent newspaper clipping that shows how common this problem is with cooling systems in hotels and con-

Chilly rooms anger people at conferences, social events

Updated 8/6/2008 9:46 PM | Comments 63 |

Event attendees have found themselves shivering at meetings this summer despite the heat outdoors. Too-cold rooms have been the main complaint at conferences and other gatherings.

By Elizabeth Weise, USA TODAY

As summer-clad office workers, conference attendees and event guests stream into meeting rooms and reception halls nationwide, many will be met with a blast of Arctic air.

"Over-chilling" is one of the top complaints of meeting planners, says Donald Young of the International Facility Management Association in Houston.

Ezra Eichelberger, a catering professor at the Culinary Institute of America, calls it "a universal problem." He remembers the room was so cold for one banquet that women were given tablecloths to wrap around themselves.

Meeting planner Steve Kemble of Dallas sees it all the time in the surveys that conference attendees mail in. "If two or three weeks later they're *still* thinking about the temperature, then that's an issue." He hates cold rooms. It's so frustrating "to plan a fabulous program, with a speaker you paid $100,000 for, and all you hear from people leaving is 'Brrr, it was freezing in there!' "

Folks in charge of thermostats say this: You have to pre-chill so you don't overheat.

"We always make it as cold as possible prior to the event starting, because it's a better experience," says Eric Whitson, director of sales at the National Conference Center in Lansdowne, Va. That means about 65 degrees. "At first its going to be chilly to most people who enter, but the temperature goes up as those numbers build." It should end up at about 70 degrees, he says.

The engineers weigh in

There's only one problem with this logic, says Kent Peterson, past president of the American Society of Heating, Refrigerating and Air-Conditioning Engineers: It's flat-out wrong. The whole idea of getting a room cold before the people come flies in the face of ASHRAE's 114 years of experience, Peterson says.

If you have to pre-cool a room, it's because something's seriously out of whack with the air-conditioning system, he says. "You should start out at 74 degrees, and as the temperature starts to rise up to 78, the air conditioning system will turn on and keep the temperature down."

And no excuses about the addition of all those heat-producing bodies. That should have been taken into consideration when the system was designed, he says. "Cooling load calculations" are based on how many people will be in the space and what activities they'll be engaged in, whether it's dancing or sitting. "Our industry is very good at doing these calculations," he says.

Peterson says many rooms are chilled to dehumidify the air. "Instead of implementing an energy-efficient design that has a dedicated system to dehumidify the outside air, they're running their air-conditioning system at full load in order to reduce the amount of moisture."

Fig. 14.6

Failure to control and dry ventilation air

Controlling both the volume and the dew point of the ventilation air allows the thermostat setting to remain at comfortable levels, rather than being forced down to remove humidity.

ference centers. Cold conference rooms turns out to be *the number one complaint* of conference planners.[12]

The problem is perceived as one of temperature control. But it actually originates with high dehumidification loads. In densely-occupied spaces like conference rooms and classrooms, high DH loads are generated by people in the room, combined with the DH loads carried into the room by the ventilation air needed to dilute the contaminants they generate.

In designing the cooling system, there are two approaches to avoiding this common problem. One can either remove the DH loads by drying the ventilation air below the desired room dew point *before* the ventilation air enters the space, or the designer can remove both the people load and the ventilation DH load *after* those loads mix into air in the space.

Pre-drying the ventilation air is the more effective alternative. This can be done with either a dedicated ventilation system or by providing dry air from a central source such as a variable air volume system. By relieving the room's cooling system of those DH loads, the equipment can be sized for just the sensible cooling load, which usually reduces its installed costs. In conference rooms, which tend to be interior rooms without many windows, the sensible loads may be very low. And of course the room occupancy is seldom at its maximum. So a smaller sensible cooling system is likely to provide better comfort for more of the time than a large system, which might frequently overcool the room.

Dry ventilation air from another system can be metered into the room in response to the room's actual occupancy. The ventilation air volume can be controlled either by using time clocks or schedules in the building automation system for regularly-defined occupancies, or controlled by CO_2 sensors for less predictable occupancies. Adjusting the ventilation air volume to fit the occupancy saves energy without compromising the indoor air quality.

If the ventilation air is not pre-dried, then the designer will need to specify cooling equipment that also has a large (and well-defined) dehumidification capability. The loads in a conference room or classroom are mostly DH loads rather than sensible cooling loads. As discussed earlier in this chapter, it's best to consult with the manufacturer to specify the lbs or kg of water vapor that the cooling system will need to remove every hour, rather than assuming a DH capacity based on catalog values of the equipment's steady-state sensible heat ratio. Manufacturers which provide dehumidifiers (or cooling equipment with DH capacity) will be able to provide the number of lbs or kg of water vapor the equipment will remove when there is no sensible heat load in the space.

Thermostat Settings Which Provide Equal Comfort
(Within 3 ft. [1m] of a glass window wall)

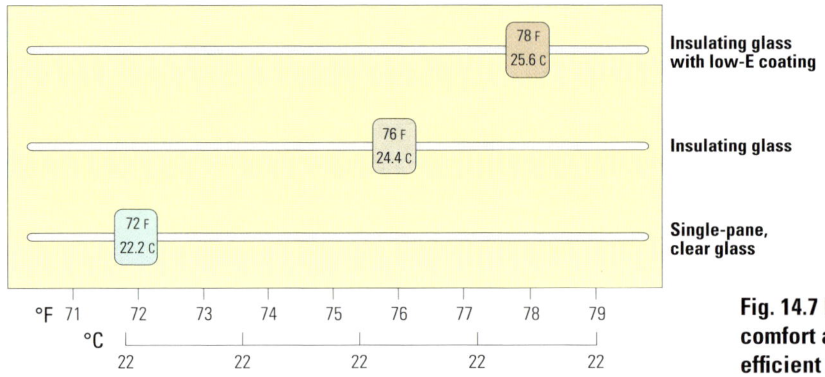

Assumptions:
- Seated occupants
- CLO = 1.0
- Air velocity less than 40 fpm [0.2 m/s]
- Solar insolation: 240 Btu/ft2/h [757 W/m]

Fig. 14.7 Better glass provides comfort at higher, more energy-efficient thermostat settings

Low-cost glazing allows so much radiant heat into the building that occupants near the windows will demand lower thermostat settings. Any comfort problems caused by the glass selection is usually blamed on the HVAC system. So the wise cooling system designer will become informed about modern glass alternatives, and will participate actively in the early glass decisions.

Focus carefully on the exterior glass–it often sets the sensible cooling loads

In hot and humid climates, the enclosure loads—the solar heat gain through windows and the cooling loads through the roof and walls—often strongly dominate the sensible cooling loads. That is not true in all occupancies, of course. In data centers, the internally-generated cooling loads from computers or telecom equipment will be greater than the enclosure loads.

But for nearly all other occupancies, it's the exterior glass which really generates the largest loads.[13] In particular, note the importance of the glass on the westerly exposures of the building. That glazing lets in heat at the end of the day, after the accumulated cooling loads from other sources have reached also their maximum.

This fact is covered in Chapter 12 (Estimating Cooling Loads). But it also has important implications for the design of the system itself. If the glass has a high solar heat gain coefficient (SHGC), or if its surface has a high thermal emissivity, then the occupants near that glass are likely to need more cooling than occupants in other parts of the building because they will be feeling the high radiant heat from the glass. Figure 14.7 shows the thermostat adjustments that would be necessary to maintain constant thermal comfort with rising amounts of radiant heat from nearby windows, when all other variables are held constant (e.g.: clothing coverage and air velocity over the skin).[14]

What this means is that the amount of cooling the designer needs *near the widows* depends heavily on the indoor surface temperature of that glass, and its emissivity. The higher its temperature and its emissivity, the more cooling will need to be supplied to the nearby occupants, independent of the dimension of the other sensible cooling loads in that same space. In other words, near poorly-performing exterior glass, the occupants might need low temperature air or more of it, at a time when occupants just a bit further away from that glass may be feeling much too cold! Poor glazing is one of the many reasons for "thermostat wars" between occupants of the same space.

To avoid this common problem (which inevitably gets blamed on the AC system rather than the architect's glass decisions) it's important for the HVAC designer to have a clear understanding of what glass will really be installed in the building. In particular, what will be the inside surface temperature and the surface emissivity of the glass selected by the architect? The glazing supplier will be able to provide these numbers.

The glass decisions often change mid-way through design documents or after initial bids, when the owner and architect really realize that high-performing glass is not cheap. At that awful moment when the glass goes over budget, snap decisions can make it very difficult for the HVAC designer to recover comfort by redesigning the system and resizing the equipment.

So the wise cooling system designer will:

- Make it his business to become very knowledgeable about window design and glass alternatives.[15,16,17]
- Participate fully in the very early conferences which define the size, type, shading and exact location of the exterior glazing.
- Keep track of any changes in the glazing specification and the window design which results in a higher indoor glass surface temperature or a higher thermal emissivity.
- Make sure the owner and architectural designer understand that when glazing specification are *changed*, they will need advice and perhaps HVAC redesign or rebudgeting to avoid thermal discomfort near windows—such as in places like those attractive offices on the corners of glass buildings which are occupied by the Senior Executives who may have paid for the building's construction. It's always a bad day when the Chairman of the Board is uncomfortable.

Design air systems which are really air-tight

All air conditioning systems move air either through air handling equipment or through ducts, or through both. It's very important, when designing a cooling system for a hot and humid climate, to make sure that *all* the connections, seams and joints in air handling equipment and duct work are sealed up, *air tight*.

Air-tight systems and equipment save energy, provide better comfort and reduce mold risk compared to systems and equipment which leak air. Leaky equipment and duct connections are responsible for about a 30% increase in annual energy consumption, and a greatly-increased mold risk, as explained in detail in Chapter 2 (Reducing Energy Consumption) and Chapter 5 (Avoiding Bugs, Mold & Rot).

These facts may be familiar to many HVAC designers. What may not be as well-known is the importance of sealing up:

- Any return air or supply air plenums, especially where the walls and ceilings or the walls and floors of such plenums meet.
- Any vertical plumbing chases or similar building cavities used as return air ducts, supply air ducts or toilet exhaust ducts.
- All exhaust air duct connections—all the way from the exhaust air grill to where the air leaves the building.
- All connections of return air grills, supply air diffusers and exhaust air grills *to the wall surfaces* they penetrate in addition to the ducts they connect to.
- The joints where packaged rooftop units rest on the roof curbs which support them.

Chapter 16 will discuss air tightness in more detail. But air tightness is also very important to keep in mind during the process of designing cooling systems. Air that leaks out of the system means the system's cooling capacity is wasted, and it means that more air will have to be cooled and moved through that system, to make up for the waste.[18,19]

On the return air side of your system, inward leakage from the outdoors is common. It means the system will ultimately be pulling untreated outdoor air into the building from the weather, greatly increasing the cooling loads while also adding moisture to the building materials to support mold growth.[20] The same suction problems happen with leaking exhaust air ducts.

So for *all* types of ducts and building cavities which carry air, it's important for the HVAC system designer to specify that the connections must be sealed up using mastic (as opposed to tape, which does not last long in a hot and humid environment).

That modest investment in mastic covering all air-side joints and longitudinal seams pays off with about a 30% energy savings over a typical year.[18,19] The investment also makes the building much more resistant to mold problems. Sealing duct connections only adds between 3% and 5% to the cost of the sheet metal portion of the installation contract.

The careful designer could also specify that the supply, return and exhaust duct systems must be leak-tested, and that the amount of air leakage must be recorded after all air-side components have been installed. That would be a powerful incentive to better installation of the sealant around the duct connections. But such testing will cost far more than the cost of sealing itself. So measured tests might be most applicable to higher-quality buildings, or those with strict air separation requirements, such as hospitals and medical facilities. For most commercial occupancies, perhaps the photo-documented application of mastic to each joint would be adequate to ensure that leakage has been minimized. Lots of photos, to be sure. But such a specification requirement would get the point across to the installers that the designer is serious about air sealing, without the cost of a complete test.

Cautions for buildings with operable windows

In many countries around the world as well as in some code jurisdictions in North America and Europe, ventilation is not a function of the mechanical system because ventilation air is not required when operable windows are provided. The building occupants simply open the windows when they feel the need for outdoor air. This is quite common in residential occupancies such as hotels and resorts, high-rise apartments and condominiums, or military and university dormitories, and even in some eldercare facilities and hospital patient rooms.

For example, a common pattern in South Asia and the Middle East is to cool the building with wall-mounted "mini-split" or "multi-split" DX cooling equipment.[21, 22] This type of equipment sometimes has very little provision for ventilation air. Often, occupants prefer to

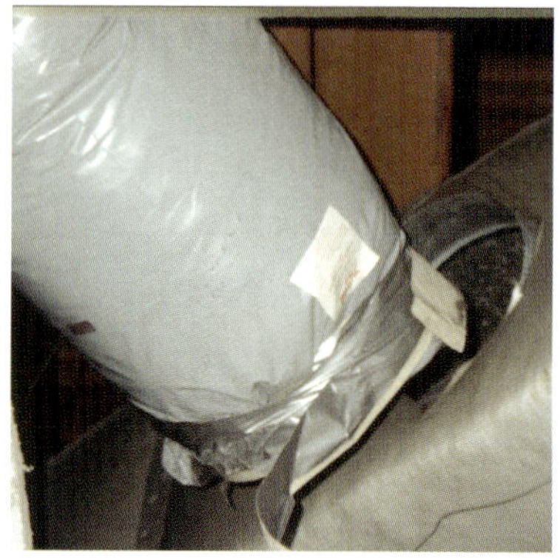

Fig. 14.8 Duct tape vs. mechanical fastening and mastic
The upper photo shows an example of why duct tape over slip joints is not an effective sealing system over time in hot and humid climates. To make cooling systems effective and energy-efficient both at installation and over time, specify that all duct connections—especially to air handlers—be mechanically fastened and then sealed air-tight using mastic and reinforcing tape, as shown in the photo above.

open windows to provide both cross-flow ventilation and additional cooling during the evening and night time hours.

The problem is that, as discussed in more depth in Chapter 11 (Estimating Dehumidification Loads), the early evening is just when the dew point of the outdoor air is likely to be at it's peak. The humid outdoor air flows into the building through the open windows, where it meets surfaces of walls, ceilings and building cavities which have been chilled down below the outdoor air's dew point. The cool surfaces then absorb humidity from the incoming humid outdoor air. Some surfaces will even be cold enough to produce condensation and dripping water. That moisture supports mold growth, especially in closets and closed building cavities where the dry air from the AC equipment may not reach later, to dry out the surfaces. Moisture absorption and condensation are what lead to disastrous mold growth.

There are no easy, low-cost solutions to this problem. But it can be mitigated in two steps, depending on the available budget. First, the designer can interlock the thermostat which controls the AC unit with window-actuated switches. When any window opens, it will turn off the AC unit. That way, at least the AC units won't continue to chill the building when they have no hope of drying the incoming outdoor air.

Next, for a much more reliable solution, add dry ventilation air to each room with a dedicated variable air volume system for ventilation air only. The supply of dry ventilation air to each room is interlocked with the window sash, just like the thermostat. In other words, the large sensible cooling loads are removed by the room cooling units on demand, as long as the windows are closed. And whenever the windows are closed, the rooms are constantly flushed with a small amount of dry ventilation air. This lets the walls, ceilings and furnishings dry out whenever the windows are closed, even if the AC unit is turned off to save energy when the room is unoccupied.

Cautions for comfort in hot and humid climates

Designers who do not live in hot and humid climates themselves may not fully appreciate that to many people, typical North American indoor temperatures of 72° to 75°F [22 to 24°C] are very uncomfortable, especially in residential occupancies.[23, 24, 25] A western-dressed office worker might indeed prefer cold temperatures. But when that same person goes home at night, he or she often sheds formal business attire and dresses in more climate-appropriate clothing. In many cultures, even in offices there will be a strong preference for increasing air movement with slow-rotating fans rather than dropping the air temperature to typical North American levels. These facts are discussed in more depth in Chapter 2 (Improving Comfort).

The designer of cooling systems for hot and humid climates should keep in mind there is often a strong preference for warm-but-dry conditions adjusted by personal fans—as opposed to the traditional brute-force North American approach of "adding some extra tons and some extra supply air to make sure it *never* gets too hot." Design the cooling system so it does not overcool the space when the occupants prefer a warmer, drier environment. This will save energy, in addition to pleasing the occupants. Just make sure the overall HVAC system can keep the building below a 55°F dew point [12.8°C] *without* overcooling the spaces, as discussed often in this chapter and in Chapter 13 (Designing Dehumidification Systems).

References

1. Gatley, Donald P., *Mold and condensation behind vinyl wall covering*. 1990. Gatley & Associates, Atlanta, GA.
2. Harriman, Lewis G, III and Thurston, Steven, *Mold in Hotels and Motels—Survey Results.* 1991. American Hotel & Lodging Association. Washington, DC.
3. Shakun, Wallace. "A review of water migration at selected Florida hotel/motel sites." Proceedings of the biennial symposium on improving building practices in hot & humid climates. October 1990. Texas A&M University, College Station, TX.
4. Murphy, John and Bradley, Brenda. *Dehumidification in HVAC Systems*. 2002. Trane Commercial Systems Division, Ingersol-Rand, Inc. LaCrosse, WI. Trane Applications Engineering Manual SYS-APM004-EN
5. McMillan, Hugh and Block, Jim. "Lesson in curing mold problems." *ASHRAE Journal,* May 2005. pp 32-37.
6. Shirey, Don B. III and Henderson, Hugh. 2004. "Dehumidification at part-load." ASHRAE Journal, April, 2004. pp. 42-47.
7. Table 1. Columns 12a, b, c. "Climatic Design Information." Chapter 28, ASHRAE Handbook—Fundamentals 2005. ASHRAE, Atlanta, GA.
8. ASHRAE Standard 62.1-2007 (Ventilation for Acceptable Indoor Air Quality) ASHRAE, Atlanta, GA www.ashrae.org and...

   62.1 User's Manual ASHRAE/ANSI Standard 62.1 (Ventilation for Acceptable Indoor Air Quality) 2005 ASHRAE, Atlanta, GA www.ashrae.org ISBN 1-93862-80-X
9. Komor, Paul. 1997. Space cooling demands from office plug loads. ASHRAE Journal, December 1997. pp.41-44. www.ashrae.org
10. Wilkins, Christopher and Hosni, M.H., 2000. "Heat gain from office equipment." ASHRAE Journal, June 2000 pp.33-39 www.ASHRAE.org
11. Pratt, Robert G, "Errors in audit predictions of commercial lighting and equipment loads and their impacts on heating and cooling load estimates." ASHRAE Transactions, 1990. AT-90-11-2 pp.994-1003 www.ashrae.org
12. Weise, Elizabeth. 2008. "Chilly rooms anger people at conferences, social events." USA TODAY, August 6th, 2008. Gannett Publishing, Arlington, VA
13. Huang, Joe & Franconi, Ellen. *Commercial Heating and Cooling Component Loads Analysis* 1999. Report LBL-37208 Building Technologies Department, Lawrence Berkeley National Laboratory. Berkeley, CA 94720
14. Larson, James. 2008. "Thermostat correction factors for constant thermal comfort as window surface temperature varies." Spreadsheet file. Cardinal Corporation, Eden Prairie, MN. www.cardinalcorp.com
15. McGowan, Alex. 2008. Introduction to green window design and performance. Journal of Green Building, Volume 3, Number 2, Spring 2008 pp.3-12 www.collegepublishingus/journal.htm
16. O'Connor, Jennifer, Lee, E. Rubenstein, F. & Selkowitz, Stephen, *Tips for Daylighting with Windows - The Integrated Approach* Report no. LBNL-39945 1997. Building Technologies Program. E.O. Lawrence Berkeley National Laboratory, Berkeley, CA

17. Carmody, John; Selkowitz, Steven; Lee, Eleanor; Arasteh, Dariush and Wilmert, Todd. *Window Systems for High Performance Buildings* 2004. Norton & Company, 500 5th Avenue, New York, NY. 10110 ISBN 0-393-73121-9

18. Cummings, James B., Withers, C. R. Withers, N. Moyer et al. 1996. *Uncontrolled air flow in non-residential buildings.* Final report. FSEC-CR-878-96. April 15th, 1996. Florida Solar Energy Center, Cocoa, FL

19. Wray, Craig. *Energy impacts of leakage in thermal distribution systems.* 2006. Report to the California Energy Commission. Lawrence Berkeley National Laboratory. Berkeley, CA. http://epb.lbl.gov/ Report no: PIER II #500-98-026

20. Harriman, Lewis, Lstiburek, Joseph and Kittler, Reinhold. 2000. "Improving humidity control in commercial buildings." ASHRAE Journal, November 2000. pp. 24-30.

21. Sekhar, S. Chandra. 2004. "Enhancement of ventilation performance of a residential split-system air-conditioning unit." *Energy and Buildings.* 36 (2004) 273-279. Elsevier B.V., www.elsevier.com/locate/enbuild

22. Sekhar, S.Chandra and Lim, A.H., 2003. "Indoor air quality and energy issues of refrigerant modulating air-conditioning systems in the tropics" *Building and Environment* 38 (2003) 815-825 Pergamon Press, www.elsevier.com/locate/buildenv

23. Sekhar, S. Chandra, K.W. Tham and K.W. Cheong. 2003. "Indoor air quality and energy performance of air-conditioned office buildings in Singapore" *Indoor Air* 13 (2003) 315-331 Blackwell Munksgaard, www.blackwellpublishing.com/ina

24. Jitkhajornwanich, Kitchai et al. 1998. "Thermal comfort in transitional spaces in the cool season of Bangkok" ASHRAE Transactions, Volume 104, Part 1.

25. Chan, Daniel, et al. "A large-scale survey of thermal comfort in offices in Hong Kong" ASHRAE Transactions, Vol 104, Part 1.

Chapter 15

Designing Ventilation Air Systems

By Lew Harriman

Fig. 15.1 Ventilation air is a precious resource

Note the size of the outdoor air intake, and also the size of the air handler and utilities needed to clean and dry that ventilation air. The HVAC designer can avoid massive costs and energy waste by **measuring and controlling** *the ventilation air flow, instead of simply assuming that, if enough air is injected into the building, it will somehow end up in the right places, at the right times.*

Key Points

ASHRAE recommends that buildings be ventilated with outdoor air to dilute the concentration of indoor pollutants, so that occupants can be comfortable. It follows that the principal tasks for a designer, installer and operator of a ventilation air system are to ensure that:

- The appropriate amount of outdoor air reaches the breathing zones of occupants (rather than sending ventilation air to other places, and rather than ventilating where no people are present).
- When ventilation air is breathed in by occupants, it must be clean, dry and free from pollutants.

To achieve these goals, designers and owners of buildings might wish to follow these suggestions:

1. Size the ventilation air flows and design the overall system with the assistance of ASHRAE Standard 62.1 - Ventilation for Acceptable Indoor Air Quality.[1] For detailed assistance, the 62.1 User's Manual[2] expands and explains the logic behind the standard's provisions.
2. Dry all of the incoming ventilation air below a 55°F dew point [12.8°C dew point], whenever the outdoor air dew point is above that level. This provides better comfort, and avoids condensation and potential mold growth.
3. Install a MERV-8 filter on any incoming outdoor air stream to remove excess particulate. If the building is located in an area with high levels of ozone and particulate, install a MERV-11 filter to remove particulates, and also install a carbon filter or similar air cleaner to reduce ozone.
4. To reduce energy costs and to provide better air quality in spaces where occupancy varies, modulate the amount of ventilation air using either time clocks which change the flows according to anticipated changes in occupancy, or by CO_2 or motion sensors which estimate the actual occupancy in each space.
5. Recover waste energy to reduce energy consumption, especially whenever the ventilation air flows are over 5,000 cfm [2355 l/s].

Ventilation Costs Money, So It's Controversial

Conditioning ventilation air properly is expensive, especially in hot and humid climates. It costs money from the owner's operating budget to clean outdoor air, dry it, cool it and push it into the breathing zone. And the fans, filters, dehumidifier, duct system and controls which deliver outdoor air require money from his construction budget.

So it's understandable that the amount and quality of ventilation air needed to provide acceptable indoor air quality is controversial. In fact, the question of exactly how much ventilation air should be provided for buildings may be the most enduring controversy in HVAC engineering.

In January 1895, during the very first meeting of what is now ASHRAE, Edward Bates, the Society's first President declared: "Ventilation comes next to Godliness. Every family has a right to an abundance of good fresh air, even if it is not aware of its rights. I hearby suggest that this be one of the first problems which we handle."

At its next meeting in January of 1896, the Society reported that: "A very little work sufficed to show the committee that there was a wide divergence of opinion among authorities as to the proper amount of air to be supplied per minute per person to crowded rooms." The Society went on to recommend that buildings occupied principally by those under 15 years of age be ventilated at 30 cfm/person; buildings occupied by persons over 15 years of age be ventilated at 33 cfm/person and buildings illuminated by open gas flames should be ventilated at 50 cfm/person. [14.1 l/s/person, 15.6 l/s/person and 23.6 l/s/person].[3]

These amounts are similar to recommendations from some research, even today. But other research suggests considerably lower or higher values per person. Somewhat surprisingly, more than a century later, our understanding of the health and human perception issues

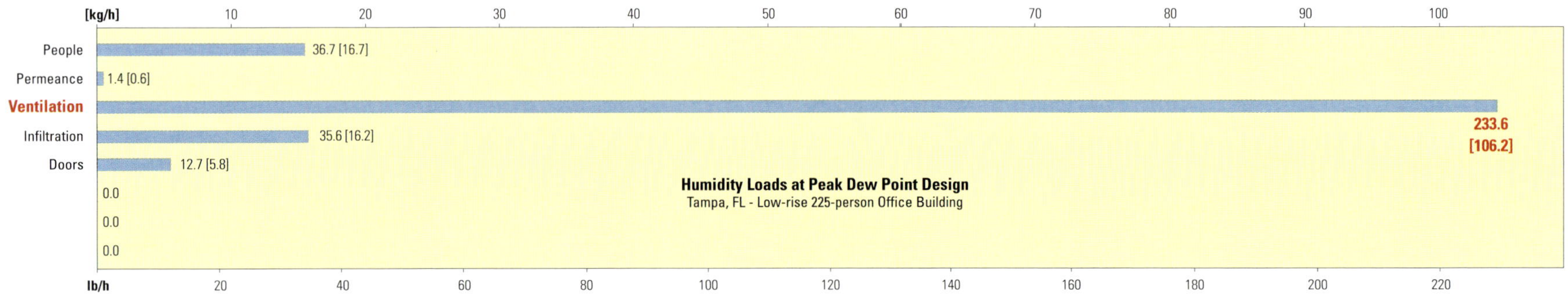

has not advanced to universally-accepted conclusions. There are still ardent debates among researchers, practitioners, occupants, government regulators and building owners about the amount of ventilation air which is appropriate for different occupancies.

The current recommendations of ASHRAE Standard 62.1 (Ventilation for Acceptable Indoor Air Quality) are much more elaborate and detailed than the Society's first recommendations of 1896. Also, the recommendations are based on a rigorous consensus process. But the enduring uncertainty among experts regarding the appropriate quantities for different occupancies, and the means by which these are calculated and delivered, all seem to suggest that much remains to be understood about "a good ventilation system." One hopes that, before another 100 years have elapsed, we will have conducted the research needed to provide simple, actionable and unambiguous guidance for means of controlling, verifying and delivering ventilation air to the breathing zone, and that we will have verifiable standards for the purity and humidity of that air—standards which can be firmly supported by a broad consensus of real-world occupant experiences and perceptions as well as by laboratory research.

In the mean time, one must recognize that ventilation air quantities, delivery systems and levels of purity and humidity remain the subject of some debate. This chapter will provide a summary of key issues, and will make suggestions to assist ventilation decisions in hot and humid climates.

Ventilation Dehumidification and Air Cleaning

It's not enough to simply bring outdoor air into a building to dilute indoor pollutants. Ventilation air offers a risk as well as a benefit to buildings and its occupants. Incoming air must be cleaned and dried. Otherwise—especially in a hot and humid climate—ventilation air can make the indoor air quality worse rather than better, and it can even damage the building.

Drying ventilation air

Ventilation air generates the building's largest single humidity load, by far. In most building types, the ventilation air accounts for more than 60% of the total peak humidity load (the latent heat load). Figure 15.2 shows estimated peak humidity loads for a small office building if it were located in Tampa, Florida. Note that ventilation air accounts for more than 73% of the total humidity load.

More importantly, unlike cool or mixed climates, in hot and humid climates the ventilation humidity load *stays high* nearly all year long. Figure 15.3 shows hourly weather observations for a typical year in Tampa. Note that even during the "dry winter months", the outdoor humidity is nearly always above a desirable level for an indoor space (below a 55° dew point [12.8°C dew point]).

The fact that the ventilation humidity load is both high and nearly continuous is the first of several good reasons to dedicate a separate air handling system to drying the ventilation air. In recent years, the

Fig. 15.2 - Ventilation is the largest humidity load

The graph shows humidity loads for a 3-story, 225-person office building located in Tampa, FL. Note that ventilation accounts for more than 73% of the total load. That's why it's so important to dry ventilation air in a hot and humid climate—otherwise, indoor humidity stays high enough to allow mold and bacteria to grow near or on damp surfaces.

Indoor vs. Outdoor Air Dew Points
In hot and humid humid climates, the outdoor air dew point can be very high even during "winter." So the system must dry the ventilation air during **all** seasons.

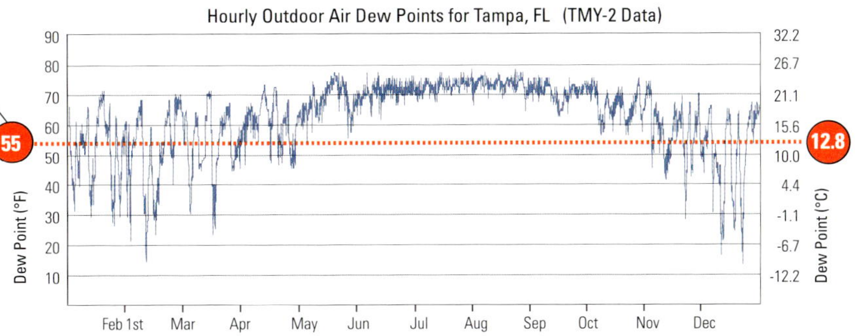

Fig. 15.3 - High humidity all year long
This graph shows hourly dew points for a typical meteorological year (TMY-2 data). Dehumidification loads occur all year long in hot and humid climates like Tampa, FL.

costly problems caused by indoor moisture and high indoor humidity have made the dedicated ventilation dehumidification system a popular solution.

In the United States, for example, since 2003 all new federal buildings have been required to have separate, dedicated ventilation dehumidification systems to keep excessively humid air entirely out of the building. The Federal Facility Standard[4] requires drying all ventilation air below a 50°F dew point [10°C dew point] at all times. This level of dryness allows the ventilation air to "act as a sponge," absorbing the humidity loads generated by office workers inside the building and therefore keeping the indoor dew point below 55°F [12.8°C].

If the designer elects not to install dedicated ventilation dehumidification systems, humidity control becomes a task for the cooling system. When that is the designer's choice, the cooling system must be designed very differently from most cooling systems—it must have an effective dehumidification component that will *dry the air even when that air does not need cooling*. The designer must take special care to question the cooling equipment suppliers regarding the equipment's measured dehumidification capacity at "part-load" conditions—its dehumidification performance when the outdoor environment is at the peak outdoor dew point, not when it is operating at the peak outdoor dry bulb temperature.

The classic problem with most constant-volume cooling equipment is that when cooling loads are low, the equipment operates for such short periods or at such relatively high coil-leaving temperatures that the system has nearly zero dehumidification effect. The humidity load from ventilation remains high, so occupants feel cold and clammy as the humidity load builds up in the space. This leads to discomfort for most of the operating hours of the year and exposes the building to moisture accumulation in large and risky amounts.

Without dedicated ventilation dehumidification, the designer will need to install some form of effective dehumidification equipment at some other place in the system. Most importantly, the designer must make sure that equipment *will really remove the humidity load* anytime the dew point of the mixed ventilation and return air is above the dew point desired in the space. Under those circumstances, one

1. Outdoor Air
Hot and humid, it must be cleaned and dried.

2. Ventilation Dehumidifier
Dries the incoming air to a condition below the desired humidity dew point, removing the internally-generated humidity loads

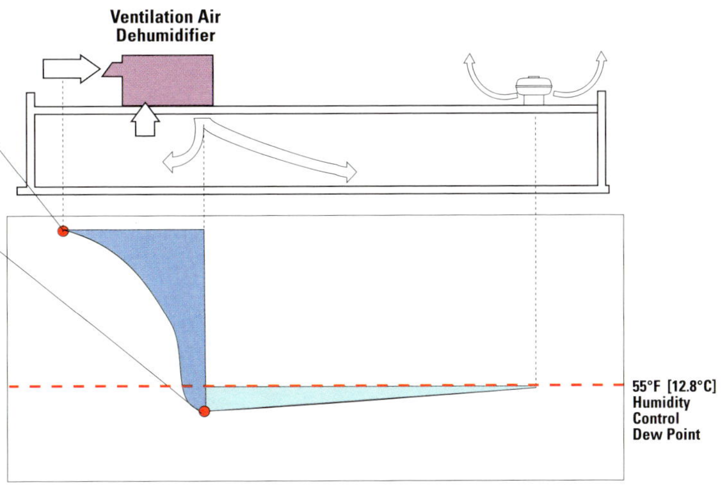

Fig. 15.4 - Ventilation Dehumidification
The ventilation air can be dried in many ways—but it must be dried. For that, the system will need a dehumidification component controlled by the indoor dew point signal, not the room temperature. This diagram shows one approach—a dedicated ventilation dehumidification system, sometimes call a dedicated outdoor air system (DOAS).

Fig. 15.5 Dedicated outdoor air system (DOAS)
The photo shows an example of a desiccant-based dedicated outdoor air system designed to dry the ventilation air deeply, so that it can absorb part of the internal dehumidification loads as shown in figure 15.4.

good alternative is to use a variable-air-volume (VAV) cooling system. These systems cool the supply air to a low temperature and a low dew point constantly, and vary only the amount of cold air supplied to each space as sensible heat loads change.

By constantly cooling the supply air to a low dew point, the variable air volume system dries the combined return and ventilation air without the need to provide a separate system for drying the incoming humid ventilation air. But because the coil-leaving temperature is so low (to ensure dehumidification), the challenge for designers is to reduce the flow low enough, or to reheat the supply air warm enough so it does not overcool the spaces which have low cooling loads. For example, early in the morning all the cooling loads are still low—but the air flow from a VAV system will have to stay high enough to provide adequate ventilation and adequate dehumidification for each space. In that situation, reheat may be necessary.

To reduce reheat costs, and to comply with energy codes and the provisions of ASHRAE Std 90.1 (Energy Standard for Buildings), first reduce air flows to the minimum needed for adequate ventilation (often, 30% of full flow still provides enough ventilation air in the mix). After flow reduction, the reheat energy should come from either waste heat or from renewable energy, such as condenser heat or solar-heated hot water. It's also very important not to give in to the temptation to "save energy" by resetting the cooling coil to a higher leaving-air temperature—one which would not be cold enough to remove the humidity loads from the ventilation air and from internal sources. Raising the VAV coil-leaving air temperature to levels above 55°F [above 12.8°C] may indeed save energy—but when outdoor air is humid, those savings are achieved at the expense of discomfort for occupants and at the risk of condensation, corrosion and mold growth in the building enclosure.

The importance of effective dehumidification of ventilation air cannot be overemphasized for buildings in hot and humid climates.

If the designer chooses not to dry the ventilation air before it enters the system, then he must somehow ensure that the internal system components will really remove the humidity from the mixture of ventilation and return air, even when the thermostats are not calling for cooling. Ask the equipment supplier to submit the moisture removal performance of his proposed system when the building is operating at the peak outdoor dew point condition (when little or no cooling is required). Otherwise, one can expect mold problems.

Filtering particles

Ventilation air needs to be filtered before it is brought through the HVAC system. Otherwise, the particles it carries will clog cooling coils, increase energy use, provide a growth medium for mold and bacteria inside duct work, and eventually coat the indoor surfaces (and the occupants) with unhealthy[6] dust.

At the risk of oversimplifying a complex subject, if the outdoor air is filtered by a MERV-8 filter, more than 70% of the larger particles carried by the ventilation air (3 microns and larger) will be removed before that air enters the HVAC system. This level of cleanliness should be sufficient for most commercial and institutional buildings in most

ASHRAE Standard 52.2 - Particle Removal

MERV	3-10 microns	1-3 µm	0.3-1 µm
1	<20%	-	-
2	<20%	-	-
3	<20%	-	-
4	<20%	-	-
5	20-30%	-	-
6	35-50%	-	-
7	50-70%	-	-
8	>70%	-	-
9	>85%	<50%	-
10	>85%	50-65%	-
11	>85%	65-80%	-
12	>90%	>80%	-
13	>90%	>90%	<75%
14	>90%	>90%	75-85%
15	>90%	>90%	85-95%
16	>95%	>95%	>95%
17	>99%	>99%	>99%
18	>99%	>99%	>99%
19	>99%	>99%	>99%

Std 52.1 - Dust Spot

Removal Rating	Particle Size
<20%	>10 µm
<20%	3.0-10 µm
20-25%	3.0-10 µm
25-30%	3.0-10 µm
40-45%	1.0-3.0 µm
50-55%	1.0-3.0 µm
60-65%	1.0-3.0 µm
70-75%	1.0-3.0 µm
80-85%	0.3-1.0 µm
90-95%	0.3-1.0 µm
~95%	0.3-1.0 µm
>95%	0.3-1.0 µm

MERV 8 — Better than a "30%" filter → 25-30%

MERV 11 — Similar to a "65%" filter → 60-65%

Fig. 15.6 ASHRAE MERV Ratings

Standard 52.2 clearly defines particulate removal in discrete size ranges. MERV classifications provide more useful detail for decision-making than did the older dust-spot rating system described by Standard 52.1.

MERV ratings are now the basis of the ventilation filtration recommendations described in Standard 62.1.

locations. On the other hand, higher MERV ratings would be appropriate in more polluted areas, or for more sensitive occupancies, such as health care facilities. High concentrations of small particulates (2.5 microns and smaller) are a health concern.[6]

MERV ratings

ASHRAE Standard 52.2 establishes particulate removal criteria for 17 Minimum Efficiency Reporting Values (MERV ratings) for air filters. The higher the MERV rating, the more particles the filter will remove. And the higher the MERV number, the greater the percent removal in three different ranges of particle sizes, as shown in Figure 15.6.

For example, a filter with a MERV rating of 4 or lower is likely to remove less than 20% of the easiest-to-remove, larger particles (those with a diameter between 3 and 10 microns). And that filter will not predictably remove particles in the two smaller and more difficult size ranges (1-3 microns and 0.3-1 microns).

In contrast, a filter with a MERV rating of 17 is likely to remove 99% of all particles in all three size ranges.

Since the removal efficiencies of different filters depend on which size particles are being removed, the MERV rating system quantifies performance in all three particle size ranges. Quantified performance data allows those who are concerned with health issues to make more informed decisions about filter selection.

But for many designers and building owners, the MERV classifications are less familiar than ASHRAE's older method of rating filters based on arrestance and dust-spot tests. To help make the transition to understanding the more useful MERV ratings, one can think of a traditional "30%" dust-spot filter as being roughly similar to the performance of MERV-8. And a "65%" dust-spot result is similar to MERV-11 performance.

More particulate? - Use filters with higher MERV ratings

The MERV ratings indicate the *percentage* of particles removed by a given filter. So when the ambient air is more highly polluted, one would want to remove a higher percentage of that larger number of particles, to keep the *net* air cleanliness at the same level as in less polluted areas.

Urban areas usually have higher particulate loadings, especially near highways. Particles are generated from automotive exhaust. In addition, traffic stirs up and aerosolizes dust that lands on the highway. Also agricultural activity generates particles which must be removed from ventilation air. Figure 15.7 shows annual average particulate concentrations for particles with a diameter of 2.5 microns or smaller, in selected locations throughout 48 US states during 2003.[6] It is a popular misconception that only congested urban locations have high airborne particulate levels. As seen in Figure 15.7, many less urban locations and agricultural areas have quite elevated levels.

Each geographic location will have different annual particle loadings and different particle size distributions. But it's also useful to keep in mind that these levels will often vary widely over the space of a few hundred meters in the exact same location. An air inlet which faces an active construction site or an elevated highway may have a far higher particulate loading than an air inlet on the other side of that same building which faces a wide expanse of lush green lawn.

In locations where the outdoor particle count is higher, it would be prudent to select a ventilation air filter with a higher efficiency. For example, a MERV-11 filter removes 85% of the larger particles, and also more than 65% of the smallest particles—removal rates which will not be achieved by the MERV-8.

Particulate loads and MERV ratings—US EPA Clean Air Standards and ASHRAE Standard 62.1 recommendations

In the United States, the Federal Environmental Protection Agency (EPA) has defined several levels of concern for outdoor pollutants. Each county in the nation has been defined either as having "attained" a clean air level, or as being in a "non-attainment" zone—an area which has too many contaminants to meet federal clean air standards.

It is easy to find the current status of any location in the US. The current levels of particles, ozone and five other contaminants are monitored, recorded and displayed every 15 minutes for the entire country. At the EPA's website, one can quickly learn both the attainment classification and the current air cleanliness level for any county in the United States, in near-real time (www.EPA.gov/air).

The newest edition of ASHRAE Standard 62.1 uses these classifications along with the MERV ratings to assist designers and owners in selecting particulate filters for ventilation air. But in fact, the particulate filter recommendations are based on EPA standards for ozone. Ozone catalyzes reactions with particulates, generating other pollutants which have adverse health consequences.[6] And high levels of ozone often occur in locations which also have high levels of particulate.

For ozone, the EPA has segmented their "non-attainment" classification into six levels of pollution: marginal, moderate, serious, severe-15, severe-17 and extreme. (The numbers signify the years allowed—after non-attainment designation—for the community to attain clean air.)

ASHRAE Standard 62.1 recommends a MERV-8 filter as the minimum for ventilation air in clean air areas, and for areas which are marginal or moderate non-attainment zones for ozone. For areas which are categorized as either serious, severe or extreme, MERV-11 is the recommended minimum.

To assist those who wish to follow ASHRAE Std 62.1 recommendations in other countries, Figure 15.8 shows the maximum annual 8-hour average ozone concentrations which correspond to the US EPA non-attainment classifications.

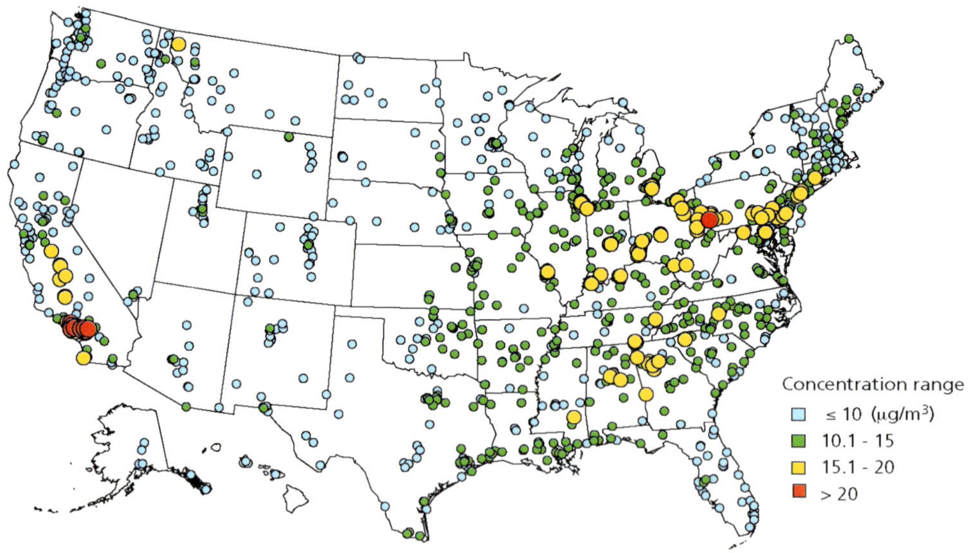

Fig. 15.7 PM$_{2.5}$ concentrations

The US Environmental Protection Agency tracks the particulate concentration in the United States to monitor compliance with the Clean Air Act. These averages from the year 2003 show that air in agricultural areas can be just as heavily-loaded with small particles as urban areas. The EPA website displays current particle concentration in near-real-time for the US.[6]

Fig. 15.8

Ozone concentration sets the filtration recommendations of Standard 62.1

When ozone concentration is high, its reactions with particles are especially damaging to health. At high ozone concentrations, ASHRAE recommends better filtration for particles, and also removal of ozone from ventilation air.

ASHRAE STD 62.1 Guidance	U.S. EPA Ozone Non-Attainment Criteria: 8-Hour Design Value Parts Per Million	Zone Classification
MERV-8	0.085 to 0.092	Marginal
MERV-8	0.092 to 0.107	Moderate
MERV-11, AIR CLEANERS	0.107 to 0.120	Serious
MERV-11, AIR CLEANERS	0.120 to 0.127	Severe-15
MERV-11, AIR CLEANERS	0.127 to 0.187	Severe-17
MERV-11, AIR CLEANERS	0.187 and above	Extreme

Note: "to" means "not including"

Health issues

For most commercial buildings, the ventilation air filtration is principally aimed at keeping large masses of particles out of the HVAC system. So filtration at the MERV-8 and MERV-11 levels are typical, depending on ambient particulate loading.

But when health is a principal concern, the designer and owner should focus on also removing the "respirable" particles which slip through MERV-8 and MERV-10 filters. Particles with a diameter of less than 2.5 microns are likely to enter the deepest reaches of the lungs, bypassing the body's protective mechanisms for dealing with particulate-related health hazards.

For example, health care facilities often include two banks of filters in series for the ventilation air. Designs for hospitals might include a MERV-8 prefilter for the larger mass of big particles, followed by a MERV-13 filter to catch a defined percentage of the more health-risky small particles.

Filtering gaseous pollutants - emphasizing ozone

Extensive medical research supporting the US EPA's clean air standards show that, in addition to particulates, four gaseous pollutants of outdoor air represent a health risk. These include carbon monoxide, nitrogen dioxide, oxides of sulfur and ozone. The limits for all pollutants currently regulated by the EPA are shown in Figure 15.9.

Of these gases, ozone currently receives the most attention in the US and Canada. This is not because the others are less harmful, but because the others are usually present in lower concentrations than in the past. In recent years regulations have reduced emissions from many commercial, residential and industrial sources, and the economies of these two countries have moved away from heavy industry, leaving motor vehicles and electrical power plants as the largest sources. In the more rapidly-industrializing countries, all five of these gaseous pollutants are likely to be of concern. But as in the US and Canada, ozone is particularly harmful, and is common in all urban areas because of automotive emissions.

Further, sunshine generates ozone. Especially in hot and humid climates the long and intense solar exposure helps generate ground-level ozone from products of combustion, principally emissions of nitrogen oxides from automobiles and power plants.[6] And heavy solar exposure greatly accelerates the production of harmful *by-products* of ozone reactions with particulates and other gasses. So ozone is a particular concern for designers of ventilation systems in any hot and humid climate, and especially for any building in an urban area.

Ozone is the principal ingredient of smog, which causes shortness of breath in nearly all people, and often more serious health effects for vulnerable individuals. According to documents produced by the US EPA, ozone, even when inhaled at low levels, can cause severe respiratory problems, including triggering asthma attacks.[6] Children are most at risk from both ozone and particles, because they breathe in a higher volume of air per unit of body mass than adults. And 15 to 20% of all summertime respiratory-related hospital visits in the Northeast US are caused by high ground-level ozone concentrations.[6]

ASHRAE Standard 62.1 calls for air cleaners or carbon filters to remove ozone from ventilation air when the outdoor concentration is in the serious, severe or extreme categories—values of 0.107 ppm and above when measured as an eight-hour average. (As mentioned earlier, both current and 8-hour classification values for any US location are available from the US EPA's website: www.epa.gov/air.)

United States National Air Quality Standards

Pollutant	Maximum Concentration	Averaging Time
Carbon Monoxide	9 ppm [10 mg/m³]	8 hours [a]
	35 ppm [40 mg/m³]	1 Hour [a]
Lead	1.5 µg/m³	Quarterly Average
Nitrogen Dioxide	0.053 ppm [100µg/m³]	Annual Arithmetic Mean
Particulate ($PM_{2.5}$)	15 µg/m³	Annual Arithmetic Mean
	35 µg/m³	24 Hours
Ozone	0.08 ppm	8 Hours
	0.12 ppm	1 Hour
Sulfur Oxides	0.03 ppm	Annual Arithmetic Mean
	0.14 ppm	24 Hours [a]

Notes:
a. Not to be exceeded more than once per calendar year (Not a rolling 12 month period)
b. mg/m3 = **Milli**gram per cubic meter (1/1000 gram)
c. µg/m3 = **Micro**gram per cubic meter (1/1,000,000 gram)
d. $PM_{2.5}$ = Number of particles which have a maximum diameter of 2.5 microns or smaller

Fig. 15.9 - US Air quality standards

ASHRAE recommendations in Standard 62.1 depend on understanding the US clean air standards. For those in other countries who wish to follow ASHRAE recommendations, this table will be helpful.

When pollutant concentrations are above these levels, the area is classified as a "non-attainment" zone by the US EPA. In those areas, ASHRAE recommends more comprehensive filtration of ventilation air, as shown in Figure 15.8.

Carbon filters, formerly a costly technology used principally for industrial applications and for museums, are now more convenient and economical. They are increasingly common in commercial and residential buildings because the allowable air velocities through modern carbon-impregnated filter media are much higher than for the granular media which is used for very high levels of ozone removal. Higher velocity allows pleated carbon filters (which look like the familiar "30%" particle filters) to use conventional flat filter frames rather than custom-built "V" or "W" shaped frames which require much more space in the direction of air flow.

Effective Ventilation Air Distribution

In a perfect system, a small amount of clean, dry ventilation air—the absolute minimum needed for health and comfort—would be injected through tubes, directly into the nostrils of each occupant. And those tubes would be absolutely air tight, so that none of the valuable clean dry air would be wasted by leaking out into ceiling spaces or wall cavities, where there are no people to breathe it.

In real-world systems, there are many reasons that adequate amounts of ventilation air might not actually get to the breathing zones of occupants (ineffective ventilation). Common examples include: air leaks from duct connections, no actual measurement or control of outdoor air volume, VAV systems without the ability to increase the percentage of outdoor air as overall supply airflow reduces, or overly-diluted mixtures of ventilation air into return air in constant-volume systems.

Specify durably air-tight duct connections

If the designer and installer allow ventilation air to blow into the spaces above the ceilings and inside the walls through leaky duct work, then there will not be enough clean dry air delivered to the occupants' breathing zones. With leaky duct connections there are only two choices: inadequate ventilation—or adding more fan capacity, more dehumidification capacity, more filtration, and more cooling to bring larger amounts of air into the building to compensate for the air lost to building cavities. Sealing up duct connections and seams

Fig. 15.10
Mastic for duct connections
The upper photo shows an example of why most duct tape is simply not durable enough to provide sealed duct connections over time. The lower photo shows duct connections durably sealed with mastic and glass fiber tape before insulation is applied.

There's not much point in spilling expensive clean and dry ventilation air into building cavities through duct leaks. Sealing up ducts with reliable methods like mastic over reinforcing tape prevents energy losses of 20 to 30% per year, and gets the ventilation air to the people who need it, rather than to the stud cavities and mechanical spaces, which don't.

is far less expensive than adding larger equipment to compensate for leaks, and it helps avoid condensation in hidden spaces.

The simple way to specify tight duct work which will stay tight over time is to require that all seams and *all connections* be sealed with mastic, reinforced with glass fiber tape as shown in Figure 15.10. When the duct work is rigid metal fastened with screws or bolts, then the mastic probably does not need glass fiber tape—glass fibers mixed into the mastic itself will probably be sufficient to keep the mastic from cracking and leaking over time.

Avoid using a single constant-volume cooling system to ventilate many zones

If a separate, dedicated ventilation system is installed in the building, and if that system has independent duct work which delivers ventilation air directly to each occupied space, it's easy to ensure that adequate amounts of outdoor air gets to occupants' breathing zones.

Shortages of ventilation air usually arise when all the ventilation air is blended into a single, unified duct system which carries a constant volume of air used for cooling the spaces in addition to ventilating them. These systems are very common. Indeed they are the rule rather than the exception in North American building practice.

The problem is that these small spaces may have widely-varying heat loads and different occupancy schedules. So in fact, *neither the cooling capacity nor the ventilation air volume should be constant*. As one example, consider office buildings with a single system serving the reception areas, all the closed offices, all the open-plan office cubicles and all the conference rooms. Another typical example is a wing of a school, with several classrooms, a common room and two or three offices. In these situations, a single constant-volume system will be delivering too much or too little ventilation air to all spaces, all the time. Their occupancies vary widely day-to-day and hour-to-hour, throughout the 8760 hours of the year.

To meet codes at the lowest-possible construction cost, the designer usually decides in favor of adequate ventilation during periods of peak occupancy, and so designs the total amount of ventilation air for what he estimates as the probable simultaneous peak occupancy for all the spaces served by that single system. The ventilation air flow is fixed at that value initially, and it probably stays at that high level for its entire operating life.

Since the proportion of ventilation air to recirculated air is fixed, all of the spaces are always either under-ventilated or over-ventilated as the number of people goes from zero on weekends to full capacity for a few hours in the middle of a weekday.

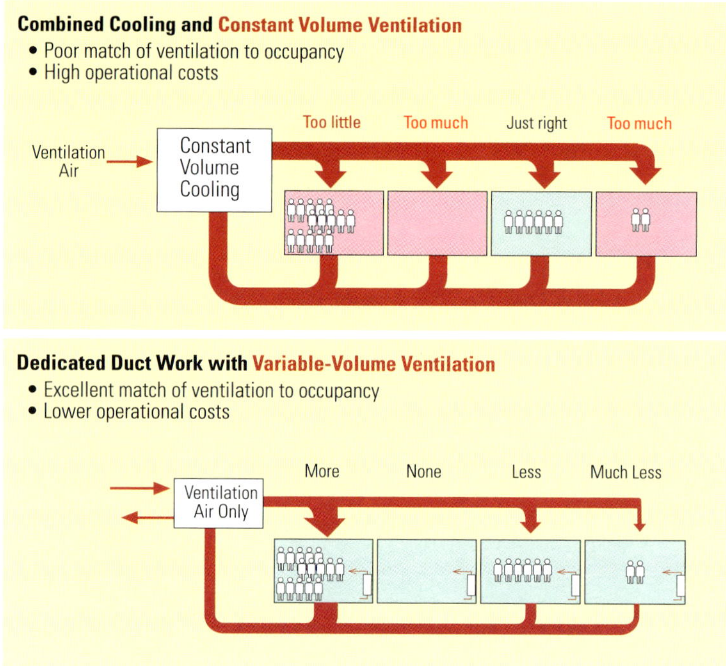

Fig. 15.11 Ventilation Air Distribution

The most certain method for ensuring adequate ventilation is to install dedicated ventilation duct work to each space.

Otherwise, when ventilation air blends into the supply air, the spaces are usually under-ventilated or over-ventilated as their occupancy changes.

Dedicated duct work and dedicated ventilation dehumidification add up-front costs. But they provide more effective ventilation at lower operational cost for the life of the building, because ventilation air can be more easily reduced or increased as occupancy changes.

Such constant-volume systems save on construction costs and probably comply with building codes in many jurisdictions. But *operating* a school or office served by these systems in a hot and humid climate—in a way which ensures adequate ventilation air—is expensive. Also, such systems can generate a mold problem in an unoccupied building.

Unless they are equipped with ventilation control dampers, constant-volume systems will bring in ventilation air any time their fans operate. Given the intermittent operation typical of unoccupied periods, the cooling coil rarely cools the air for *long enough* to dry out the humid mixture of ventilation and supply air. This leads to mold, as described in more detail in Chapter 5 (Avoiding Bugs, Mold & Rot).

Among the alternatives are variable volume systems, which provide ventilation air in proportion to the cooling load, or the dedicated ventilation system, which provides ventilation independently from any cooling or heating load. These cost more money initially, but they can deliver adequate ventilation without energy waste. Assuming they also dry that ventilation air to a dew point of less than 55°F [12.8°C], they will also reduce mold risk for the life of the building.

Reducing The Cost Of Ventilation

Ventilation in hot and humid climates is seen as expensive. It costs a considerable amount to clean and dry incoming outdoor air.

There are several ways to reduce the annual cost of ventilation by more than 50%—but this takes more thought on the part of the designer, and a slightly larger construction budget from the owner. We'll begin with the least-cost, highest-benefit suggestions, and then move on to suggestions with larger construction cost implications.

Measure the ventilation air volume, then set it accurately

In spite of the widespread and accurate perception that ventilation air is expensive, designers and owners seldom take the most essential step towards reducing that cost—measuring and controlling the amount of ventilation air fed to the building.

Field research consistently shows that buildings tend to be highly over-ventilated or under-ventilated. Seldom is the amount of ventilation air appropriate to the occupancy. For example, in field measurements of five federal courthouses in Florida, the actual ventilation rates were found to be between 500 and 2,000 cfm/person—as opposed to the ASHRAE standard (at that time) of 20 cfm/person, which was probably closer to the designer's intention.[7] Field measurements of ventilation rates in offices[8] and schools[9] have similarly discouraging results.

To avoid this needless waste, measure and set the ventilation air flows to the values intended by the designer. However, to make that *possible* the designer must ensure that:

- Dampers or perforated plates are installed on all ventilation air intakes to allow technicians to set the air flow.

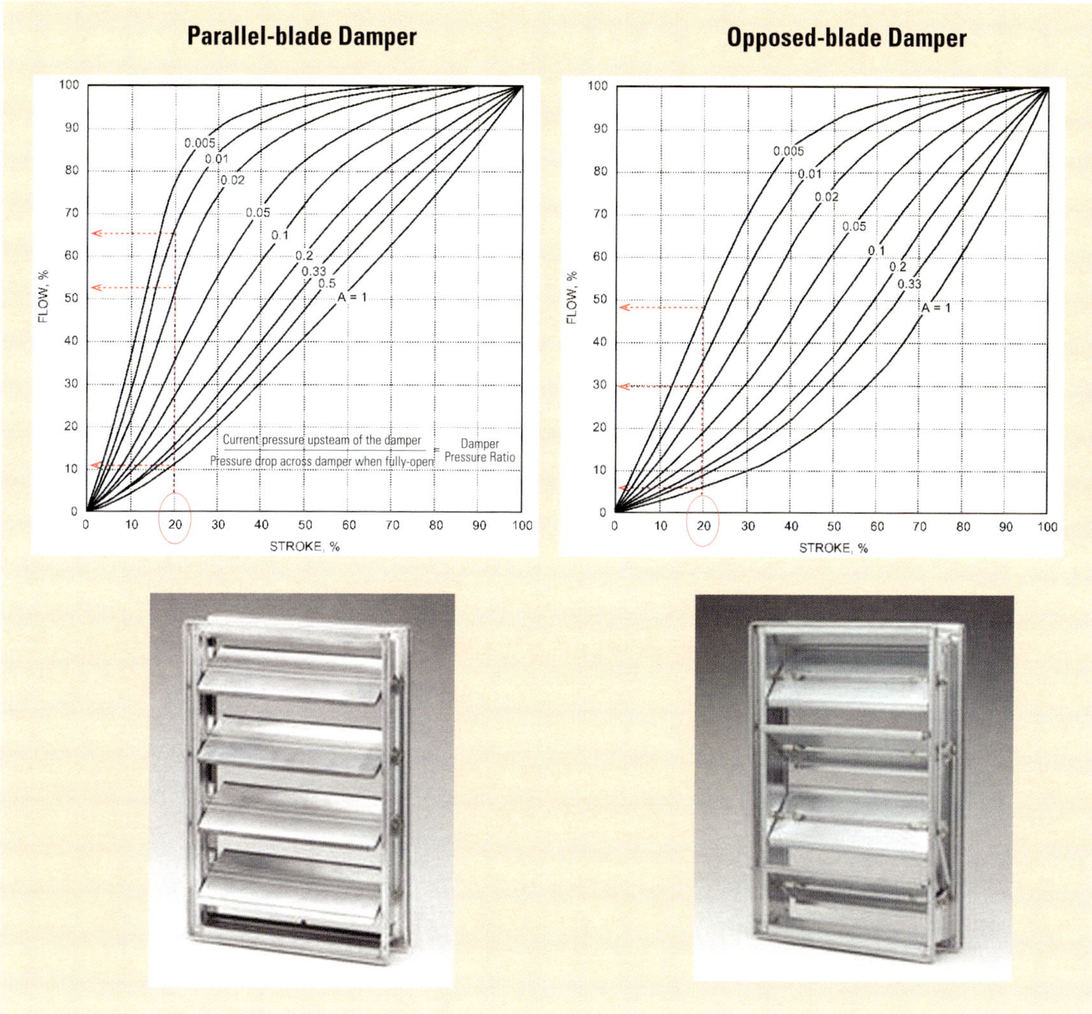

Fig. 15.12 Damper position controls—but is not the same as—the air flow through that damper

System pressures greatly affect air flow through a damper. The damper stroke position (its percent open area) is not a reliable indicator of the amount of air flowing through it. Pressure upstream of an outdoor air damper changes with wind pressure, constantly. Instead of relying on damper stroke position, measure the ventilation air flow with some device designed specifically for measurement, such as a flow station for air flow or a CO_2 sensor for net ventilation effectiveness—then use the damper to modulate the ventilation air flow.

Source: ASHRAE Handbook–Fundamentals 2005, Chapter 17, figures 13 and 14.

- The specification requires some specific responsible party to measure the flow of each incoming ventilation air stream, set it to its correct value and document those actions, in writing, to the owner.

An important caution is that damper position is not linear with air flow volume. That is to say, a damper which is set to an angle of 20% open does not pass 20% of the full volume. It may pass far more, or much less than 20% of full flow. As shown in Figure 15.12, the volume varies according to the ratio between the current upstream air pressure, and the pressure drop through the damper when it is fully-open. This value is called either the "damper pressure ratio" or more recently the "Authority" of the damper-pressure relationship. To avoid waste, measure ventilation flow with a device designed for that purpose—then control that flow by modulating a damper. Don't misuse a conventional damper's stroke position as the outdoor air flow measurement.

Also, it's important to understand that damper linkages corrode and slip out of adjustment, and building needs change over time. Certainly every few years or on change of ownership, remeasuring and resetting ventilation air flows has three big benefits. First and most important, the ventilation air quantity is appropriate to assuring good indoor air quality for the comfort and health of occupants. This is not only the right thing to do, but it also greatly reduces the potential for complaints. Secondly, energy costs are usually reduced, saving more than enough to pay for the cost of measuring and setting air flows. Finally, the overall system becomes more responsive, improving comfort and reducing calls for maintenance troubleshooting.

Remeasuring and resetting ventilation air flows on a regular basis is the least expensive way to improve indoor air quality and reduce operating costs at the same time.

Time clocks to reset ventilation air volumes

Ventilation is for people, and the number of people in a building varies widely over the 8760 hours in a year. Therefore, to avoid the high costs

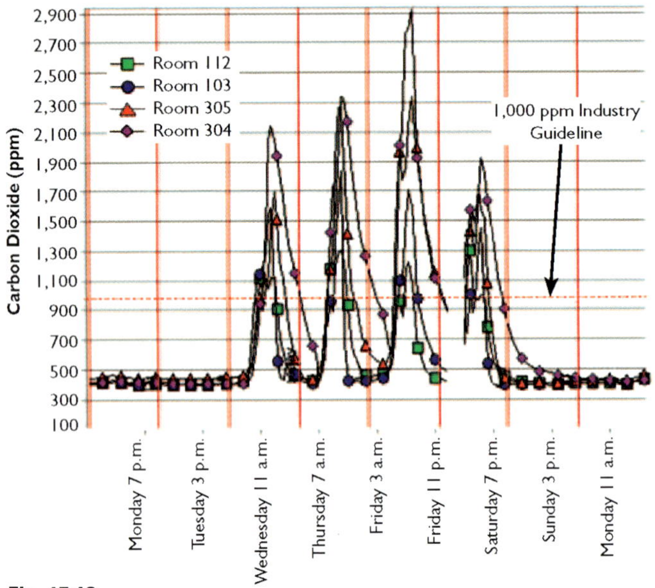

Fig. 15.13
CO_2 concentration is an excellent indicator of human occupancy[11]

of wasted energy, the ventilation air should vary as well, controlled according to the number of people actually in the building.

One low-cost addition to a ventilation system is a time clock controlling an actuator and two-position damper. The damper reduces ventilation air flow to its minimum, based on the assumed occupancy after hours, overnight and during vacations.

To ensure that ventilation air flow is both adequate and not wasteful, each ventilation inlet will need to be measured for air flow in the field under installed and operating conditions. *Then* the two damper positions can be set to achieve the flows defined by the designer for both occupied and partly-occupied modes.

An improvement on the time clock is to install a motion sensor in each space. The motion sensor signals the damper actuator to close, shutting off the ventilation to that space when the sensor indicates the space is no longer occupied.

Time clocks and motion sensors are relatively simple, inexpensive and don't require much maintenance. And for smaller spaces occupied by relatively few people on relatively predictable schedules, like classrooms, small offices and small conference rooms, such simple devices provide a reasonable and low-cost means of avoiding greatly over- or underventilated rooms. But in large spaces such as auditoriums, gymnasiums, theaters and conference center rooms, the population for any given event will very widely. And the space will be occupied at many different times which are not really practical to anticipate and schedule with a time clock.

So for a much closer match between occupancy and ventilation airflow in large spaces with widely varying populations, a sensor can measure the concentration of carbon dioxide (CO_2) in the occupied spaces. Then modulating dampers can vary the amount of ventilation air to each space in proportion to its actual occupancy.

Variable-volume (CO_2 demand-controlled) ventilation

Indoors, carbon dioxide (CO_2) is basically a tracer gas, not a pollutant. It has no known adverse health effects until it rises to extremely high levels (Inside a submarine, for example, concentrations rise to 11,000 ppm without noticeable adverse health effects.)[10] But CO_2 is an excellent indicator of human metabolic activity, and therefore of the number of people occupying a building.[21]

As the human body digests and then metabolizes food, it breaks down organic material into its constituents. As one of those "products of combustion," carbon dioxide is exhausted from our lungs in proportion to the number of calories metabolized. When the concentration of CO_2 rises in a space, it's an indication of either greater physical exertion, or more occupancy, or both.

In either case, the pollutants and odors generated by those occupants are also increasing, so more ventilation is needed. Conversely, when the CO_2 concentration falls, it's an indication that the amount of ventilation air can probably be reduced without significantly affecting the quality of the indoor air.

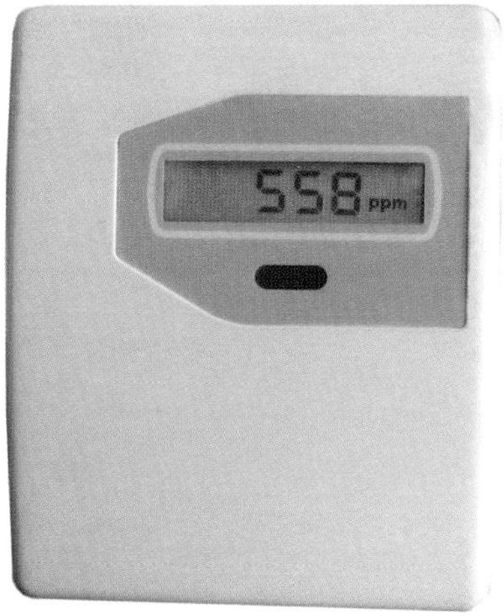

Fig. 15.14 CO_2 sensor-transmitter
As CO_2 concentration rises , the sensor can call for more ventilation air. Conversely, the sensor can signal for a reduction in the ventilation air flow when occupancy falls, saving a great deal of money each year, while helping to ensure good indoor air quality.

In the past, an approximate target concentration for CO_2 has been "700 ppm above the CO_2 concentration in the local outdoor air." (An increase limited to 700 ppm indicates that about 15 cfm [7 l/s] of outdoor air is being provided to the space, assuming the occupants' metabolic rates are typical of classroom and office activities.) In most locations other than deep forests or near auto exhausts or smokestacks, background CO_2 concentration is usually between 300 and 500 ppm. So traditionally, the target limit for occupied spaces has usually been set somewhere between 1,000 and 1,200 ppm. The current edition of Std 62.1 (2007) now has an appendix which shows how to more precisely set target limits for CO_2 concentration. Accurate occupant definition, along with detailed metabolic calculations may allow mixed-use occupancies to have a rise of greater than 700 ppm above the outdoor CO_2 concentration.[2,22]

CO_2 sensors can provide a very clear indication of the match between ventilation airflow and space occupancy. For example, Figure 15.13 shows the CO_2 concentration in a school classroom.[11] Note how the concentration in room 304 rises above 1,900 ppm after it fills with students, and then falls back to about 400 ppm at the end of the day after the students leave. At that point, it would be useful to reduce the ventilation air volume to the bare minimum needed to flush the building of contaminants generated from interior furnishings and finishes. Also, it's clear from the CO_2 concentration peaks that the system might be better arranged if it provided *more* ventilation air when those classrooms are heavily occupied.

Tracing the CO_2 concentration and adjusting the amount of ventilation air accordingly can save a great deal of money compared to simply "setting and forgetting" the ventilation air volumes. This is particularly true in buildings, like schools, which have highly variable occupancies over the typical day and school year. The cost consequences of continuous ventilation based on *assumed peak* occupancy rather than *current*, *measured occupancy* using CO_2 concentration are often the reason that adequate ventilation has a reputation for being expensive.

For example, the schools partly described by Figure 15.13 are located in a cold climate.[11] One of the eight schools in that study had been recently renovated, to eliminate indoor air quality problems caused by poor ventilation. Its total annual heating costs *doubled* after the ventilation improvements. The design was based on an assumed peak occupancy, and the ventilation air flow was constant—it was not reduced at low occupancy (many thousands of hours per year). One assumes that the school district's construction planner was either not informed of the probable increase, or chose not to invest in CO_2- controlled ventilation or other means of avoiding that doubling of annual heating costs.

There are two useful cautions about CO_2- controlled ventilation. The first concerns pollutants generated by sources other than people, and the second concerns calibration and maintenance.

CO_2- controlled ventilation is allowed under the guidelines in Standard 62.1, *provided* that other pollutants do not exceed acceptable levels. Such a problem could occur, for example, in a small room with only one person, but which contains many photocopiers or laser printers. Often, these emit ozone and volatile organic vapors. So the designer must remain conscious of possible pollutant sources such as office machinery, and establish higher minimum ventilation air volumes for spaces where large pollutant sources are expected. The CO_2 sensors will then automatically raise ventilation air flow upwards from that higher baseline, when more people occupy the space.

Finally, some of the early CO_2 sensors had a well-deserved reputation for inaccuracy and calibration drifting widely over time. And currently, as more lower-cost sensors enter the market and make their use more economically attractive, the calibration and accuracy of CO_2 sensors over time continues to be an issue. From a designer's perspective, it's useful to question the sensor supplier on these issues, to obtain assurance of exactly what sort of accuracy can be expected, and what sort of annual maintenance has proven to be necessary in similar installations.

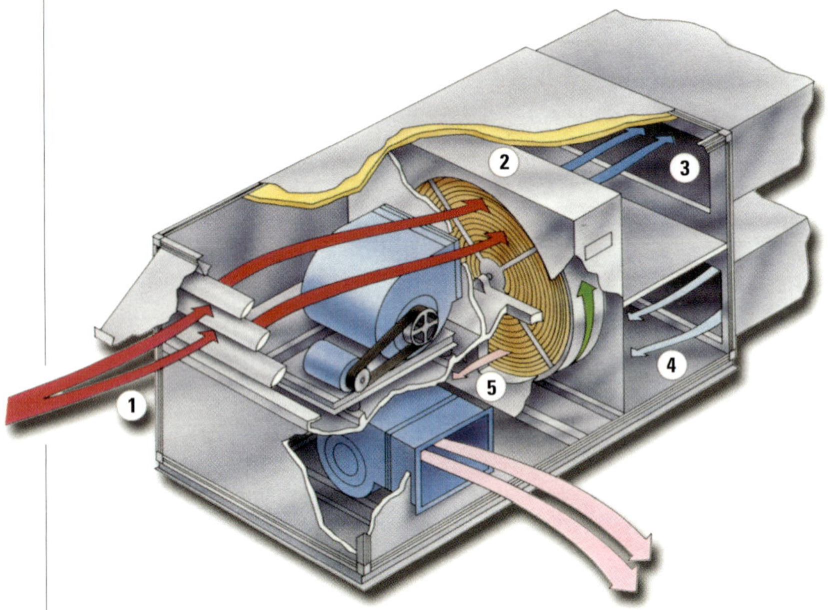

1. Hot, humid outdoor air enters the unit.
2. The air gives up part of its heat and moisture as it flows through the passive desiccant wheel.
3. Leaving the wheel, the ventilation air carries a much smaller heat and humidity load.
4. Relatively cool and dry exhaust air leaving the building strips heat and moisture from the passive desiccant wheel.
5. The exhaust air, now warm and moist, leaves the system, exhausted to the outdoors.

Fig. 15.15 Enthalpy heat exchanger

When air flows are large and operating hours are long, enthalpy heat exchangers can save money in the operating budget.

But actually, the biggest cost reduction comes during construction, when the reduced load allows downsizing of cooling and dehumidification equipment.

Note that air filters are essential. These are located upstream of the wheel on both the supply and exhaust air sides. They have been omitted from this diagram, to show other components more clearly.

To some extent this caution is similar to that for humidity sensors, which display relative humidity to two decimal places, even though their measurement tolerance is ±3% rh. The accuracy of the device is perfectly adequate for the purpose, as long as you don't get too excited about the accuracy *implied* by the display, which is closer to marketing promotion than it is to a reliable field measurement.

In other words, it would not be wise to set the ventilation controls based on the assumption that CO_2 concentrations can be reliably measured from a single sensor for all points in the breathing zone to an accuracy of ten parts per million. That's not necessary and is technically impractical in any case. Tolerances of plus or minus 50 or (even 100) parts per million may be more practical decision points for adding or subtracting ventilation air in most commercial and institutional buildings.

Enthalpy heat exchangers can reduce ventilation costs

In a hot and humid climate, exhaust air is a very valuable asset. All of the exhaust air (except from kitchens and laundries) is much cooler and drier than the outdoor air which will shortly have to replace it. Cleaning and drying the replacement air—the makeup air—is going to cost a lot of money every year. Between 30 and 50% of this annual cost can be avoided. Also, the installed cost of the ventilation drying and filtering equipment can be cut by as much as 30% by installing enthalpy heat exchangers in the ventilation system.

That's why ASHRAE Standard 90.1 (Energy Standard for Buildings Except for Low-Rise Residential Buildings) requires energy recovery for any ventilation air systems larger than 5,000 cfm [2360 l/s] in which outdoor air makes up 70% or more of the design supply airflow.[2]

These heat exchangers remove heat and moisture from the incoming ventilation air by transferring that energy to the outgoing exhaust air. Figure 15.15 shows one of these devices—the popular rotary enthalpy heat exchanger, often called a heat wheel or total heat wheel. The effectiveness of an enthalpy heat exchanger varies between 80 and 30%, depending on how much of the exhaust air can be brought back to where the ventilation air enters the rest of the system.

The cost to operate a rotating enthalpy heat exchanger includes the cost of the fan power needed to push the exhaust and ventilation air streams through the wheel (usually less than 0.5"wg for each air stream [less than 125 Pa]), plus the much smaller cost to run the fractional horsepower motor which rotates the wheel.

Historically (before ASHRAE Standard 90.1 required some form of energy recovery), enthalpy heat exchangers were often installed because using them allowed the designer to cut the size of the cooling and dehumidification equipment by as much as 50%. In buildings with very large make-up air requirements such as hospitals, those equipment cost savings are sometimes large enough to reduce the net installed cost of the entire system, even after accounting for the cost of the heat exchanger and its associated duct work and controls.

Also in the past, when enthalpy heat exchangers were *not* used, it was often because the exhaust air left the building so far away from the point where the ventilation air enters the building that either the cost or the difficulty of bringing duct work back to that location was prohibitive. Another common reason for owners and designers to ignore their large potential cost savings is that enthalpy wheels do leak some air between the incoming and outgoing air streams. And they also require occasional attention from the maintenance staff to make sure the wheel assembly does not loosen and begin to wobble. Finally, some designers and owners have been under the misimpression that an enthalpy wheel is the only dehumidification device needed for ventilation air. If the design and operation of a system is based on that misimpression, humidity goes out of control during part-load hours, and the enthalpy wheel is blamed for the problem.

To avoid these potential problems and gain the maximum benefits from rotary enthalpy heat exchangers, the designer should keep in mind several important points.

Dry exhaust air is essential

An enthalpy heat exchanger is not a *substitute* for a dehumidifier. It *only reduces the load for a dehumidifier.* The heat exchanger removes heat and moisture from ventilation air only when the exhaust air is cool and dry. If there is no dedicated dehumidification component somewhere in the system, or if it is not working, humidity will build up in the space. Then the exhaust air will become too humid to absorb excess humidity from the incoming ventilation air, no matter how efficient the heat exchanger.

In hot and humid climates, the largest benefit of an enthalpy heat exchanger is the load reduction during peak sensible load hours.

Downsize cooling and dehumidification equipment accordingly. Otherwise, installing an enthalpy heat exchanger adds to the cost of construction rather than reducing it.

Infrequent operation or low air flow rates prevent operational savings

Annual savings depend on number of hours of operation, multiplied by the air flow and by the enthalpy difference between indoors and outdoors, and by the heat exchanger efficiency.

In a hot and humid climate, the enthalpy difference is large for most hours in the year. But if the ventilation system *does not operate* for many hours during a year, as is typical for an auditorium or a place of worship, then the annual savings will be quite small. Likewise, adding enthalpy recovery to a single intermittent toilet exhaust with a flow of 60 cfm [28 l/s], is not going to save much energy each year. But in both cases, the load reduction at the peak design condition may still provide a significant cost reduction in the construction budget, even if the annual energy savings are negligible.

Install ducts and dampers to bypass both sides of the wheel

The cost of the wheel's pressure drop is not balanced by savings when the weather outdoors is cool and dry. For example, there are many climates like that of Northern Florida or Northeast Texas which are certainly hot and humid, but which still have low cooling and dehumidification loads during several months. During those months, the wheel does no good, but still costs money because air must be forced through it.

To avoid this cost, install bypass ducts and dampers to allow both air streams to flow around the wheel when the outdoor dew point falls below 55°F [12.8°C]. The air should bypass the wheel until the outdoor air becomes cold enough to require heating before it enters the building. Then the bypass can close, and the wheel can provide heating energy savings.

To provide savings, rotors and seals must be kept tight

Although very cost-effective, an enthalpy heat exchanger is not a free lunch. The wheel (and therefore the entire system) will fail if the bolts fastening the rotor assembly are not tightened every year, or if the air seals are missing, or if the wheel wobbles as it rotates, forcing the seals back and generating excessive leakage. Historically, an enthalpy wheel requires so little annual attention that the device is often forgotten until it fails. Annual maintenance is just as important for an enthalpy wheel as it is for a chiller of similar capacity.

How Much Air & Where - ASHRAE Std 62.1

The air outdoor volume has a strong influence on the construction cost of the HVAC system and on its annual cost of operation. ASHRAE Standard 62.1 provides extensive guidance in this area, which will not be repeated here (except for its current ventilation air flow requirements which are shown at the end of this chapter as Figure 15.23). Instead of repeating the entire standard, we will discuss several "big picture" aspects of the standard which are useful for the architect, owner, HVAC designer and operator to keep in mind.

Std 62.1 now makes demands on Architects and owners

Thirty years ago, Standard 62 was largely limited to setting ventilation rates for different occupancies. Its underlying assumption was

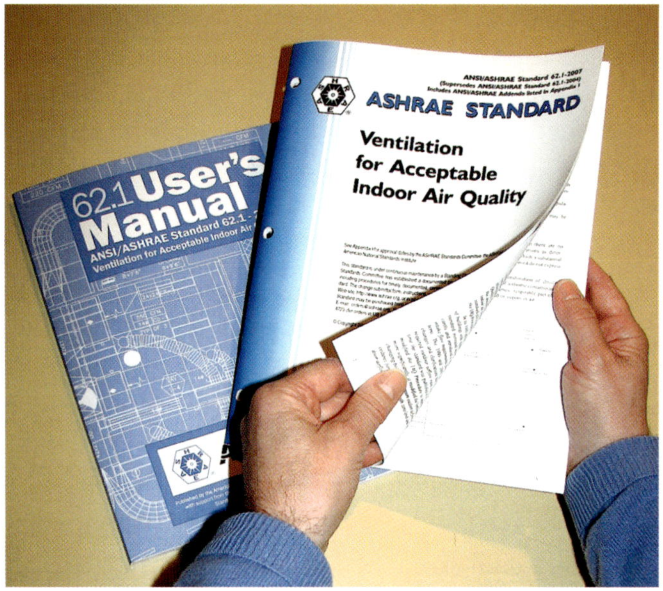

Fig. 15.16
ASHRAE Standard 62.1 with its User's Manual

that if enough outdoor air was brought into the building, the indoor air quality would be adequate for comfort and health. Consequently, architects and owners believed that assuring indoor air quality was the province of the HVAC designer, and that providing adequate amounts of ventilation air would accomplish that goal.

But in recent years, owners, building scientists and HVAC designers have come to understand that assuring adequate indoor air quality involves more variables than simply injecting untreated outdoor air into the occupied spaces. In particular, one must keep the building from growing mold and bacteria, and these risks remain high when humid outdoor air floods into a cool building.

After tens of thousands of painful and costly indoor air quality investigations, there have been major expansions in the scope of Standard 62.1. It has expanded greatly, requiring practices which avoid what we now know to be sources of indoor air quality problems.

Many of these requirements are not under the exclusive control of the HVAC designer. But that's because the standard does more than

Fig. 15.17 Access for maintenance

Contorted equipment squeezed into small spaces make it very difficult if not impossible to maintain that equipment, which means the indoor air quality will suffer, through the entire life of the building.

*That's why providing adequate access to all equipment for maintenance is now a requirement of ASHRAE Standard 62.1. This requirement will need emphasis during **early** discussions with the owner and architectural designer.*

Length

The length should match the width of the aisle required for maintenance access. In most buildings, several sticks with different lengths will be required.

Normal wear & tear

Often used during design conferences, an Andy Stick should be softer than the conference table surface.

Digital accuracy

Appropriate precision is important for any quality-assurance tool

Emphatic Advocate

Any tool is only as effective as the person who uses it. Like this young engineer, many HVAC designers and operations personnel prefer the two-handed grip. This helps them emphasize the importance of maintenance access when the architectural designer and owner decide how much space to allow for mechanical rooms.

Fig. 15.18 The "Andy Stick"

An effective tool for quality assurance during on-site inspections, an Andy Stick[12] quickly measures compliance with the aisle width specified for maintenance access. Adequate access is now a requirement of Std 62.1

reducing symptoms by using dilution air. Now, the standard is aimed at *preventing* the problems which *cause* those symptoms.

For example, the HVAC designer must now think long and hard about operation and maintenance. He is required to take specific steps in the design to ensure that the system can actually be maintained and kept clean. In the current standard, it is no longer acceptable for the designer to pack the equipment into closets or above dropped ceilings and hope the operations staff can somehow clean and maintain the duct work, coils, compressors, fans, sensors and controls despite any actual access to that equipment.

The requirement for adequate access, in turn, makes what may be unfamiliar demands on the owner and the architect. If the owner wants to comply with Standard 62.1, the owner and architect must allow the HVAC designer adequate mechanical space for maintenance access. Figure 15.17 shows an example of design and installation contortions needed to fit ducts into inadequate space. Figure 15.18 shows a low-cost tool which has sometimes been used during design conferences to obtain space for adequate access.[12]

Further, the Standard requires that some competent authority start the system up, balance the flows, test the drain pans, document the system and provide an operating and maintenance manual. Documentation must now include the design criteria and assumptions, the system design narrative, final design drawings, control sequences and the startup and air balance report. These services and documents can now be expected by the owner when he requires (or when local building codes require) compliance with Standard 62.1.

Also, the owner will need to pay for these services, and will need to define who, specifically, will provide them. In the past, startup, commissioning and documentation were not often provided by the HVAC designer. In the future, the owner will need to be clear about what he means, exactly, when he requires that the HVAC designer must "comply with ASHRAE Standard 62.1" Contracts between the owner,

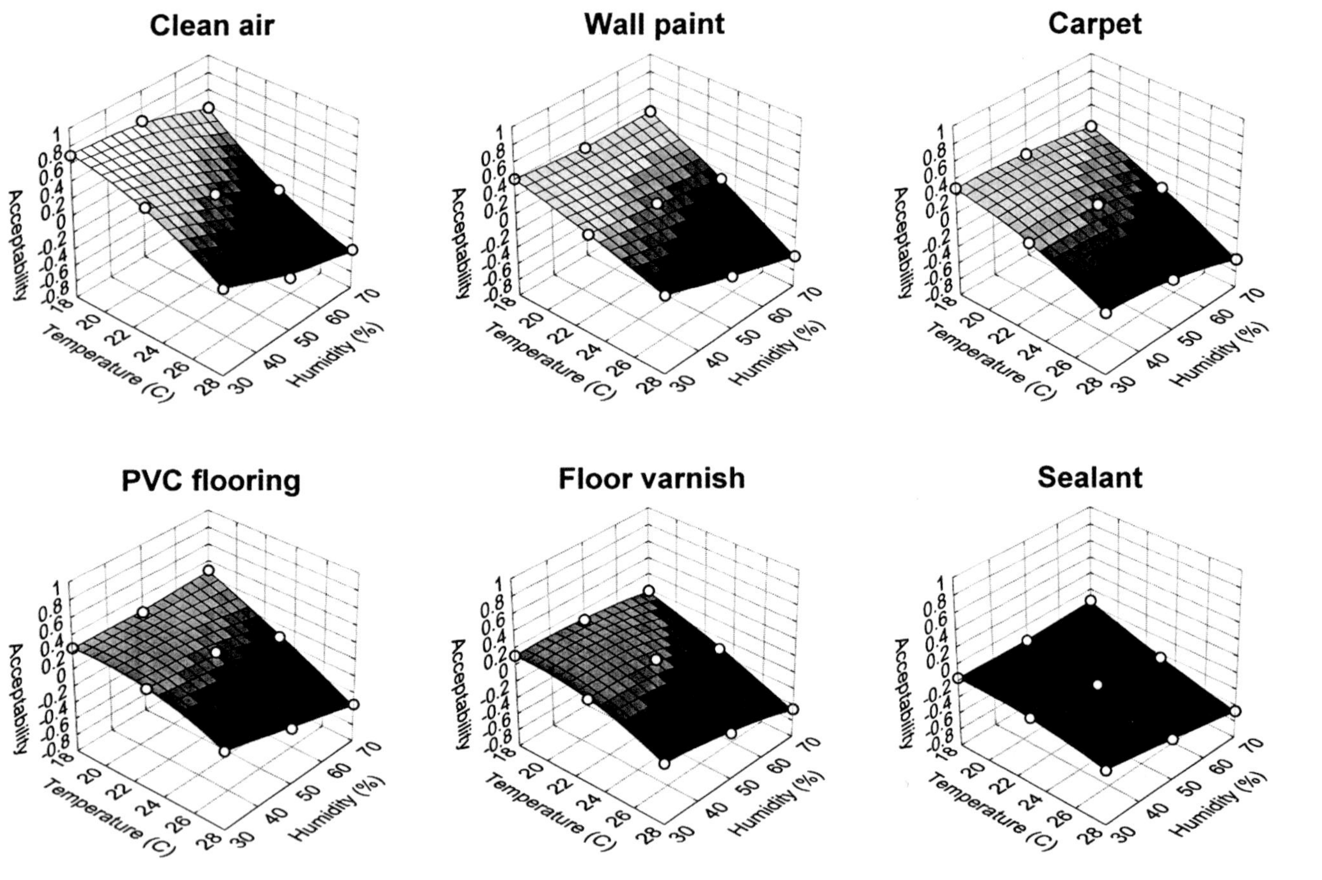

Fig. 15.19

When indoor air is polluted, occupants find the environment less thermally acceptable

Studies performed at the Technical University of Denmark show that when indoor air is more polluted, occupants ask for colder temperatures. [15]

Adequate ventilation dilutes these pollutants, allowing warmer temperatures to be comfortable, which saves energy and reduces the potential for condensation and mold.

architect, HVAC designer and builder will need to be clear on who is responsible for which aspects of Standard 62.1 compliance.

For example, the standard also now requires that the ventilation system be maintained in specific ways, on a schedule which is also defined by the standard. Dehumidification devices must be inspected and cleaned of any microbial growth at least once every year. Drain pans must be cleaned on the same schedule. Outdoor air dampers must be inspected every three months, and the airflows they allow must be verified on that schedule. These and several other perpetual maintenance requirements are part of Standard 62.1; and are usually the province of the owner rather than the HVAC designer.

In summary, Standard 62.1 is no longer just guidance which sets ventilation air quantities. Compliance requires participation and specific actions from each member of the delivery team, and also from the owner and operator of the building. The HVAC designer is encouraged to make this fact known in productive ways, at appropriate times, to other members of the team who may not be aware of the cross-cutting requirements of this important international standard.

Ventilation rates depend on occupancy + floor area

In the 2004 edition of Standard 62.1, the society split the recommended ventilation rate into two components: air to dilute contaminants generated by people, and air needed to flush out contaminants generated by the building and its furnishings.[20]

This recognizes the fact that many furnishings and interior finishes out-gas volatile organic vapors (pollutants) which, in sufficient concentrations, are either irritants or hazards to heath. Dividing the ventilation rate into two components covers the cases where occupant density may be low, but the surface area emitting pollutants may still be high.

For example, even when a classroom is not fully-occupied, it still has many surface feet of flooring, furnishings and finishes which emit airborne pollutants. To ensure that the air quality will remain acceptable, Standard 62.1 calls for providing 10 cfm per person, *but also 0.12 cfm per square foot of classroom floor space* no matter how few people occupy that classroom. [5 l/s per person and 0.06 l/s/m^2] That ventilation rate per unit of floor space is required until that particular classroom is completely vacant. Only then can its ventilation air be closed off entirely.

The ventilation air flow recommendations of the 2007 edition of Standard 62.1 are included at the end of this chapter in Figure 15.23. However, recommended rates are subject to change. So the designer should consult the current edition of Standard 62.1 for all the rates for different types of buildings and different use categories.

But basically, looking at Figure 15.25 one can see that the recommended rates go higher in areas where the physical activity (the metabolic rate of occupants) is higher. And rates also go up for buildings with a great deal of floor space, or for buildings which contain strong pollutant emitters.

For example, health clubs should be ventilated at 20 cfm/person [10 l/s/person]. This is twice the rate per person compared to a school classroom, where students are sedentary. On the other hand, in swimming pool areas, no ventilation is required *per person*—but a constant ventilation rate of 0.48 cfm *per square foot* is required to dilute the pollutants emitted by the pool water [2.4 l/s/m2].

Similarly, in a beauty shop one expects hair sprays and cleaning fluids which emit high rates of volatile organics. So the rate per person is 20 cfm in spite of the low level of physical activity, while the rate per square foot is 0.12—the same rate per square foot as in a classroom [10 l/s/person and 0.3 l/s/m2].

Recommended rates are *minimums*, not maximums

As noted earlier, ventilation rates are controversial technically, socially and economically. There is constant pressure from building owners to reduce ventilation rates, and constant pressure from occupants to increase them. In the US, many building codes require compliance with ventilation rates set by ASHRAE Standard 62.1. But a few building code jurisdictions sometimes decide that the rates are simply unaffordable and therefore compliance is optional.

So it's useful to keep in mind that ASHRAE has no regulatory authority, unless the local authority having jurisdiction requires compliance. Also, the rates represent a consensus—a complex compromise between competing economic interests and competing technical research. The rates are the recommended minimums.

Each owner may wish to consider whether his building should really be built for these minimum rates, or whether higher rates might be beneficial to achieving the building's purposes. For example, research has documented productivity improvements which require more than the ventilation rate recommended by Std 62.1 for offices.[13] Also, owners may not realize that minimum rates are based on the perception of indoor air quality by "adapted" occupants—*not* those who have recently entered the building.

Minimum rates do not eliminate odors

The minimum ventilation rates are based on the fact that newly-arrived people entering heavily occupied spaces may indeed find the odor objectionable—but after about six minutes, most people (80%) will become "adapted" to the environment, and therefore are not likely to complain about the smell.[14]

Another way of looking at this criterion is that when the designer supplies only the recommended minimum ventilation rates, the owners should expect that 20% of the population, when asked, may indeed wish to complain about the smell, even after their noses have become acclimated to those odors.

On the face of it, this criterion seems insufficient for buildings which depend on an attractive, inviting environment to achieve their purposes. Retail stores, for example, have financial consequences if customers enter the store, perceive unpleasant odors, and then leave without buying, or if they don't linger to buy more than a minimum amount of merchandise. In the merchandising business, much has been written about the strong correlation between time-in-store and money spent per visit.

On the other hand, for buildings such as hotels, the consumer's "buying decision" is based primarily on location and price. And for elementary schools, the occupants may not have any choice at all about which building they occupy. For these and many other buildings, the perception of pleasant indoor air is not really a factor when occupants "choose the building." So perhaps in those cases the minimum ventilation rate is more economically rational from the owner's perspective, and the prospect of having 20 out of every one hundred occupants dissatisfied with the indoor environment is not a significant concern. To those who may be approaching this subject for the first time, it may seem odd that ASHRAE standards expect a significant number of unsatisfied adapted occupants. But one must consider the alternatives and economic consequences, and the biological facts about humans and odors.

First, upon initial contact with an unfamiliar scent, human perception of odors is quite acute. One suspects that this has been useful throughout human history—making quick judgements about what is safe to eat, and what environment is safe to breathe. Secondly, after some time (usually between 6 and 20 minutes) the olfactory sensors in the nose "saturate", and perception of the odors becomes less acute. Perhaps over human history, this too has been an advantage, allowing us to occupy smelly spaces shared with animals, unwashed companions and not-always-well-preserved food supplies.

The basic fact is that humans have a very different perception of odors on entering a space than after they have become adapted a few minutes later. Also, women will perceive many odors at lower concentrations than do men. And odor perception peaks in middle age, and then declines in older people.[14] Finally, the more humid the environment, the more sensitive humans become to odors. Figure 15.19 shows that as humidity rises, the thermal acceptability of different polluted environments goes down. In other words, people perceive odors as discomfort, as the indoor humidity rises above 50% rh.[15]

So there is a wide variation in what people perceive as acceptable indoor air quality, and the perception of air quality goes down as the humidity and temperature go up. The amount of ventilation air needed to avoid *all* perception of *all* odors for *all* people in *all* types of buildings under *all* circumstances would certainly be very expensive, and probably impractical from the perspectives of both sustainability and public policy.

The bottom line is that when a building owner wants a greater percentage of the occupants to perceive *high* indoor air quality, he should consider a higher ventilation rate than the Std 62 minimums for the occupied spaces, and keep the humidity much lower than the Std 62 maximum limit of 65% rh. A maximum limit of a 55°F dew point [12.8°C] would be a great improvement for the perception of indoor air quality and thermal comfort, and it would also greatly reduce the amount of indoor condensation and mold growth, as discussed in more depth in Chapter 5 (Avoiding Bugs, Mold and Rot).

The building must either exclude water, or tolerate it

When water leaks into a building, it allows the building materials to become food for mold and bacteria. These grow, and then emit unpleasant odors and particles which trigger allergies and asthma.[16]

For many years, such "sick building" symptoms were thought to be problems caused by the ventilation system. But over the last 20 years, tens of thousands of investigations in the US and in other countries have consistently shown that the really dramatic problems of microbiological growth are caused more by water leaks than by insufficient ventilation air. (Although to be fair, these investigations also show that when ventilation air is not dried, it *contributes* to these severe problems, even if it is seldom their principal cause.)

In any case, that's why Standard 62.1—the indoor air quality standard—now requires that the building not leak water, and that its materials must tolerate incidental water penetration without damage to the building enclosure (microbial growth or corrosion).

This requirement of Std 62.1 is clearly in the province of the owner, the architectural designer and the building contractor rather than the HVAC designer. Since those team members are not usually familiar with the details of ASHRAE standards, the HVAC designer would do well to make them aware of the fact that if the building leaks water, indoor air quality problems will result, and that these cannot be remedied by adjusting or redesigning the ventilation system.

Much more is said about this issue in Chapter 5 (Avoiding Bugs, Mold & Rot). But briefly, most problems are caused by water leaks around and through windows, and by leaks where different building materials come together. The basic advice is, in the words of one building scientist: "If you want to save cash—flash."[17]

Access for HVAC maintenance is now a requirement

Investigations of indoor air quality problems consistently show that a "lack of maintenance" is a frequent contributor to these problems. By this unhelpful generality, the investigator often means that cooling coils are dirty and the drain pans under them do not actually drain water away. The combination of dirt and moisture on cooling coils and standing water in drain pans can lead to odors.

But often, the reason that coils and the interior of duct work get dirty and that condensate drains get plugged is that filters are not changed regularly. And that is often because the designer has simply not allowed any space to access and inspect the coil, nor enough space on the side or in front of the HVAC equipment to pull out the filters and replace them. Frequently, the designer does not have enough space to access and change components, sometimes because the architect and owner have decided that such valuable space must be used for other purposes.

The size and location of mechanical rooms and mechanical spaces is a constant negotiation between owner, architect and HVAC designer. But the important point made by the current edition of ASHRAE Std 62.1 is that, if the owner wants good indoor air quality, the system and its components must be maintainable. And for that, they must be accessible. The HVAC designer must be allowed enough clear space beside, in front of and behind the components, for doors and windows for cleaning and inspection, and enough space to pull out filters and replace them. (See figure 15.18 for a useful tool to aid in obtaining this space.)

Manufacturers must also take note of this requirement. To comply with Std 62, the ventilation-related equipment, such as air handlers with cooling coils or desiccant wheels, must be equipped with access doors and panels before and after cooling coils, filters, desiccant wheels, fans and air cleaners. That's a lot of access, and not all manufacturers provide it in their equipment. So the HVAC designer should question the manufacturer on this point in particular.

Use the peak dew point for dehumidification calculations

The peak humidity load occurs when the outdoor temperature is moderate, not when it is hot. In hot and humid climates, the humidity load is 30 to 100% larger under moderate temperatures than it is when the outdoor air dry bulb temperature is at its peak.

Figure 15.20 shows the difference between humidity load at the peak dew point compared to the humidity at the peak dry bulb temperature. At the peak dew point, the ventilation humidity load is 320 lbs/hr/1000 cfm. In contrast, at the peak dry bulb condition, the

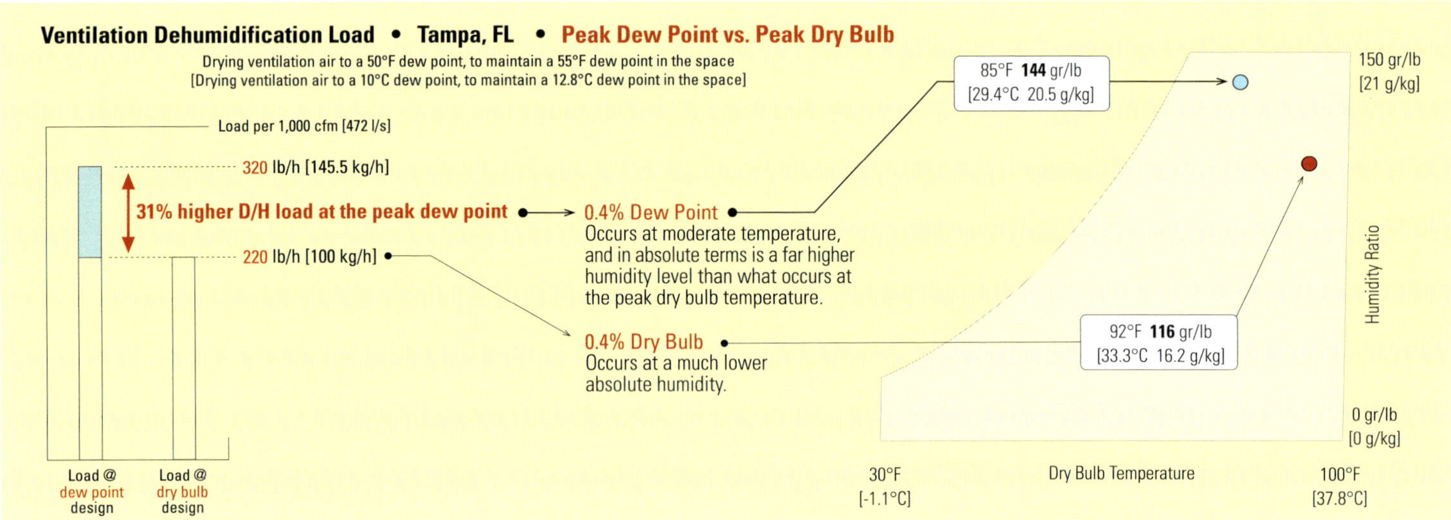

Fig. 15.20

Calculate ventilation humidity loads using the peak dew point

Standard 62 calls for calculating the dehumidification load in ventilation air using the peak dew point design data, not the peak dry bulb conditions. The humidity load is much higher at the peak dew point.

ventilation humidity load is more than 30% lower at 220 lbs/hr/1,000 cfm. [145 kg/h/500l/s vs.100 kg/h/500l/s].

This large difference is the reason that Std 62.1 requires that the ventilation dehumidification load calculation must be based on the peak outdoor dew point conditions and not the peak outdoor dry bulb conditions. Both of those sets of data are available in printed form in the 1997 and 2001 editions of the Fundamentals volume of the ASHRAE Handbook. For the 2004 edition, the climate design data expanded to more than 4400 stations. These data are not in the printed volume, but are available in electronic form on the CD which is attached to the inside back cover of the volume. A partial set of the printed climatic design data is expected to return to the printed volume in the 2009 edition, and data for more than 5,000 stations will be on the CD which accompanies that printed volume.

65% rh upper limit - a 55°F dew point is a better one

Standard 62 establishes an upper limit for indoor humidity at 65% rh. But if mold prevention is the goal, a better maximum upper limit for humidity would be a 55°F dew point [12.8°C dew point].

The logic for the relatively high 65% rh limit is not entirely clear. It was probably influenced by the fact that a few species of mold will grow on some building materials when they have an equilibrium moisture content of 70% rh.[18] Keeping the relative humidity below 65% may have been intended as a prudent margin of safety with respect to 70% equilibrium moisture content, estimated to be the lowest limit for mold growth.

The difficulty with a 65% rh limit is that the *surface* relative humidity varies widely inside a building. And the surface relative humidity governs sorption, and therefore governs moisture content.

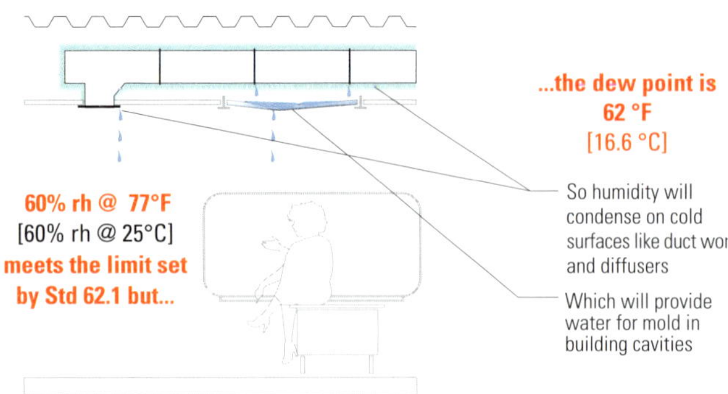

Fig. 15.21 Control the dew point rather than the relative humidity

When the indoor dew point is high, moisture condenses on materials which are cool or cold. The rh does not clearly indicate the condensation potential. The dew point does.

Fig. 15.22

High indoor dew point leads to condensation and moisture accumulation

These photos show an example of the problems that result from focusing on rh rather than on the dew point. This thermal image show an area of suspected moisture accumulation under a chilled-water fan-coil unit. The temperature of the suspect area is cool, suggesting either cold air—or evaporating moisture.

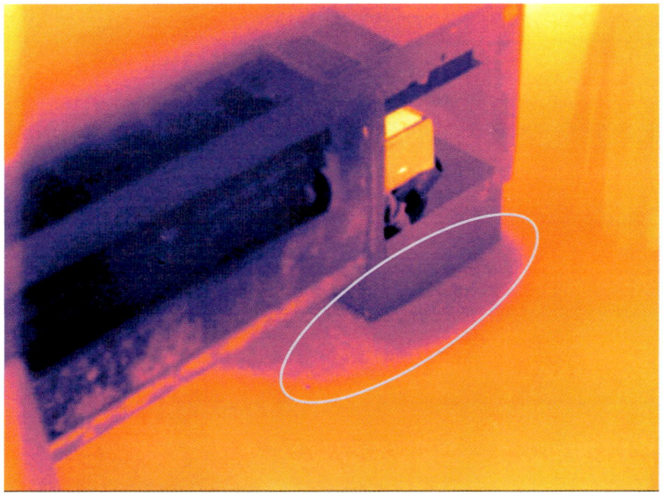

The moisture meter confirms elevated moisture in the carpet and sub-flooring.

The last photo shows the cause of the problem: constant drips from the surface of the cold condensate drain line. The moisture did not come from water leaks, but rather from surface condensation. The rh in the room is well below 65%—but the **dew point** *is over 60°F [15.5°C]. So moisture condenses on the cold surface of the condensate line. That water drips down into the carpet, then through the flooring, growing mold and eventually ruining the ceiling of the room below.*

Keeping the air in the middle of the room at 65% rh will not prevent moisture from being absorbed into a cool surface such as the area around a cold supply air diffuser. At that colder location, the surface relative humidity may be well above 80 or even 90% rh, which allows enough moisture to be absorbed into sensitive materials to grow mold. There are many examples of buildings in which relative humidity measured in the air was held below 65% rh, but which still suffered prolific mold growth in walls, ceilings and furnishings.

The logic for a clearly-defined upper limit—a 55°F [12.8°C] dew point—is that, even in an air conditioned building, there are very few surfaces with a temperature below that level. Therefore condensation and prolonged high surface relative humidity are very unlikely. So if the dew point is kept below a 55°F [12.8°C], if any condensation happens at all, there will not be very much of it, and it probably won't occur for long periods. So it's unlikely that sensitive materials will stay wet enough—for long enough—to grow mold. They'll dry out before they can grow mold, if the dew point stays low.

Commissioning, documentation & maintenance are required

Unless the ventilation system is set up and operated the way the designer intended, there's an increased risk to indoor air quality when that system operates, because in a hot and humid climate there is a high potential for microbial growth in damp or humid materials.

To avoid this risk, Standard 62.1 now requires that the system be commissioned, documented and then maintained—all in very specific ways, so that it will provide the result intended by the designer, for the energy investment assumed by the designer.

Yet again, these provisions of Standard 62.1 are prompted by field investigations of buildings, which tend to be either highly over- or under-ventilated.[8]

For example, consider one study of 140 buildings equipped with small, packaged rooftop units.[19] Among the findings were:

- 70% of the units with air-side economizers were either wired backwards (no ventilation air when outdoor conditions were moderate, and 100% outdoor air when outdoor conditions were extreme), or not wired at all, or open constantly, or closed constantly.

- 45% of the units were set up so that fans operated during unoccupied periods when the units should have shut down entirely.
- 12% of the units had less total air flow than what the designer intended and the manufacturer provided as the unit's capacity.
- 8% of the units were set up to provide no outdoor air at all, under any circumstances.
- 7% were set up so that during the summer, the units actually heated the air first—then cooled it back down to air conditioning temperatures.

For another example, consider a study of 510 randomly-selected US office buildings performed for the Environmental Protection Agency by the National Institute of Standards and Technology.[8] Among many other values, the field investigators measured the actual vs. the intended ventilation rates, and also compared the measured ventilation flow rates to the number of actual occupants as opposed to the number of occupants assumed by the owner and designer before the systems were constructed. Figure 15.23 shows the results. Against the designers' probable intent of providing 20 cfm of outdoor air person, within one standard deviation, the actual measured values were between 0 and 158 cfm/person, with 63 cfm/person being the most typical and 117 cfm/person being the average of all 510 observations. [9.4 l/s/person intended, but 29.6 l/s/person typically delivered].

In other words, the ventilation for those 510 buildings was costing about three to five times more to operate than what was necessary to satisfy ASHRAE Standard 62.1 minimum requirements. It's doubtful that's what the building owner wanted to spend. And it's doubtful that three to five times the minimum ventilation air quantity resulted in a worthwhile improvement for the occupants.

These facts of life in the real world are what prompted Standard 62's requirements for system start-up. This includes measuring and setting air flows, testing outdoor air dampers and ensuring that drain pans actually drain. Also, so that the system can be set up properly and maintained, the system must be documented. That documentation must include:

- An operating and maintenance manual describing the equipment, its operation, and its maintenance schedule.
- Controls description, diagrams, schematics, sequence of operation narrative and their maintenance and calibration requirements.
- A copy of the air balance report required during the start-up phase.
- Construction drawings of record, control drawings and final drawings.
- System design criteria and assumptions.

What the standard *does not* establish is *who* must be responsible for these tasks, exactly. So the owner, Architect, HVAC designer, commissioning agent and contractor will have to agree and document who is required to start up and document the system, if the owner wants to ensure that the building complies with Standard 62.1.

Since the other team members are not likely to be familiar with this standard, the HVAC designer might wish to point out the value of this requirement in keeping the owner's costs to a minimums and in assuring indoor air quality, and the fact that the owner must decide who will be responsible (and therefore who will be compensated) for these tasks.

Key Maintenance Aspects Of Ventilation

When ventilation systems are not maintained, they create indoor air quality problems instead of solving them.

To improve indoor air quality instead of making it worse, ventilation systems must dry and clean the incoming air. They must also make sure that the right amount of air is delivered instead of too little or too much, and they must not contribute to moisture accumulation inside the equipment or its duct work.

Standard 62.1 discusses minimum maintenance requirements extensively. That guidance is expanded and explained in more detail in the Std 62.1 User's Manual. A few of the most important maintenance issues are discussed here.

Replace outdoor air filters every month

Clean filters are a very important element in maintaining ventilation air quality, and in maintaining the reliability and capacity of the HVAC equipment.

If dirt gets into the system, mold and bacteria will grow. And if coils are dirty, they won't cool and dehumidify to the levels needed in the building. And when outdoor air filters load up with particulate, they restrict the makeup air flow, forcing the building to "go negative" and start pulling dirty, humid outdoor air through the walls, where it supports mold growth. But the question is: *how often* should the outdoor-air filters be replaced?

Standard 62.1 says: "Maintain filters and air cleaning devices according to the operating and maintenance manual." And usually, O & M manuals for HVAC equipment (and advice from filter manufacturers) is roughly: " Change filters as required to maintain indoor air quality, but in general, the outdoor air filters will need more frequent replacement than the supply air filters."

This advice is not helpful. But it stems from the fact that particulate loadings in the outdoor air really do vary widely between different locations, and between different times of the year and even different times of the day. Also, hospitals need cleaner outdoor air than office buildings or swimming pools. So it's difficult to be ***sure*** how often outdoor air filters will need to be replaced.

That said, an informal poll conducted by the author suggests that facility managers and service technicians agree that monthly replacement is a prudent interval for outdoor air filters.

In dusty areas and during construction, more frequent replacement will be needed. And in a few cases, the system operates for so few hours that its air filters can be changed less frequently. But for planning purposes, once a month is a good minimum.

Observe position and operation of outdoor air dampers every three months

After ensuring clean filters, the next most important maintenance item is ensuring that neither too little nor too much outdoor air is being brought into the building. Standard 62.1 requires that outdoor air

Fig. 15.23 Measuring and setting ventilation flows avoid waste

This study of 510 buildings shows how much money and energy is wasted when ventilation air flows are not adjusted to the actual building occupancy.[8]

*The target flows were 20 cfm/ person [9.4 l/s]. The real-world flows averaged six times higher! This is one reason why adequate ventilation gets an undeserved reputation for being expensive, and why ASHRAE Standard 62.1 now requires **measuring** and **setting** air flows at start-up and again each year.*

Measured Ventilation Air Flow - 510 Office Buildings

ASHRAE Standard Requirement (At the time of measurement)
Excess ventilation

	Per Person cfm/person [l/s/person]	Per Unit Floor Area cfm/ft2 [l/s.m2]	Air Changes Per Hour
Mean (Average)	117 [55]	0.4 [2.04]	2.00
Median (Most Common)	63 [30]	0.2 [1.03]	0.98
Standard Deviation	158 [74]	0.5 [2.54]	2.45
Std 62.1-1989	**20 [9]**	0.07 [0.36]	0.34

dampers and actuators be visually observed or remotely monitored for correct function every three months. The studies referenced earlier in this section are ample evidence of why this frequency is necessary and how it benefits the owner. Figure 15.23 shows how far ventilation airflows can be out of adjustment. Eliminating the massive energy waste of excessive ventilation is one of the easiest ways to save operating costs. This is done by measuring and resetting the ventilation air flows to match actual current requirements, rather than the requirements envisioned by the owner and HVAC designer years earlier, during design conferences.

Recalibrate CO_2 sensors & outdoor air rh sensors each year

As discussed earlier, many types of CO_2 sensors drift in calibration over time. When accuracy closer than ±100 ppm is important, the sensor should be recalibrated regularly. In most cases, annual recalibration should be adequate, but the manufacturer's recommendations will provide more certain guidance.

With respect to humidity, the sensors which control outdoor air economizers are the chief concern. Usually, these measure relative humidity rather than dew point. Condensation or near condensation on the outdoor air sensor is so common that calibration can drift from ±5 %rh to ±15% rh in a matter of weeks. Then the control system can make very poor decisions about how much outdoor air should be brought into the building, and when.

To avoid the energy waste and the excessive indoor humidity which result, outdoor air relative humidity sensors should be checked every six months and recalibrated or replaced if necessary.

Clean coils, drain pans and damp duct interiors once a year

When dirt collects on damp surfaces, it feeds and promotes the growth of mold and bacteria. The "dirty socks" smell familiar to technicians who service residential air conditioning systems results from bacteria growing in the dirt that accumulates in the middle of the wet cooling coil when the air is not well filtered.

Similarly, odors are generated from the bacteria which grow in standing water of wet drain pans, and on the dirt which usually lines the inside of duct work downstream of cooling coils. That's where moisture and high humidity make dirt particles more adhesive and a better growth medium for mold.

If filters are not replaced regularly, they can clog and blow out of their frames. Then large "clots" of dirt, insects, leaves and bird feathers can flow into the system to plug up condensate drain lines. Clogged drains lead to standing water in the drain pan, and then to all manner of unhealthy and smelly microbiological growth on cooling coils and in drain pans.

The whole point of frequent filter replacement is to avoid the accumulation of dirt inside the system. So when the outdoor air filters are changed monthly, the need for frequent cleaning is not so great. But still, at least once each year, the wet surfaces need cleaning.

Summary

Ventilation is certainly a critical element in assuring good indoor air quality for building occupants. Yet in some ways ventilation air in hot and humid climates is like strong medicine: if handled incorrectly, the cure can be worse than the disease.

Ventilation air is for people. So don't bother ventilating when nobody is in the building. And especially in hot and humid climates, the ventilation air *must be cleaned and dried, and the system must be set up properly and maintained* to avoid causing more problems than it solves.

Table 15.24
Ventilation air flow rates - Std. 62.1-2007

GENERAL NOTES

1. **Not identical to Table 6-1:** The rates in this table are based on table 6-1 of Standard 62.1-2007. HOWEVER, THE TABLE HEADINGS AND NOTES SHOWN HERE ARE NOT THE SAME. THESE WERE MODIFIED FOR CLARITY, IN COMPENSATION FOR THE ABSENCE OF THE COMPLETE TEXT OF THE STANDARD. FOR FULL GUIDANCE, CONSULT THE STANDARD ITSELF.
2. **Related requirements:** The rates in this table are based on all other applicable requirements of Standard 62.1-2007 being met.
3. **Smoking:** This table applies to non-smoking areas. Rates for smoking areas must be determined by other methods. See section 6.2.9 for ventilation requirements for smoking areas.
4. **Air density:** Volumetric airflow rates are based on an air density of 0.075 lb_{da}/ft^3 [1.2 kg_{da}/m^3], which corresponds to dry air at a barometric pressure of 1 atm [101.3 kPa] at an air temperature of 70°F [21°C]. Rates may be adjusted for actual density, but such adjustment is not required for compliance with this standard.
5. **Default occupant density:** The default occupant density shall be used when the actual occupant density is not known.
6. **Default assumptions:** These rates are based on the assumed minimum occupant densities. ASHRAE Standard 62.1-2007 states that these assumed minimum densities *shall* be used whenever the actual occupancy is not known. The rates in these columns include the ventilation air required to dilute contaminants emitted by people (at that assumed density), plus the air needed to dilute contaminants emitted by the materials and contents of the building itself. For occupancy categories without an assumed minimum occupant density, refer to the columns labeled "...outdoor air per unit of floor area" to calculate the minimum amount of outdoor air required for the space in question.
7. **Unlisted occupancies:** If the occupancy category for the proposed space is not listed, the requirements for the occupancy category most similar to the proposed use in terms of occupant density, activities and building construction shall be used.
8. **Health-care facilities:** Rates shown here reflect the information provided in ASHRAE Std 6.1-2007, Appendix E. They have been chosen to dilute human bioeffluents and other contaminants with an adequate margin of safety and to account for health variations between different people and activity levels.
9. **Occupancy-specific requirements:** Notes A - K provide additional clarification of outdoor air requirements shown in this table.

MINIMUM VENTILATION RATES IN BREATHING ZONE (Based on—but not identical to—Table 6-1 of ASHRAE/ANSI Standard 62.1 - 2007)

Occupancy category	Outdoor air per occupant, plus...		...outdoor air per unit of floor area		Notes	Default Assumptions (For use when the actual occupancy is not known)			Air Class
						Occupants per 1000ft² or 100m²	Combined minimum outdoor air[5]		
	cfm/person	L/s • person	cfm/ft²	L/s • m²			cfm/person	L/s • person	
Correctional Facilities									
Cells	5	2.5	0.12	0.6		25	10	4.9	2
Dayrooms	5	2.5	0.06	0.3		30	7	3.5	1
Guard Stations	5	2.5	0.06	0.3		15	9	4.5	1
Booking/waiting rooms	7.5	3.8	0.06	0.3		50	9	4.4	2
Educational Facilities									
Daycare (though age 4)	10	5	0.18	0.9		25	17	8.6	2
Daycare sickroom	10	5	0.18	0.9		25	17	8.6	3
Classrooms (ages 5-8)	10	5	0.12	0.6		25	15	7.4	1
Classrooms (ages 9 & older)	10	5	0.12	0.6		35	13	6.7	1
Lecture classroom	7.5	3.8	0.06	0.3		65	8	4.3	1
Lecture hall (Fixed seats)	7.5	3.8	0.06	0.3		150	8	4.0	1
Art classroom	10	5	0.18	0.9		20	19	9.5	2
Science laboratories	10	5	0.18	0.9		25	17	8.6	2
University/College laboratories	10	5	0.18	0.9		25	17	8.6	2
Wood/metalworking shop	10	5	0.18	0.9		20	19	9.5	2
Computer lab	10	5	0.12	0.6		25	15	7.4	1
Media center	10	5	0.12	0.6	A	25	15	7.4	1
Music/theater/dance	10	5	0.06	0.3		35	12	5.9	1
Multi-use assembly	7.5	3.8	0.06	0.3		100	8	4.1	1
Food & Beverage Service									
Restaurant dining rooms	7.5	3.8	0.18	0.9		70	10	5.1	2
Cafeterias/quick-service dining	7.5	3.8	0.18	0.9		100	9	4.7	2
Bars/cocktail lounges	7.5	3.8	0.18	0.9		100	9	4.7	2

Important note: These data reflect the recommendations of ASHRAE Standard 62.1-2007 as of July, 2008. However, the standard is in continuous maintenance, which means it's recommendations are likely to change more frequently than this book can be updated. When compliance with ASHRAE Std 62.1 is important, the reader should consult the most current edition of that standard (along with any amendments).

MINIMUM VENTILATION RATES IN BREATHING ZONE (Based on—but not identical to—Table 6-1 of ASHRAE/ANSI Standard 62.1 - 2007)

Occupancy category	Outdoor air per occupant, plus...		...outdoor air per unit of floor area		Notes	Default Assumptions (For use when the actual occupancy is not known)			Air Class
						Occupants per 1000ft² or 100m²	Combined minimum outdoor air[5]		
	cfm/person	L/s • person	cfm/ft²	L/s • m²			cfm/person	L/s • person	
General									
Break rooms	5	2.5	0.06	0.3		25	10	5.1	1
Coffee stations	5	2.5	0.06	0.3		20	11	5.5	1
Conference rooms/meeting rooms	5	2.5	0.06	0.3		50	6	3.1	1
Corridors	-	-	0.06	0.3		-	See note K	See note K	1
Storage rooms	-	-	0.12	0.6	B	-	See note K	See note K	1
Hotels, Motels, Resorts, Barracks & Dormitories									
Bedroom/sleeping area	5	2.5	0.06	0.3		10	11	5.5	1
Barracks sleeping areas	5	2.5	0.06	0.3		20	8	4.0	1
Laundry rooms (central)	5	2.5	0.12	0.6		10	17	8.5	2
Laundry rooms in dwelling units	5	2.5	0.12	0.6		10	17	8.5	1
Lobbies/pre-function areas	7.5	3.8	0.06	0.3		30	6	2.8	1
Multipurpose assembly areas	5	2.5	0.06	0.3		120	6	2.8	1
Office Buildings									
Office space	5	2.5	0.06	0.3		5	17	8.5	1
Reception areas	5	2.5	0.06	0.3		30	6	3.0	1
Call center/data entry clusters	5	2.5	0.06	0.3		60	6	3.0	1
Main entry lobbies	5	2.5	0.06	0.3		10	11	5.5	1
Miscellaneous Spaces									
Bank vaults/safe deposit vaults	5	2.5	0.06	0.3		5	17	8.5	2
Computer rooms (no printers)	5	2.5	0.06	0.3		4	20	10.0	1
Electrical equipment rooms	-	-	0.06	0.3	B	-	See note K	See note K	1
Elevator machine rooms	-	-	0.12	0.6	B	-	See note K	See note K	1
Pharmacy prep area	5	2.5	0.18	0.9		10	23	11.5	2
Photo studios	5	2.5	0.12	0.6		10	17	8.5	1
Shipping/receiving areas	-	-	0.12	0.6	B	-	See note K	See note K	1
Telecom closets	-	-	0.00	0.0		-	See note K	See note K	1
Transportation waiting areas	7.5	3.8	0.06	0.3		100	8	4.1	1
Warehouses	-	-	0.06	0.3	B	-	See note K	See note K	2

Important note: These data reflect the recommendations of ASHRAE Standard 62.1-2007 as of July, 2008. However, the standard is in continuous maintenance, which means it's recommendations are likely to change more frequently than this book can be updated. When compliance with ASHRAE Std 62.1 is important, the reader should consult the most current edition of that standard (along with any amendments).

OCCUPANCY-SPECIFIC NOTES

A. For high school and college libraries, use the values shown for public assembly spaces-libraries.

B. Rates may not be sufficient when stored materials have potentially-harmful emissions.

C. Rate does not allow for humidity control. Additional dehumidification may be required to keep the indoor dew point low enough to prevent structural damage to the building enclosure.

D. Rate does not include dilution and exhaust of pollutants from special effects such as dry ice vapor (CO_2) or theatrical smoke.

E. When combustion equipment is used on the playing surface (such as ice-resurfacing vehicles) additional ventilation and/or source control shall be provided beyond the rates shown in this table.

F. Default occupancy for dwelling units shall be two people for studio and one-bedroom units, with one additional person for each additional bedroom.

G. Air from one residential dwelling unit shall not be recirculated or transferred to any other space outside of that dwelling unit.

H. Floor area for estimated maximum occupancy for health care facilities is based on the net occupiable area rather than the gross floor area.

I. Special requirements or codes or required air pressure relationships between adjacent spaces in health care facilities may determine ventilation rates and filter efficiencies which are different from the values shown in this table. Also, medical or other procedures which generate contaminants may require higher rates than those shown in this table.

J. Air shall not be recirculated from autopsy rooms into other spaces.

K. ASHRAE Standard 62.1-2007 has not provided a minimum assumed occupancy for this space. However, outdoor air remains a requirement, in order to dilute contaminants generated by the building itself and it's contents. Refer to the columns labeled "... outdoor air per unit of floor area" to calculate the minimum outdoor air requirement for this space.

OCCUPANCY-SPECIFIC NOTES

A. For high school and college libraries, use the values shown for public assembly spaces-libraries.
B. Rates may not be sufficient when stored materials have potentially-harmful emissions.
C. Rate does not allow for humidity control. Additional dehumidification may be required to keep the indoor dew point low enough to prevent structural damage to the building enclosure.
D. Rate does not include dilution and exhaust of pollutants from special effects such as dry ice vapor (CO_2) or theatrical smoke.
E. When combustion equipment is used on the playing surface (such as ice-resurfacing vehicles) additional ventilation and/or source control shall be provided beyond the rates shown in this table.
F. Default occupancy for dwelling units shall be two people for studio and one-bedroom units, with one additional person for each additional bedroom.
G. Air from one residential dwelling unit shall not be recirculated or transferred to any other space outside of that dwelling unit.
H. Floor area for estimated maximum occupancy for health care facilities is based on the net occupiable area rather than the gross floor area.
I. Special requirements or codes or required air pressure relationships between adjacent spaces in health care facilities may determine ventilation rates and filter efficiencies which are different from the values shown in this table. Also, medical or other procedures which generate contaminants may require higher rates than those shown in this table.
J. Air shall not be recirculated from autopsy rooms into other spaces.
K. ASHRAE Standard 62.1-2007 has not provided a minimum assumed occupancy for this space. However, outdoor air remains a requirement, in order to dilute contaminants generated by the building itself and it's contents. Refer to the columns labeled "... outdoor air per unit of floor area" to calculate the minimum outdoor air requirement for this space.

MINIMUM VENTILATION RATES IN BREATHING ZONE (Based on—but not identical to—Table 6-1 of ASHRAE/ANSI Standard 62.1 - 2007)

Occupancy category	Outdoor air per occupant, plus...		...outdoor air per unit of floor area		Notes	Default Assumptions (For use when the actual occupancy is not known)			Air Class
						Occupants per 1000ft² or 100m²	Combined minimum outdoor air[5]		
	cfm/person	L/s • person	cfm/ft²	L/s • m²			cfm/person	L/s • person	
Public Assembly Spaces									
Auditorium seating area	5	2.5	0.06	0.3		150	5	2.7	1
Places of religious worship	5	2.5	0.06	0.3		120	6	2.8	1
Courtrooms	5	2.5	0.06	0.3		70	6	2.9	1
Legislative chambers	5	2.5	0.06	0.3		50	6	3.1	1
Libraries	5	2.5	0.12	0.6		10	17	8.5	1
Museums (children's)	7.5	3.8	0.12	0.6		40	11	5.3	1
Museums/galleries	7.5	3.8	0.06	0.3		40	9	4.6	1
Residential									
Dwelling unit	5	2.5	0.06	0.3	F,G	See note F	See note F	See note F	1
Common corridors	-	-	0.06	0.3		-	See note K	See note K	1
Retail									
Sales (except as below)	7.5	3.8	0.12	0.6		15	16	7.8	2
Shopping mall common areas	7.5	3.8	0.06	0.3		40	9	4.6	1
Barbershop	7.5	3.8	0.06	0.3		25	10	5.0	2
Beauty & nail salons	20	10	0.12	0.6		25	25	12.4	2
Pet shops (animal areas)	7.5	3.8	0.18	0.9		10	26	12.8	2
Supermarket	7.5	3.8	0.06	0.3		8	15	7.6	1
Coin-operated laundries	7.5	3.8	0.06	0.3		20	11	5.3	2
Sports and Entertainment									
Sports arena (playing area)	-	-	0.3	1.5	E	-	See note K	See note K	1
Gymn/stadium (playing area)	-	-	0.3	1.5	K	30	See note K	See note K	2
Spectator areas	7.5	3.8	0.06	0.3		150	8	4.0	1
Swimming pool (pool and deck)	-	-	0.48	2.4	C	-	See note K	See note K	2
Dance area	20	10	0.06	0.3		100	21	10.3	1
Health club/aerobics room	20	10	0.06	0.3		40	22	10.8	2
Health club/weight room	20	10	0.06	0.3		10	26	13.0	2
Bowling alley (seating)	10	5	0.12	0.6		40	13	6.5	1
Gambling casinos	7.5	3.8	0.18	0.9		120	9	4.6	1

Important note: These data reflect the recommendations of ASHRAE Standard 62.1-2007 as of July, 2008. However, the standard is in continuous maintenance, which means it's recommendations are likely to change more frequently than this book can be updated. When compliance with ASHRAE Std 62.1 is important, the reader should consult the most current edition of that standard (along with any amendments).

MINIMUM VENTILATION RATES IN BREATHING ZONE (Based on—but not identical to—Table 6-1 of ASHRAE/ANSI Standard 62.1 - 2007)									
Occupancy category	Outdoor air per occupant, plus...		...outdoor air per unit of floor area		Notes	Default Assumptions (For use when the actual occupancy is not known)			Air Class
						Occupants per 1000ft² or 100m²	Combined minimum outdoor air[5]		
	cfm/person	L/s • person	cfm/ft²	L/s • m²			cfm/person	L/s • person	
Sports & Entertainment (Continued)									
Game arcades	7.5	3.8	0.18	0.9		20	17	8.3	1
Stages, studios	10	5	0.06	0.3	D	70	11	5.4	1

Health Care Facilities (Summarizing Appendix E - ASHRAE Standard 62.1-2007 - See general note 7 and occupancy-specific note H)									
Patient rooms	-	-	-	-	I	10	25	13	-
Medical procedure	-	-	-	-	I	20	30	15	-
Operating rooms	-	-	-	-	I	20	30	8	-
Recovery and ICU	-	-	-	-	I	20	15	8	-
Autopsy rooms	-	-	0.5	2.5	J	20	See note K	See note K	-
Physical therapy	-	-	-	-	I	20	15	8	-

Important note: These data reflect the recommendations of ASHRAE Standard 62.1-2007 as of July, 2008. However, the standard is in continuous maintenance, which means it's recommendations are likely to change more frequently than this book can be updated. When compliance with ASHRAE Std 62.1 is important, the reader should consult the most current edition of that standard (along with any amendments).

OCCUPANCY-SPECIFIC NOTES

A. For high school and college libraries, use the values shown for public assembly spaces-libraries.

B. Rates may not be sufficient when stored materials have potentially-harmful emissions.

C. Rate does not allow for humidity control. Additional dehumidification may be required to keep the indoor dew point low enough to prevent structural damage to the building enclosure.

D. Rate does not include dilution and exhaust of pollutants from special effects such as dry ice vapor (CO_2) or theatrical smoke.

E. When combustion equipment is used on the playing surface (such as ice-resurfacing vehicles) additional ventilation and/or source control shall be provided beyond the rates shown in this table.

F. Default occupancy for dwelling units shall be two people for studio and one-bedroom units, with one additional person for each additional bedroom.

G. Air from one residential dwelling unit shall not be recirculated or transferred to any other space outside of that dwelling unit.

H. Floor area for estimated maximum occupancy for health care facilities is based on the net occupiable area rather than the gross floor area.

I. Special requirements or codes or required air pressure relationships between adjacent spaces in health care facilities may determine ventilation rates and filter efficiencies which are different from the values shown in this table. Also, medical or other procedures which generate contaminants may require higher rates than those shown in this table.

J. Air shall not be recirculated from autopsy rooms into other spaces.

K. ASHRAE Standard 62.1-2007 has not provided a minimum assumed occupancy for this space. However, outdoor air remains a requirement, in order to dilute contaminants generated by the building itself and it's contents. Refer to the columns labeled "...outdoor air per unit of floor area" to calculate the minimum outdoor air requirement for this space.

References

1. ASHRAE Standard 62.1-2007 (Ventilation for Acceptable Indoor Air Quality) ASHRAE, Atlanta, GA www.ashrae.org
2. 62.1 User's Manual ASHRAE/ANSI Standard 62.1 (Ventilation for Acceptable Indoor Air Quality) 2005 ASHRAE, Atlanta, GA www.ashrae.org ISBN 1-93862-80-X
3. *Proclaiming The Truth - An Illustrated History of the American Society of Heating, Refrigerating and Air Conditioning Engineers*. 1995. ASHRAE, Atlanta, GA ISBN 1-883413-20-6
4. Chapter 5 - Mechanical Systems. *Facilities Standards for the Public Buildings Service (P100 - 2005)* Office of the Chief Architect, U.S. General Services Administration, Washington, DC.
5. McMillan, Hugh and Block, Jim. "Lesson in curing mold problems" *ASHRAE Journal*, May 2005, pp.32-37 ASHRAE, Atlanta, GA www.ashrae.org
6. United States Environmental Protection Agency "The Particle Pollution Report - Current Understanding of Air Quality and Emissions Through 2003." December, 2004. EPA 454-R-04-002 U.S. EPA Office of Air Quality Planning & Standards, Emissions, Monitoring & Analysis Division, Research Triangle Park, NC www.EPA.gov/air
7. Cummings, James. Private communication
8. Persily, Andrew; Gorfain, Josh; Brinner, Gregory. "Ventilation Design and Performance in U.S. Office Buildings." *ASHRAE Journal*, April 2005, pp.30-35 ASHRAE, Atlanta, GA www.ashrae.org
9. Äsk, Andrew. "Ventilation and Air Leakage" *ASHRAE Journal*, November 2003, pp.29-34 ASHRAE, Atlanta, GA
10. National Academy of Science "Emergency and continuous exposure guidance levels for selected submarine contaminants" 2007.

The National Academies Press, Washington, DC. Text is online at no cost at http://www.nap.edu/catalog/11170.html Printed edition: ISBN 0-309-10661-3

11. Schulte, Robert; Bridge, Barry, Grimsrud, David. "Continuous IAQ Monitoring" *ASHRAE Journal*, May 2005, pp.38-45 ASHRAE, Atlanta, GA www.ashrae.org

12. The "Andy Stick" This tool is named for Andrew Äsk P.E., the innovative young engineer who first designed, manufactured and used it. The Andy Stick can be used as a motivational device during design conferences with Architects and owners. But it is primarily used, in different lengths, as a measurement tool for testing whether adequate space has been provided to allow access to equipment for adjustments and for maintenance in mechanical rooms and above ceilings.

13. Fisk, William. "A review of health and productivity gains from better indoor air quality." 2000. LBNL-48218 Lawrence Berkeley National Laboratory, Berkeley, CA. http:repositories.cdlib.org/lbnl/LBNL-48218

14. Chapter 13 (Odors) *ASHRAE Handbook—Fundamentals, 2005.* ASHRAE, Atlanta, GA www.ashrae.org

15. Fang, L.; Clausen, G.; Fanger, P.O., "The impact of temperature and humidity on perception and emissions of indoor air pollutants." Indoor Air Conference 1996 4:349-354. Tokyo Institute of Public Health. ASHRAE, Atlanta, GA www.ashrae.org

16. National Institute of Medicine *Damp Indoor Spaces and Health* 2004. Electronic files available at no cost: http://books.nap.edu/catalog.php?record_id=11011

Printed and bound copies: National Academies Press, Washington, DC ISBN 0-309-09193-4

17. This brief catch-phase is a useful aid to memory. It has been widely used by Joseph Lstiburek, P.Eng, Ph.D. Fellow, ASHRAE, and published in the Builder's Guide Series from Building Science Corporation, Westford, MA. www.BuildingScience.com

18. Flannigan, Brian and Miller, J. David. "Microbial Growth in Indoor Environments." *Chapter 21, Microorganisms in the Home and Indoor Work Environment.* 2001. Taylor & Francis, 29 West 35th St. New York, NY ISBN 0-415-26800-1

19. Jacobs, Pete. "Small Packaged System Commissioning—How to Eliminate the Most Common Problems." November 2002 *HPAC Engineering*. pp.60-61 Penton Publishing, Cleveland, OH.

20. Carter, Randall and Zhang, Jiansun. "Definition of standard office environments for evaluating the impact of office furniture emissions on indoor VOC concentrations." 2007. ASHRAE transactions volume 113, Part 2. ASHRAE, Atlanta, GA. www.ashrae.org

21. Schell, S.C. Turner, and R.O. Shim, "Application Of CO_2 Demand-Controlled Ventilation Using ASHRAE 62: Optimizing Energy Use And Ventilation," ASHRAE Transactions, vol. 104, pt. 2, pp. 1213-1225

Image Credits

Fig. 15.10 - David Bearg, Life-Energy Associates, Concord, MA

Fig. 15.12 - Damper photos: © Kruger Corporation

Fig. 15.14 - © 2005 Air Test Technologies

Fig. 15.15 - © 1999 Semco Inc.

Fig. 15.22 - Mason-Grant Consulting, Portsmouth, NH

Chapter 16

Airtight HVAC Systems

By Lew Harriman

16.1

"Advanced technology" which saves energy and reduces mold risk

Sealing up all air connections so they don't leak air is the lowest-cost way to save 25% of annual HVAC operating costs and to avoid the most notorious causes of mold in buildings in hot and humid climates. Often, the most effective technologies are the simplest ones.

Key Points

Field investigations show that air leaks into and out of HVAC equipment and duct connections are responsible for more than 25% of annual operating expense of a typical system.[1,2,3] Also, air leaks into exhaust and return ducts are partly responsible for the expensive mold problems found in buildings located in hot and humid climates.[4,5]

Therefore, to save energy and help prevent mold, all HVAC components and duct connections should be sealed up airtight, using sealants which last for the life of the system. Suggestions include:

1. Fasten duct sections to each other and to air handler casings using mechanical fasteners like screws and clamps. Then, if the duct connections do not have airtight gaskets, seal those joints and any duct seams with mastic rather than relying on duct tape. Tape is not an effective substitute for mechanical fasteners, and tape has a very poor track record as a durable air seal.
2. Don't use building cavities as either return air plenums, supply air plenums or exhaust ducts. Building cavities are notoriously leaky. If they cannot be avoided, recognize the probable energy waste and mold risks of the air leaks. Take steps to minimize those problems by sealing up all plenum seams, penetrations and connections airtight, using spray-applied smoke seal or fire sealant.
3. Seal up all the seams and connections of ducts which carry indoor air out to exhaust fans. Exhaust fans remove air from the building. If the exhaust duct connections leak, then the fan creates suction in the building cavities where the leaking seams are located. That suction often leads to infiltration of humid outdoor air, which then supports mold growth in the building cavities. Sealing all exhaust duct connections avoids this common and difficult-to-locate source of mold problems.
4. Specify airtight air handler enclosures. Then seal the joints where an air handler connects to a duct; or where it sits on a roof curb; or where it connects to the interior face of wall board. Any joint or seam near a fan sees a larger pressure difference than joints or seams located further away from that fan. Therefore, sealing the joints and seams nearest to the fans produces the biggest beneficial effect in reducing system leakage.

Airtight Systems... Are They Necessary?

Yes, they are. Airtight air systems are the lowest-cost way to reduce the HVAC energy consumption of a building by more than 25%, and airtight connections are very important to preventing mold in hot and humid climates.

Energy consumption and leaky air systems

Leaking duct connections are responsible for a surprising amount of energy waste. That's one reason that increasingly, energy codes and best practices guides require air systems to be tightly sealed and tested for air leakage.

As of the publication of this book, sealed duct connections are required by energy codes in Canada, and in the states of Florida, California and Washington State. Also, the Air Conditioning Contractors of America (ACCA) guidelines clearly and emphatically call for sealed systems.[6] As energy becomes a bigger concern around the world, this requirement for sealed air systems, which began in Scandinavia more than 20 years ago, seems likely to become standard in all countries. Sealed air systems make economic sense. Field studies have consistently shown that air leakage into and out of duct connections costs a great deal of energy—both in the form of lost cooling, dehumidification and heating, and also in terms of lost fan energy.[1,2,3]

In the past, not all energy codes and HVAC industry standards required sealed duct connections. The logic went like this: when any air leaks from or into interior duct work, its cooling effect will not really be wasted, because that air remains inside the thermal boundary of the building. But that logic has major shortcomings which have become more apparent in recent years.

In theory, cold supply air leaking out of ducts above the ceiling or behind the interior walls still cools the interior of the building. And that is true. But the lost air does not do *enough* work in the building cavities to be *effective* in cooling the occupied spaces. When less cool air reaches the occupied rooms, people run the air conditioning longer, or drop the thermostat setting, or both. That means the cooling equipment consumes extra energy, and it means the supply air fan must push extra air into the duct system to make up for the air lost through leaks. The electrical energy of the wasted fan effort is a major reason why the theory of "all-the-cooling-energy-stays-in-the-building" breaks down when energy use is actually measured.

Mold and leaky air systems

The other reason for sealing up duct connections is to avoid mold growth behind walls and above ceilings. Exhaust ducts and return air systems are a particular concern in this regard.

In the past, exhaust duct connections were very seldom sealed. Most HVAC designers and contractors reasoned that any leakage would be *into* rather than out of the exhaust duct, so odors would not escape. That logic is correct, as far as it goes. But it misses a much more important problem. Pulling excess air into the exhaust duct creates suction in building cavities, which eventually pulls untreated air into the building from outdoors.

As long ago as 1990, a study performed for the American Hotel and Lodging Association showed a high correlation between hotels which have continuous exhaust systems and those which are most likely to have major mold problems. (Figure 16.2)[4] That survey of 1100 buildings reported that of all the factors surveyed, the combination of vinyl wall covering and continuous toilet exhaust were the factors most closely correlated with major mold problems.

The field investigation behind the photos in Figure 16.3 was performed in the mid 1990's. It provides visual evidence of the mechanism which explains the high correlation between mold growth, vinyl wall covering and unsealed HVAC ducts.[5]

In that investigation of a business hotel, the unsealed exhaust ducts created a very slight suction inside the walls and above the ceilings. Fan suction pulled outdoor air into the building cavities continuously, through construction joints in the outside wall. The outdoor air provided excess humidity which condensed behind the vinyl covering of the cool walls. The resulting mold created the musty odors so familiar in humid climates. The walls had to be pulled out and replaced, to eliminate potential health risks from mold.

This same combination of leaking-ducts-creating-suction-leading-to-condensation-behind-vinyl has been seen in hundreds of subsequent investigations. These constantly-repeated errors really annoy forensic engineers and Building Scientists, who firmly believe that, nearly 20 years after the AH&LA report, both HVAC designers and building owners must surely have heard enough about this problem to understand how to avoid it.[4,5,7,8,9,10]

On the other hand, some forensic building investigators console themselves with the generous consulting fees they earn so easily, by simply changing the building name and the photos in a preformatted report showing the same problem, over and over and over again. One Building Scientist—both frustrated and delighted by the ongoing problem has sometimes remarked: "Why should I complain? The owners who keep installing unsealed air systems and vinyl wall covering have put my son through Colorado College and my daughter through Princeton!"[11]

How Much Building Leakage Is HVAC-Driven?

When HVAC designers hear about building leakage, they tend to discount the role of the HVAC system in *generating* that leakage. After all, most HVAC systems are designed with an excess of ventilation air to *prevent* infiltration.

Fig. 16.2
Early symptoms of leaky ducts
In 1990, the Executive Engineer's Committee of the American Hotel & Motel Association surveyed 1012 facilities. The survey found that the combination of continuous toilet exhaust and vinyl wall covering was closely correlated with major mold problems. Later, field investigations (Figure 16.3) showed that leaky HVAC duct connections explain this correlation.

Fig. 16.3
Leaky ducts lead to mold
Humid outdoor air flows though the construction joints in the building's wall, drawn by suction created by the unsealed exhaust ducts. The cost of sealing that duct system would have been very small, compared to the millions of dollars spent to remove and rebuild the moldy walls.

It's easy (and also appropriate) to turn to the architectural designer and the builders to ask why buildings leak so much air. Holes, open joints and unsealed construction seams are certainly beyond the control of the HVAC designer. To be sure, buildings should be built tighter than they have been, so that they will waste less energy conditioning the air leakage which enters through holes.

On the other hand, the HVAC system creates large air pressure differences as a matter of course. And if those pressure differences "escape" the HVAC system, they will pull in outdoor air through *localized* suction and wind pressures, even at the same time the building as a whole has an *average* positive pressure.

Measured wind-driven building air leakage

Buildings do leak a great deal of air. Much more than one might expect. A field investigation of 70 buildings in Florida will help the designer visualize the magnitude of these forces in typical low-rise commercial buildings. (Reference 1 - Cummings et al. 1996)

To investigate the potential for wind-driven infiltration, each building was pressurized internally with a blower, so that the average pressure difference across the exterior wall was 0.2" w.c. [50 Pa]. A pressure difference of 0.2" w.c. [50 Pa] is created whenever the wind blows against the outside wall at a velocity of 25 m.p.h. [11.2 m/s].

Once that pressure difference was achieved, the air flow through the blower into the building was measured with a calibrated venturi nozzle. The air leakage from 70 low-rise buildings showed that it does not take hurricane-force winds to create massive air infiltration. At 50 Pa positive pressure, some buildings leaked more than 50 complete air changes per hour. Others leaked less than 5 air changes per hour. But the average leakage for all buildings was about 20 air changes.

As we will discuss shortly, outdoor wind pressure is never uniform around the whole building. But even allowing for lower average pressure differences, the test results show that typical light commercial construction leaks a great deal of air. That's why the HVAC designer and contractor must take care not to create internal suction near the building wall with leaking air duct connections.

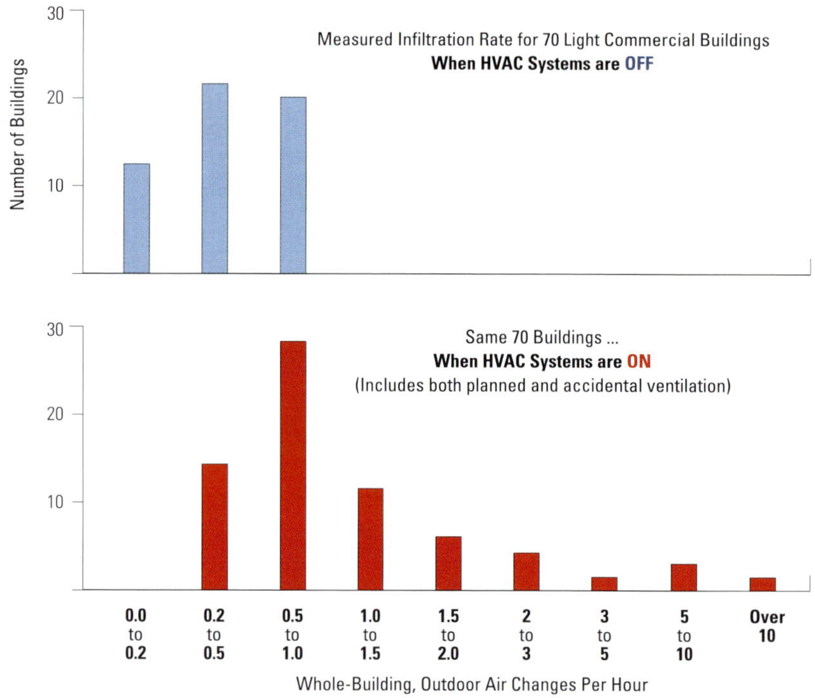

Fig. 16.4

Leaky ducts drive infiltration

This field study measured 70 light commercial buildings in Florida.[1] Note the huge increase in air infiltration rate when the HVAC systems were turned on. Leaking duct connections are responsible for this increased load, and therefore responsible for the systems' reduced cooling effectiveness. The reduced effectiveness in turn leads to needlessly high energy costs, as occupants force down the thermostat in an effort to gain comfort in spite of the humid air infiltration. Well-sealed HVAC duct connections avoid these problems.

Measured HVAC-driven building leakage

The same field investigation measured the outdoor air quantities that entered the buildings *with* and *without* the HVAC system in operation. This measurement was accomplished with tracer gas released inside the building. A faster decay in the tracer gas concentration means more outdoor air is flowing into the building.

The results are shown in Figure 16.4. At rest, the 70 buildings leaked at an average rate of 0.4 air changes per hour. But then investigators turned on the HVAC systems. With systems operating, the buildings pull in outdoor air at an average rate of 0.9 ac/h. This is far in excess of the intended amount of ventilation air. The excess comes in the form of accidental leakage caused by suction produced when fans pull air into ducts through joints that are not sealed. That suction acts on the building wall, pulling outdoor air into the building.

These field measurements stand in sharp contrast to duct system leakage values shown by standard industry references. For example, the SMACNA Duct System Inspection Guide, suggests that unsealed duct work should be expected to leak about 1.2 cfm per 100 ft^2 of duct surface.[12] Given the amount of duct work in typical commercial buildings, the total system leakage would then be about 5 cfm per 1,000 ft^2 of gross floor space. However, field measurements on 70 completed air handling systems showed they actually leak at a rate of 341 cfm per 1,000 ft^2 of gross floor space. In other words, *typical unsealed air handling systems leak at a rate 68 times larger than what the industry expects* based on leakage tests in the laboratory.

[Unsealed duct work might be expected to leak about 0.063 l/s per square meter of duct surface. Given the amount of duct work in typical commercial buildings, the total system leakage would then be about 0.027 l/s • m^2 of gross floor space. However, field measurements on 70 completed air handling systems show they actually leak at a rate

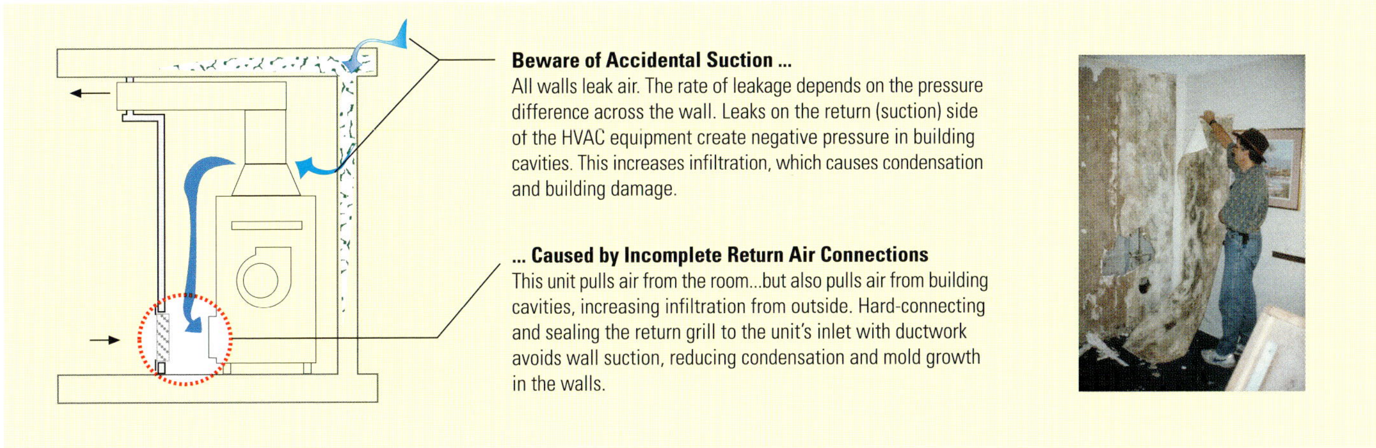

Fig. 16.5
Seal up return air connections
When return air connections are not sealed, the unit's fan will pull air from the wall cavity and eventually from outdoors. This leads to mold, as shown in the photo. To avoid the problem, seal the return air duct opening to the room-side of the wall surface.

of 1.82 l/s • m² of gross floor space. In other words, *typical unsealed air handling systems leak at a rate 68 times larger than what the industry expects* based on leakage tests in the laboratory.]

Some of the unexpected leakage comes from air handlers that are not well-sealed. Additional leakage comes from gaps between the duct inlets and the walls or ceilings. If the return grills are not sealed to the wall or ceiling, air will be pulled not from the space, but instead from the area behind the wall or ceiling. Also, while long duct runs may be well-sealed, the installing contractor may neglect to seal the straight runs to the transitions that draw air from the occupied space. Once again, that leaking joint pulls air from the building cavities.

Another classic leakage point is the casing surrounding the fan in a cooling unit mounted inside a building wall—as in packaged terminal air conditioners (PTACs) and fan/coil units set partly into the wall. If those casings leak, their fans pull air from the wall cavity, creating suction that pulls outdoor air into the exterior wall. The same problem can occur even when the casing itself is tight, if the installing contractor has not sealed the return air inlet of the unit to the inside wall surface. Then air is pulled into the unit from the wall cavity, just as if the casing itself were poorly sealed. (As in Figure 16.5)

Infiltration driven by leaking exhaust ducts is quite significant. In the Cummings study described earlier,[1] the field investigators measured the air flows entering and leaving one unsealed central toilet exhaust system in a nursing home. A single fan on the roof drew air from exhaust grills in 40 bathrooms. The total of air flows entering exhaust grills was 1324 cfm. The air flow leaving the fan was 2799 cfm. [622 vs. 1316 l/s] In other words, the leaking exhaust ducts pulled more air from building cavities than from bathrooms. (Fig. 16.6)

This leakage represents a large dehumidification load. In central Florida, the annual dehumidification load from one scfm of untreated infiltrating air is 203 lbs per year.[13] So the leaking exhaust duct system in the system described by Figure 16.6 brings a total of 299,425 pounds of excess water vapor through the hotel walls over a year's time. [An annual flow of 43 kg in every liter per second, totalling 135,639 kg/yr for this case]. That amount of excess water vapor flowing through a building certainly helps explain the mold and mildew problems of hotels in humid climates, as seen in the hotel shown in Figure 16.3.

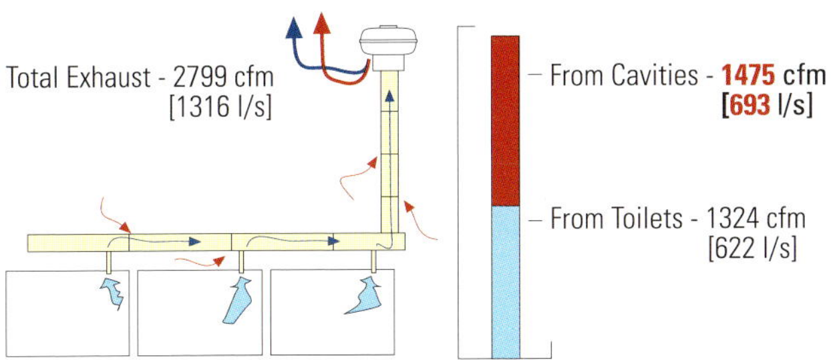

Fig. 16.6
Unsealed exhaust ducts generate large dehumidification loads and mold
In this test of a nursing home toilet exhaust system, measurements showed that more air was pulled from the cavities than from the bathrooms. The air in the cavities is eventually replaced by air from outdoors, generating a large, unplanned dehumidification load. Sealing the exhaust ducts avoids this problem.

"Positive Pressure" Alone Is Not Sufficient

To limit outdoor air infiltration, standard HVAC design practice has long been to provide a slight excess of outdoor air beyond the total volume exhausted from the building. With that slight excess of internal dry air, there's likely to more dry air leaking *outward* than humid air leaking *inward*.

To be sure, an average positive pressure is essential. If the building is "negative" (under an average negative pressure) then the outdoor air infiltration will be much worse. But while providing an excess of dried ventilation air is certainly wise, field investigations have consistently shown that an average positive pressure—by itself—is not enough to prevent large amounts of air infiltration when the air systems are not sealed up. The reasons are not obvious, and at first they seem to go against technical intuition.

In elementary physics, we were taught that air pressure is equal throughout all parts of an enclosed container like a bottle or storage tank. From this fact, which is quite correct, we often make the assumption that the same is true for a building, which seems like an enclosed pressure vessel. Unfortunately, buildings are not like bottles or storage tanks.

Buildings are highly complex assemblies of connected chambers. None of these are hermetically sealed. They have seams, cracks and joints, and all of those leaky air containers connect to other leaky containers, each of which has a slightly different pressure level. The pressure differences—therefore the air leakage volumes and directions—move in many directions at the same time inside the building itself, and also through each different section of the building's walls and roof.[14] Contributing to further complexity, the outdoor air pressures (the wind pressures) vary greatly around a building, as shown in Figure 16.15.[15]

On the windward side of the building, the average outdoor air pressure is high compared to pressure inside the building, and it varies with wind speed. So the outdoor pressure is highest at the top of the building, which is furthest away from the ground friction that slows the wind. At the same moment, on the downwind side of the building the average outdoor air pressure is low compared to the average air pressure inside the building.

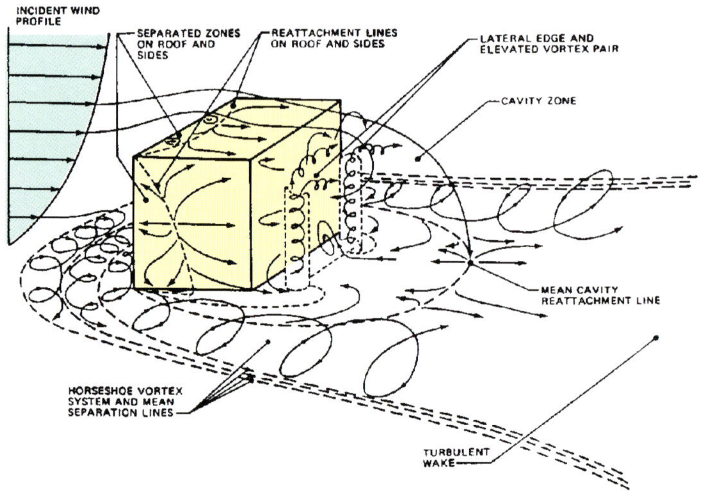

Fig. 16.7
Pressure varies constantly
It's certainly important to provide more makeup air than what is exhausted from the building. But pressure varies tremendously around the building. To avoid areas of **local** *suction, be sure to seal up the air systems which will generate areas of local suction, even in a building which is "positive"* **on average.**

These pressure differences force outdoor air though the building. Air is pushed into the building through cracks on the windward side, and pulled out through cracks on the downwind side. But as the diagram shows, the *local* pressure differences will vary widely (in both direction and magnitude) from the *overall average* pressure difference. Local suction at a single point on the downwind side could pull air in through cracks, even though the *average* downwind air pressure is below the average indoor air pressure.

From these facts, one can easily see that maintaining a positive air pressure inside the center of a building does not guarantee that the building pressure will be positive *at all points* on the exterior wall *at all times*. That's why it's important to seal all duct connections and fan casings—to avoid localized suction near the exterior face of the building, where that suction would pull in humid outdoor air.

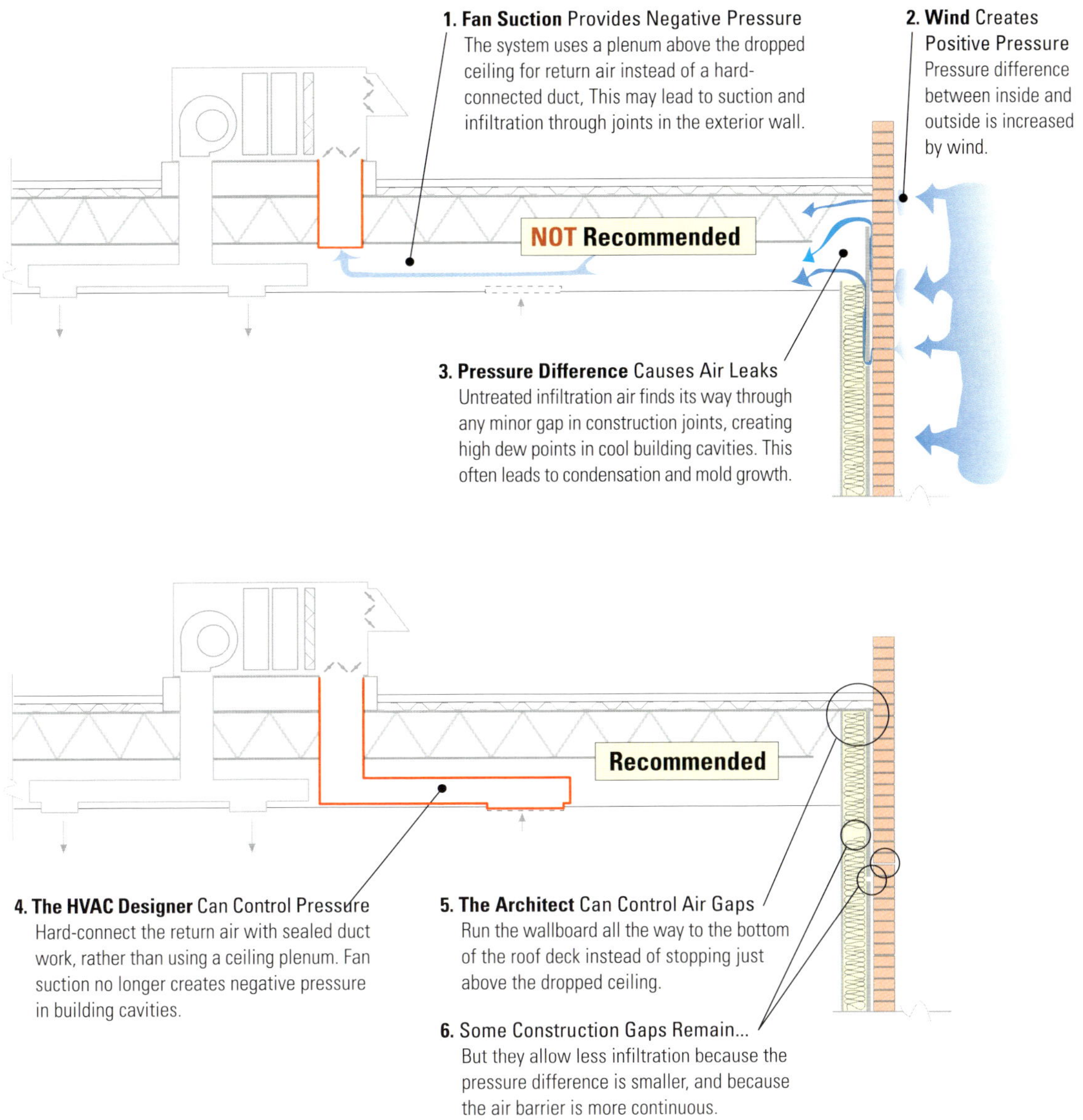

Fig. 17.8 Avoid using building cavities as air ducts—they leak too much

Designers' Guide To Limiting Air Leakage

Some aspects of system design are more critical than others. When designing to limit air leaks, focus on the locations where the pressure differences are the greatest. Also, think in terms of mechanical fasteners accompanied by mastic for every seam and every joint in the system—not just the joints between the duct sections.

Avoid return and supply air plenums

Return and supply air plenums should be avoided whenever possible. They are very difficult to seal. The cost and risks of the air leakage they create is not obvious at the time of design. And from the perspective of *sensible* cooling loads, they are not hugely problematic. It's the increased dehumidification load, increased fan energy and the increased condensation risks they create that really do the damage to the energy budget and to the building enclosure.

At a first look, return and supply air plenums save construction money. And any lost air seems to just be circulating inside the thermal boundary of the building. On closer inspection, however, one realizes that pushing and pulling air through building cavities creates suction and positive pressure, leading to leakage through the joints, seams and penetrations which are inevitable in any plenum. So not all the air reaches its intended destination. Consequently, the system must push and pull *much more* air to provide the cooling needed in the actual occupied spaces. Also, fan suction will produce local air infiltration when any negative pressure reaches the cracks and joints in the exterior wall.

The only way to avoid these problems is to *treat any plenum as a sealed duct*. But that's very difficult and expensive. The cost of duct work is usually much less than the cost of sealing up all of the joints and penetrations in a supply or return air plenum.

The leakage of unsealed air plenums is substantial. The field investigation described earlier[1] showed that buildings in strip malls have much greater air leakage than stand-alone buildings. In theory, all stores (and return air plenums) in the strip-mall are separated

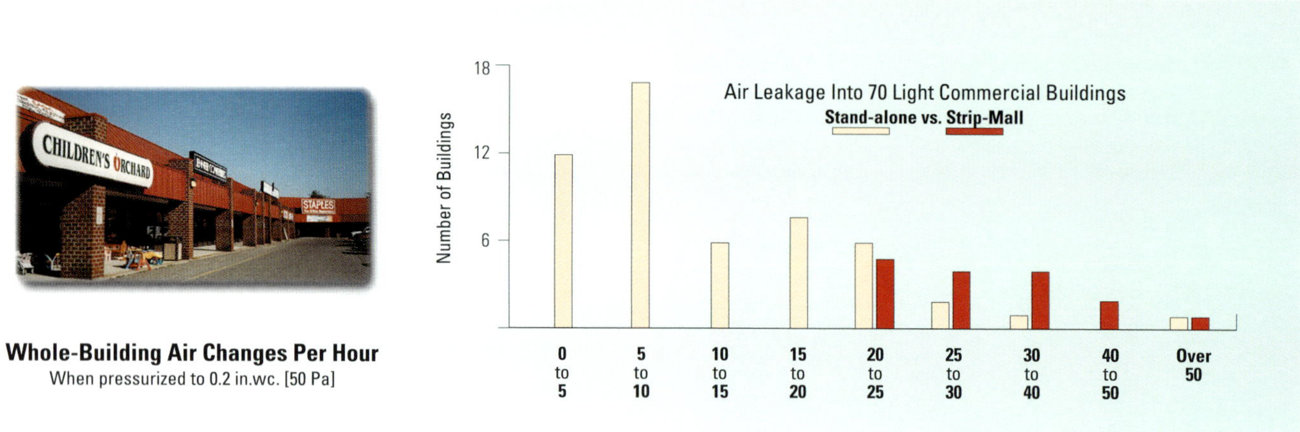

Fig. 16.9 Strip Mall Air Leakage
Strip mall building leak more than others. This may be because the owner's emphasis on costs tempts the designer to eliminate return air duct work in favor of ceiling plenum returns. These usually leak, allowing AC units to pull in outdoor air through the walls.[1]

by airtight fire walls. But the standard installation practice does not match that theory. On-site measurements showed that leaks between buildings in strip malls are very common. Suction created by a plenum return in one building pulls air from the next building. Figure 16.9 shows the impact of this leakage.

Assuming the owner has expressed a concern and identified a budget for low operating costs and better control of humidity, the HVAC designer would be prudent to invest part of that budget in well-sealed supply and return air duct work. This will provide more predictable and favorable results compared to *hoping* there will be no holes in fire walls and *hoping* that all interior finish will be tightly sealed to the roof deck and *hoping* that all wall penetrations for electrical and plumbing and telecom cables will be sealed up, airtight, after the technicians make those holes.

Spray-applied "smoke seal" for sealing up plenums

If supply or return air plenums cannot be avoided, the HVAC designer can limit the risk they create by specifying that the plenums shall be sealed up, airtight, using spray-applied "smoke seal." Smoke seal is durable, resilient and made for bridging small cracks and gaps. It is flame-spread-rated, so it can generally be used for air-side duty. But it is not fire-rated, so it is less expensive than fire-rated sealants. Smoke seal is applied by the same contractors who accomplish the fire-rated sealing of building cavities and wall and floor penetrations. An example of such spray-applied sealants is shown in Figure 16.10.

Roof curbs

Roof-mounted, packaged air conditioning equipment is supported by curbs, which hold the equipment up above the roof surface. When the supply air enters or leaves the HVAC unit through that roof curb (when the curb extends around the perimeter of the equipment) the seam between the equipment and its curb must be sealed up so that it does not leak either water or air.

Fig. 16.10 Spray-applied fire sealant
When using building cavities as air ducts cannot be avoided, they must be sealed up, airtight. This is difficult, but spray-applied fire and/or smoke sealants can be effective. Equally important, the contracting infrastructure exists for expert application, because fire safety codes require air sealing of any penetrations of fire-rated walls and floors.

Fig. 16.11
Connections to air handlers
At the connections to air handlers, pressure differences are greater than at any other point in the system. So those connections must be especially airtight.

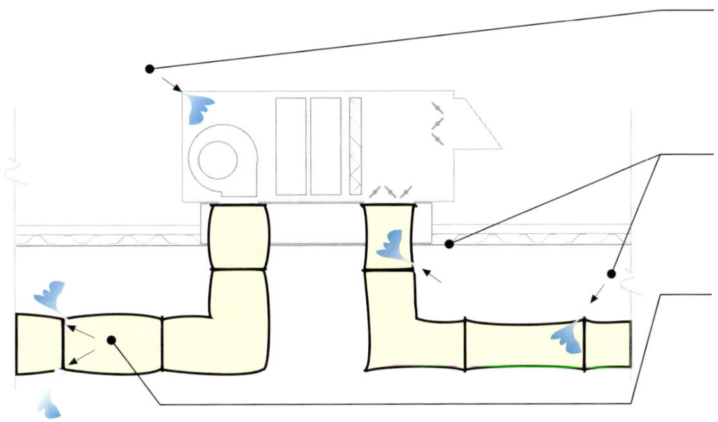

1. Eliminate Air Handler Leaks
Most commercial-grade air handlers leak a great deal of air. Seal open seams with tape and mastic, and make sure all access panels are gasketed, with all screws in place.

2. Seal Return Ducts
If return ductwork leaks, it creates low pressure in building cavities, which sucks untreated air into the wall from the outdoors. Sealing return duct work helps avoid condensation from untreated air in these cavities.

3. Seal Supply Ducts
When supply ductwork leaks in summer, cold air leaks into building cavities, cooling surfaces, which condense moisture. In winter, warm supply air carries moisture into cold building cavities, where it condenses once again. Sealing supply air ducts prevents these problems.

The fan suction is greatest at the seam between the unit and the curb. When that joint is not sealed, fan suction will pull in both humid outdoor air, wasting cooling capacity, and also pull in rain, which can drip down and soak the ceiling below the unit. Whenever air moves into and out of the unit through the curb, that seam must be sealed up so it is both airtight and watertight.

Connections to and from air handlers

Many times, duct work is installed by a sheet metal subcontractor, while the AC equipment and other HVAC components are set in place by either the mechanical contractor, or the general contractor.

In that common commercial practice, the importance of the connections between the ducts and the air handlers can be overlooked. The duct sections might be sealed up tightly, but the connections between those sections and the equipment they connect to might be quite leaky, e.g.: mastic and reinforcing tape on all the duct connections—except for the connections to and from the air handlers, which are held in place by drill screws and no sealant whatsoever.

To avoid this problem, specify clearly that all joints between ducts and equipment must be sealed up, using mastic and reinforcing glass fiber tape, by the mechanical contractor (Figure 16.11).

Seal *all* supply, return and exhaust air duct connections

If the cold supply air does not *all* arrive in the conditioned space, the fans will have to push more air through the system to make up for that loss. Also, if the cold air leaks into building cavities or semiconditioned spaces like vented attics, it will chill any surfaces near the leak site, which often leads to condensation and mold growth above ceilings and inside walls.

Also, when exhaust and return air connections leak air, they can generate suction in the cavities they pass through. Often, that suction leads to infiltration of outdoor air and then to condensation and mold growth, as described in more depth in Chapter 5 (Avoiding Bugs, Mold & Rot).

Really, there are no short cuts. To reduce mold risks and to minimize energy use, *all* duct connections must be sealed up airtight, especially where those ducts connect to other air system component such as a cooling coil, filter housing, a VAV box or any form of air handler, including in-wall packaged AC units.

In-wall packaged AC units and fan-coil units

Often, room-specific air conditioners are set partly into the wall, and draw their return air from the room through grills set into the face of that wall. Examples of such equipment include chilled water fan-coil units, packaged terminal heat pumps (PTHP's) and packaged terminal air conditioners (PTAC's).

The potential problem with this equipment was shown in Figure 16.5. If the return air inlet to the fan casing is not sealed tightly to the room-side of the finished wall surface, the fan will pull air from—and depressurize—the wall cavity. That suction encourages humid outdoor air to enter the building into the cool wall, where moisture will condense and support mold growth. To avoid this problem, specify that all return air inlets of in-wall AC equipment must be sealed to the room side of the interior wall surface.

Owners' Guide To Reducing Air Leakage

Nothing in commercial buildings and HVAC systems is perfectly airtight. Many building owners have systems and structures which—for whatever reason—leak a great deal of air. Here are a few suggestions for locating and quantifying air leakage, both on the HVAC side and for the building enclosure as a whole.

Air system leaks - Tools & techniques

The largest amount of leakage is likely to occur where the pressure difference is the greatest. That will be where the supply and return ducts meet the air handling unit, and at the doors and seams of the air handling unit itself.

Leak detection at duct connections - Puffer and pressure

Using a hand held smoke puffer, a technician can move slowly over the unit's seams and doors, and over the supply and return duct connections to pinpoint locations where air leakage is especially large. (See Figure 16.12) The smoke will be pulled rapidly into, or pushed away from, a leak point.

Where supply ducts can be visually accessed, leaking duct connections can be located by using a theatrical "fogger" to flood the system with a fine mist. Where fog escapes the duct work on its way to the supply registers, the location of a leak is relatively easy to see.

However, in most installations, the supply and return ducts will be covered by external insulation, so that leaks in long duct runs will not be visible. On the other hand, long duct runs are often well sealed, or at least do not leak a great deal of air. The evidence one way or the other should be visually clear by looking at the insulation jacket when the system is in operation. If there is a great deal of air leakage under the insulation, the return air insulation will be sucked tighter onto the duct, and the supply air insulation will be puffed-up and pushed away from the duct. Leaking sections may be identified this way.

With well-covered duct systems, the first places to inspect are all the connections between the duct work and the surface of the interior finish. Then inspect the seams where the straight runs meet the transitions to and from the space. These seams are often leaky, and sealing them may eliminate the need to strip away insulation from difficult-to-reach sections of duct further into the system.

Calibrated duct fan to measure leakage

To help the owner decide where to regain AC capacity and avoid mold by sealing leaks, technical services are available—often through test and balance contractors—to quantify the amount of air leakage in each part of the duct system. As shown by Figure 16.13, a calibrated fan is attached to a duct system. All of the other inlets and outlets of that system are temporarily sealed using sheets of cardboard and "painter's tape". The fan speed is increased until the pressure in the duct work is raised to 0.1 in.w.c. [about 25 Pa]. The amount of air passing through the fan to maintain that pressure (and therefore the amount of air leaking out) is measured by recording the pressure difference across a smooth orifice that forms the fan casing.[16,17]

Fixing leaks - Mastics & tapes

After the leaks have been located, the first step is to reconnect any detached duct connections, and then make those connections secure using mechanical fasteners such as screws, clamps, bolts, pop rivets or spot welds.

Fig. 16.12 Smoke puffer
This low-cost device shows air currents, which helps locate air leaks in duct systems and building enclosures.

Fig. 16.13 Quantifying duct leakage

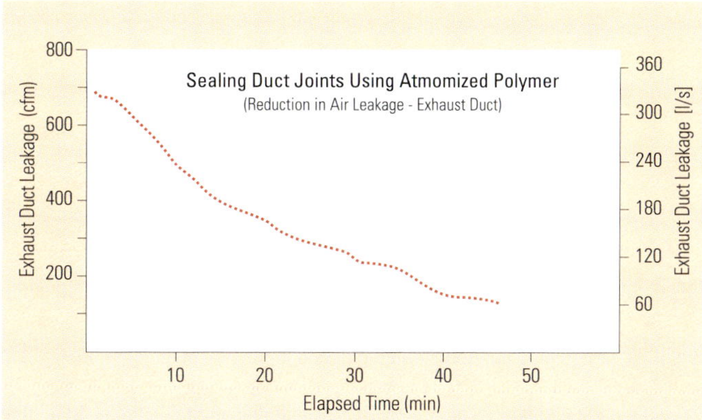

Fig. 16.14 Reducing leakage volume

An atomized polymer, injected into a leaking duct, can seal the leaks without the need to access each and every joint. This graph shows the short time needed to reduce air leakage in an unsealed exhaust duct, using this method.[18]

After the connections have been mechanically fastened, use mastic to seal the connection airtight, as shown in Figure 16.1. Note the use of a disposable glove to apply the mastic. Those who have become experts in retrofit sealing of duct work recommend putting on several layers of disposable gloves, then peeling them off as they clog with the hardening mastic. While brushes seem like the tidier application tool, the human hand can more easily reach around behind ducts, and can feel whether the seam has enough mastic on it, in the right location, to be sealed effectively.

For larger and more accessible duct connections, brush application of mastic may be quite practical. And for much larger ducts, the fire-rated spray-on mastics shown in Figure 16.10 may be more practical than applying the mastic by hand.

Also, while we have spent quite a bit of ink warning about the unreliability of tape-based duct sealing, there are some types of tape which have proven to be reliable. In the category of self-adhesive tape, products which include a layer of metal foil with butyl rubber adhesive are rated for the hot temperatures and temperature changes that defeated the less robust fabric-based tape seen in Figure 16.16.

Fixing leaks - Atomized polymer sealing system

In most large buildings with complex duct systems, there are very few practical ways to locate leaks after the obvious leaks have

Fig. 16.15 Equipment for injecting atomized polymer into a duct, to seal a leaking exhaust duct system[18]

been sealed at the air handlers and at the entry and exit points from the occupied spaces. Where hidden duct leakage remains large and spaces are not accessible, the owner may wish to simply seal the entire system internally. Work done at the US Department of Energy's Lawrence Berkeley Laboratory has shown that fogging technology can be used to seal up duct leaks from the inside, using a fine mist of suspended adhesive polymer particles.[18] As the adhesive mist escapes through any leaking connection, it quickly clogs that seam, sealing up the air leaks.

Figure 16.141 shows the measured reduction in air leakage from an exhaust duct system, as the atomized polymer finds its way to the leak points and seals them up.[19] The process requires special equipment and technical experience (Figure 16.15), but can make a big improvement in a short amount of time, and can seal up duct joints which are simply inaccessible by any other means.

Whole building leaks - Tools & techniques

A useful first step to determine whether the building as a whole is under positive or negative air pressure. When the doors are cracked open, does air leak inwards (negative internal pressure) or does air leak outwards (positive internal air pressure)? This test is very fast and inexpensive. It can be done with a strip of lightweight facial or toilet tissue paper, held near the door jamb when the outside door is cracked open slightly. The technician could also use a theatrical smoke generator or a smaller, easily available source of visible particles such as a lit cigarette or incense stick. The direction of the air leakage will be most obvious in the early morning, when the outdoor air is relatively still and not yet windy.

Another simple procedure for measuring total leakage in densely-occupied buildings is to measure the average CO_2 concentration in the return ducts when the building is occupied, and then measure that same concentration at short intervals as the building is being vacated at the end of the day. The rate of decay in average CO_2 concentration is a rough indicator of the overall ventilation rate.

Accurately quantifying whole-building leakage is an expensive and complex task. The larger the building, the more difficult it will be to locate and quantify the air leakage. For smaller buildings, the whole-building blower door test gives the quickest and most economical results. For taller, larger or more complex buildings, tracer gas tests are usually the most practical alternative.

Blower doors

As shown in Figure 16.16, a blower door includes a fan mounted in a panel that fits tightly into a doorway. After fitting the panel and sealing it with gaskets, the technician starts the fan, which pushes air into the building.[20] The fan speed is slowly raised until the pressure difference between indoors and outdoors reaches a defined limit. Standard test pressure differences are either 0.016 or 0.2 in.w.c. [4 or 50 Pa]. When the test pressure is reached, the technician reads the air flow rate being produced by the fan from the blower door's instrument panel. The blower's air flow rate is the amount of air that is leaking through the building envelope at the pressure difference selected for the test. The procedure provides a useful indication of the magnitude of the leakage at a defined pressure difference.

The total leakage number does not tell the building owner exactly where the leakage is occurring. But when a blower door puts the building under a positive pressure, technicians can find many leaks by working in pairs and using indicating smoke. One technician working inside flows smoke around the window frames and other exterior wall penetrations. The other technician, working outside, can sometimes see the smoke leaking out through cracks in the exterior walls, which locates the larger, more obvious leaks.

The basic blower door technique has also been used to quantify leakage in tall buildings.[21] Each floor is isolated by sealing up all openings other than the door to the fire stairs. That doorway is fitted with a blower, and the floor is put under positive pressure. While the procedure has been used successfully, it requires complex technique and great care to separate the air leakage through the exterior wall from the air leakage to different floors within the building. Bahnfleth concluded that tracer gas tests are usually more practical and less disruptive for quantifying and locating leaks in tall or complex buildings.

Tracer gas

Gas concentration inside a building is slowly diluted by outdoor air flowing into that building. In a tracer gas test, a known volume of an inert gas, often sulfur hexafluoride (SF_6), is released in the space and the time is noted. At periodic intervals, the technician pumps a sample of the indoor air into a vapor-tight plastic pillow and seals it securely. Later, in the laboratory, the sulfur hexafluoride concentration of each sample is measured and recorded. Since the volume of the building and the initial volume of gas is also known, the air infiltration rate can be calculated from the decline in the tracer gas concentration over the known time interval. Sulfur hexafluoride is especially useful, because it is inert and because it does not occur in high concentrations in nature. Sulfur hexafluoride requires measurement care, but its signature is usefully unique.

The ubiquitous and easy-to-handle CO_2 can also be used as a tracer gas, and its concentration can be measured in real time by

Fig. 16.16 Blower door

In smaller buildings (or sections of larger buildings) the whole-building leakage can be quantified using one or more blower doors.[20]

Fig. 16.17 Real-time, multipoint tracer gas analysis equipment

In larger buildings or in multiple spaces, multipoint tracer gas measurements can quantify air leakage more easily than using blower doors.[22]

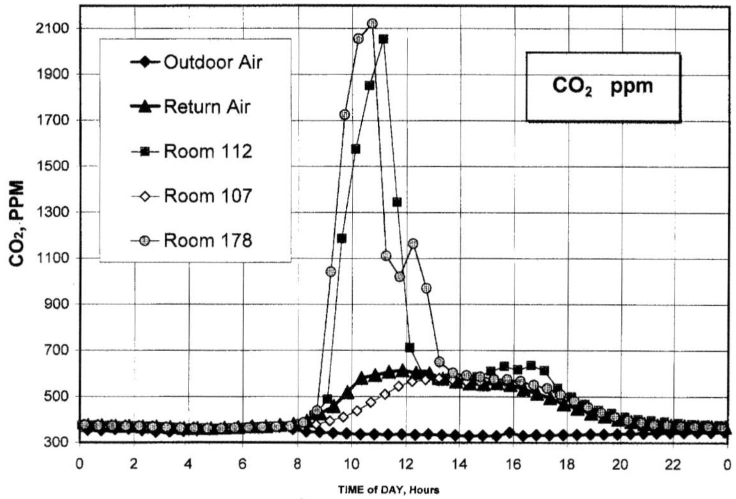

Fig. 16.18 Results of real-time, multipoint tracer gas analysis

handheld instruments. Again, the investigator releases a known quantity of CO_2 into the space and notes the time and the number of people occupying that space, as well as the concentration in the areas around the target space. The rate of decay in CO_2 concentration indicates the rate of air leakage into and out of the space, after adjusting for the CO_2 generation of the occupants.

Tracer gas techniques can be used to show air exchange rates in many isolated parts of the building over the same test interval. So compared to a blower door, tracer gas tests make it easier to quantify leakage in different parts of a building. Also, unlike blower door tests, tracer gas tests can be run both with and without the HVAC systems in operation. Comparing the results of two tests, the technician can calculate how much outdoor air infiltrates into a 'passive" building, and how much outdoor air enters when the system is in operation.

On the other hand, the tracer gas test takes time, and the sulfur hexafluoride technique requires lab equipment. To address those shortcomings, equipment has been developed to monitor tracer gas concentrations in real time from many parts of a building.[22] The equipment consists of a gas analyzer equipped with a suction pump and tube/valve manifold. Each tube runs to a different part of the building, and the suction pump draws a small sample of air through all tubes continuously. Every few seconds, the valve manifold switches, flowing a different tube's air through the gas analyzer. That way, different tracer gas concentrations from all over the building can be recorded in a matter of a few minutes. With this arrangement, the measurements are taken so frequently that transient phenomena such as door openings and short meetings can be seen in the gas concentration records.

An example of a report from a multipoint gas analyzer is shown in Figure 16.18. In that case, the issue under investigation was ventilation effectiveness. The CO_2 concentration was recorded in different parts of the building over a typical 24-hour operating day.

The graph shows how room 107 is very well ventilated, since its gas concentration is nearly the same as outdoors. In contrast, rooms 178 and 112 are considerably under-ventilated compared to their occupancy—a fact that is not apparent when measuring the average CO_2 concentration in the return air, which seems quite adequately low. This same technology is used to find and quantify outdoor air leaks in different spaces when the air system is turned off and the building is unoccupied.

Tracking down and quantifying specific leak locations

To locate leaks on the exterior walls, focus on the seams between different building components. Those are usually where the largest air gaps occur. Leak points can be located from inside or outside the building by using hand held "smoke" puffers such as those shown in Figure 16.12.

When working *inside* the building, first make sure the building is under a negative air pressure. This can be accomplished by shutting off the make-up air while allowing the exhaust fans to continue operating. Then move the smoke puffer slowly along the edges of windows and the edges of any wall penetrations, such as those made to accommodate wall-mounted air conditioning units, and those around electrical outlets. An air leak will blow the smoke away from the

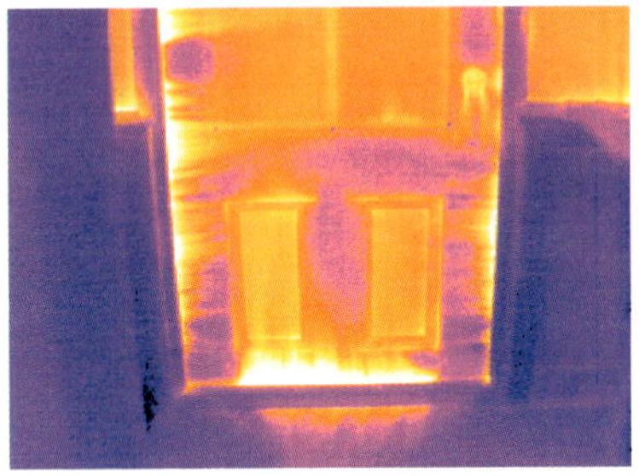

Fig. 16.19 Thermal image to locate infiltration
This thermal image, taken from inside a residential building, shows warm outdoor air flooding ***into*** *the building around the door. At this part of the building, the building pressure is negative.*

wall rapidly. This test is simple and can be performed after working hours without any great disruption to building occupants. It locates the indoor end of the worst air leaks. But to locate the outdoor end of the leak, two people will be needed.

Working *outside* the building, first put the building under a positive air pressure by increasing the amount of make-up air and reducing the exhaust air flow. Then, working in pairs an observer is stationed outside the building while the partner slowly moves a smoke puffer around window frames and other wall penetrations from inside the building. The outside partner notes any appearance of smoke leaving the building, which indicates a crack on the exterior. Locating leaks from indoors is usually much easier than working from outdoors. By the time the smoke or vapor makes its way out of the building, it's quite diluted and difficult to see, especially from a distance.

Thermal cameras for leak detection and pressure management

When there is a temperature difference between air indoors and outdoors, thermal cameras are quite useful in locating leak locations.

Figure 16.19 shows an example of a thermal image of an exterior door, seen from inside the building. The image shows a pattern of warm outdoor air entering around the door, pulled inwards by negative pressure. The image highlights the fact that a considerable amount of air is coming in around the door, so its gaskets and air seals should be replaced. Also, the suction that generated this air infiltration is extreme. The HVAC system must be re-balanced so the cleaned and dried ventilation air exceeds the exhaust air flow. When the doors show this much infiltration, it's highly probable that air is also leaking inwards at other locations which are more difficult to see.

Fig. 16.20 Thermal image to locate outward air leakage
This thermal image, taken from outside a commercial building, shows cool indoor air ***escaping out*** *of the building through construction joints. This particular building is pressurized more than it needs to be to prevent infiltration. Ventilation air can be reduced when the building is not occupied, saving money and reducing energy waste.*

Thermal cameras only see surface temperature patterns in their line-of-sight. They don't see through walls and furnishings as would happen with X-rays. So when furniture is set against the exterior wall, or when the HVAC system is leaking air behind walls, the thermal images will reflect those problems rather than any infiltrating outdoor air.

The cameras are also useful for checking the effectiveness (and any excess) of positive air pressure. The image in Figure 16.20 shows the exfiltration of air from a building being held under positive air pressure. The cool patterns at the upper corners of the windows show that indoor air is escaping at those locations.

The pattern shown in Figure 16.20 is both good news and bad news for the building operator. The good news is that the building is under positive pressure, so humid outdoor air will be mostly excluded from the building. On the other hand, the pattern is very pronounced. Probably, the amount of excess ventilation air should be reduced to avoid wasting energy by needlessly over-pressurizing the building.

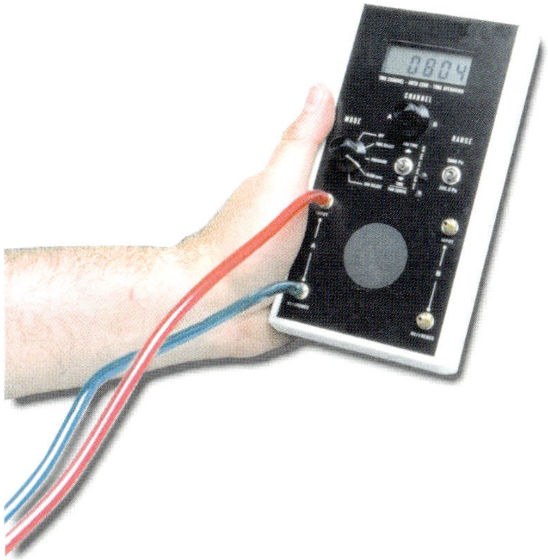

Fig. 16.21 Averaging micromanometer

*Relevant pressure differences in building investigations are so small (4-10 Pa), that the direction of pressure differences will reverse many time **per second**. Using an averaging micromanometer, one can more reliably determine both the direction and magnitude of a building's pressurization.*

Measuring and adjusting building pressure

When HVAC systems are being tested and balanced, it is often useful to quantify the local pressure difference between the room and the building cavity, and between the cavity and the outdoor ambient. The goal is to have a slight positive internal air pressure. But these pressure differences are very small. Their direction reverses many times in a single second. So a digital micromanometer with a time-averaging feature is the most useful tool for this job. (Figure 16.21)

The manometer must be very sensitive, because the pressures are so small that the process has been compared to "measuring the force of a butterfly's cough." Relevant pressure differences are likely to between 0.002 and 0.016 in.w.c. [between 2 and 5 Pa]

From the manometer, one tube is placed outside the building, and the other placed inside the wall cavity. The micromanometer must average hundreds of readings during several minutes to obtain the average pressure difference. The air handling system can be adjusted using these measurements. Also, local pressure excursions that reflect local leakage can be identified through readings taken near external wall penetrations.[23,24,25]

The averaging micromanometer can also be used for quantifying the pressure difference between the wall cavities and the conditioned space. Odom and co-workers suggest the use of a gasketed metal pie plate, taped to the inside surface of the wall so that it covers an electrical outlet.[23] One of the manometer ports is open to the room. The other is connected through a tube to the inside of the pie plate. With this arrangement, the digital micromanometer can measure the pressure difference between the wall cavity and the room, without the need to drill holes in the wall.

Measuring the pressure difference between indoors and outdoors is often difficult in commercial buildings, because windows are often sealed units. But where windows can be slightly opened, the investigator can use "painter's tape" to seal off the crack at the open window. Then punch a small hole through the tape to insert the manometer tube. Such tape is designed to adhere tightly for hours, but still come off without damage to surface finishes.

Air sealing contractors

Energy codes in many parts of North America and Europe have led to the establishment of air sealing contractors. These firms make it their business to locate and seal up air leaks in buildings and also in duct work. They use the techniques discussed here, as well as other proprietary techniques and materials.

So when the building owner or its operator would like to enjoy the energy benefits of air tightening, or wishes to avoid the mold risks of leaking air systems in existing buildings, it is possible to contract for these services rather than performing the work in-house.

Summary

Air leakage into the building envelope from both inside and from outdoors is one of the main causes of energy waste and mold growth. Sealing up the air systems—and their connections to the occupied spaces they serve—goes a long way towards reducing energy waste and avoiding mold problems, at very low cost compared to any other way of achieving those valuable benefits.

References

1. Cummings, James B, Withers, C.B, Moyer, N, Fairey, P., McKendry, B, *Uncontrolled air flow in non-residential buildings*. April 15th, 1996, Final Report of FSEC project number FSEC-CR-878-96. Florida Solar Energy Center, 1679 Clearlake Rd, Cocoa, FL 32922.
2. Henderson, Hugh; Cummings, James; Zhang, Jian Sun; Brennan, Terry. *Mitigating the Impacts of Uncontrolled Air Flow on Indoor Environmental Quality and Energy Demand in Non-Residential Buildings*. 2007. Final Report - Project # 6770. New York State Energy R & D Authority, Albany, NY

3. Wray, Craig. *Energy impacts of leakage in thermal distribution systems*. 2006. Report to the California Energy Commission. Lawrence Berkeley National Laboratory. Berkeley, CA. http://epb.lbl.gov/ Report no: PIER II #500-98-026

4. Harriman, Lewis G, III and Thurston, Steven, *Mold in Hotels and Motels—Survey Results.* 1991. American Hotel & Lodging Association. Washington, DC.

5. Shakun, Wallace. "A review of water migration at selected Florida hotel/motel sites." Proceedings of the biennial symposium on improving building practices in hot & humid climates. October 1990. Texas A&M University, College Station, TX.

6. Air Conditioning Contractors of America (ACCA). 2002. *Residential Duct Systems - Manual D*. Section 12.3 Leakage losses; Section 12.4 Leakage Loads; Section 12.5 Efficiency, operating costs and demand load; Section 12.6 Figure of merit for air distribution systems. ACCA, Arlington, VA, www.acca.org

7. Harriman, Lewis G. III, Lstiburek, Joseph and Kittler, Reinhold. 2000 "Improving humidity control for commercial buildings." ASHRAE Journal, November, 2000. pp 24-32 ASHRAE, Atlanta, GA www.ashrae.org

8. Harriman, Brundrett & Kittler, 2008. *ASHRAE Humidity Control Design Guide*, ISBN 1-883413-98-2 ASHRAE, Atlanta, GA

9. Lstiburek, Joseph, "Humidity Control in the Humid South." 1993. Proceedings of the 2nd Conference on Bugs, Mold & Rot. National Institute of Building Sciences, (NIBS) Washington, DC.

10. West, Mike and Harlos, David. 2006. "Investigating and resolving moisture problems in a Florida office building." HPAC Engineering, December, 2006. pp.30-37. Penton Publishing, Cleveland, OH. www.penton.com

11. Remarks frequently made during seminars to professionals by Joseph Lstiburek, Ph.D, P.Eng. 2003, 2004, 2005, 2006, 2007, 2008. Building Science Corporation, Westford, MA. www.BuildingScience.com

12. SMACNA. *HVAC Duct Systems Inspection Guide* (15D, 1989 The Sheet Metal Manufacturers and Air Conditioning Contractors National Association 8224 Old Courthouse Rd., Tyson's Corner, Vienna, VA. 22182 (703) 790-9890 www.smacna.org.

13. Harriman, Lewis G. Kosar, Douglas and Plager, Dean. 1997. "Dehumidification and Cooling Loads from Ventilation Air." ASHRAE Journal, November, 1997 pp.37-45. ASHRAE, Atlanta, GA. www.ashrae.org

14. Lstiburek, Joseph. "The Pressure Response of Buildings." 1999. Proceedings of the 7th Conference on Thermal Envelopes, pp:799-817. ASHRAE

15. Bearg, David W, *Indoor air quality and HVAC Systems* 1993, ISBN 0-87371-574-8, Life Energy Associates, Concord, MA 01742

16. Nelson, Gary. "Duct leakage testing." 1998. The Energy Conservatory, 2801 21st Ave. South, Suite 160 Minneapolis, MN 55407 (612) 827-1117 http://www.energyconservatory.com/duct.html

17. Sherman, Max. "The use of blower door data." 1997. Lawrence Berkeley Laboratory, Energy Performance of Buildings Group, Berkeley, CA. PDF document published at: http://epb1.lbl.gov/blowerdoor/

18. Modera, Mark. "Repairing your duct work." 1996. Aeroseal, Inc. 75 Fairview Ave. Oakland, CA 94610 (510) 601-8575 http://www.aeroseal.com/repair.htm

19. Modera, Mark. Repairing your duct work. 1996. Aeroseal, Inc. 75 Fairview Ave. Oakland, CA 94610 (510) 601-8575 http://www.aeroseal.com/repair.htm

20. Nelson, Gary. Air leakage testing of buildings. 1997. The Energy Conservatory, 2801 21st Ave. South, Suite 160 Minneapolis, MN 55407 www.energyconservatory.com/airtight.html

21. Bahnfleth, William P., Yuill, G.K., Lee, B., "Protocol for field testing of tall buildings to determine envelope air leakage rate." 1999. ASHRAE Transactions, V.105, pt 2.

22. Bearg, David W., "Monitoring for ventilation and airtightness." ASHRAE Transactions, V. 106, Pt.1, 2000.

23. Odom, J, David, III, and DuBose, George, and Fairey, P.W., "Why HVAC commissioning procedures do not work in humid climates." 1993. ASHRAE Journal, December, 1993. pp. 25-36.

24. Odom, J, David, III, and DuBose, George, Summary Report, "Indoor air quality evaluation of the Palm Beach County judicial complex". 1995. CH2M/Hill Publications, 225 East Robinson St, Suite 405. Orlando, FL 32801 (407) 423-0030.

25. Odom, J, David, III, and DuBose, George, "Preventing indoor air quality problems in hot, humid climates: Problem avoidance guidelines." 1996. CH2M/Hill Publications, 225 East Robinson St, Suite 405. Orlando, FL 32801 (407) 423-0030.

Air leakage - Further reading

26. *ASHRAE Handbook—Fundamentals. 2005*. Chapter 27, Ventilation and Infiltration, page 23.

27. Air Infiltration & Ventilation Centre. Computer database of techniques and field research regarding measuring and locating air infiltration into buildings. University of Warwick Science Park, Coventry, United Kingdom. www.aivc.org

28. Persily, Andrew K., "Myths about building envelopes." 1999. ASHRAE Journal, March, 1999, pp: 39-47.

29. Sherman, Max H. *Air Change Rate and Airtightness in Buildings*, ASTM STP 1067, 1989 (ISBN 0-8031-1451-6).

30. Treschel, Heinz , Lagus, Peter. *Measured Air leakage of Buildings*. 1986. American Society of Mechanical Engineers (ASTM STP 904, www.astm.org. ISBN 0-8031-0469-3)

31. Clarkin, Michael, and Brennan, Terry M., "Stack-driven moisture problems in a multi-family residential building." 1998. ASHRAE Transactions, V.104, Pt.2.

33. Kudder, Robert J, Lies, K.M, Hoigard, K.R, "Construction Details Affecting Wall Condensation." 1986. Proceedings of the Symposium on Air Infiltration, Ventilation and Moisture Transfer. NIBS.

34. Nelson, Gary, Nevitt, R, Tooley, J., Moyer, N., "Measured Duct Leakage, Mechanical System Induced Pressures and Infiltration in Eight Randomly Selected New Minnesota Houses." 1993. Proceedings of the 1993 Energy Efficient Buildings Association Conference. www.energyconservatory.com/articles

Image Credits

Figure 16.3 photos - © Dr. Joseph Lstiburek, www.BuildingScience.com

Figure 16.5 photo - © Dr. Joseph Lstiburek, www.BuildingScience.com

Figure 16.7 - © David Bearg, Life-Energy Associates

Figure 16.10 - © Hilti Inc, www.Hilti.com

Figure 16.12 - Mark Modera

Figure 16.12 - © Steven Winter Associates

Figures 16.15, 16.16 and 21 - © The Energy Conservatory

Figures 16.17 and 16.18 - © David Bearg and AirExpert Systems.

Figures 16.19 and 16.20 - Mason-Grant Consulting - www.masongrant.com

Special Thanks To Experts

Understanding the effects of duct leakage is one of the more important but under-researched aspects of building science. This Author is grateful for assistance supplied by many expert field investigators and design practitioners. Thanks especially to David Hales of Washington State University, Joseph Lstiburek of Building Science Corp, Andrew Persily of the National Institute of Standards and Technology, James Cummings of the Florida Solar Energy Center, Craig Wray of the Lawrence Berkeley National Laboratory, Hugh Henderson of CDH Energy, Gary Nelson of The Energy Conservatory, both J. David Odom III and George DuBose of Liberty Building Diagnostics and John Murphy of Trane Commercial Systems.

Chapter 17

Avoiding Mold By Keeping New Construction Dry

By Lew Harriman

Fig. 17.1 Construction moisture
*On job sites in hot and humid climates, rain falls **frequently**. Keeping construction dry, or drying it out after it gets wet, is the best way to avoid construction-related mold. This chapter explains how to avoid and dry out excess moisture before it can generate a mold problem.*

Key Points

When a new building smells moldy, the owner's first phone call is often to the building's HVAC designer. To most people, moldy odors suggest a need for more outdoor air ventilation. However, although an air system may spread fungal odors, in new buildings the HVAC system is seldom responsible for generating them. More frequently, mold and bacteria grow in new buildings when the building's materials never dried out, or because they became moist after construction. To avoid construction-related mold:

1. Store any moisture-absorptive materials out of the rain.
2. Make sure—through measurements—that any concrete slabs and masonry block walls in near-contact with interior wall board or flooring are dry before that wall board or flooring is installed.
3. Don't paint paper-faced gypsum wall board and don't install its wall covering until it is dried down below the moisture content limits suggested in this chapter.
4. Don't start the HVAC system early in an attempt to dry the building—it is not designed for building drying, and an premature start can easily ruin the equipment.
5. Use building drying equipment and/or services when the weather can't dry the structure quickly enough, or when water leaks or persistent rain threatens the schedule.

Construction Usually Gets Wet

Buildings are built outdoors. During construction, regular soaking of the unfinished building will occur. This is normal. But in spite of all that rain, mold does not grow to be a problem in most new buildings in hot and humid climates.

Building materials are usually quite tolerant of excess moisture. They either resist mold growth by not providing a nutrient source—as in the cases of poured concrete and masonry block; or they are able to dry out before mold grows because they are very porous, as in the cases of acoustic ceiling tile or paper-faced gypsum wall board. Mold problems usually occur when cellulosic materials such as paper-faced wall board are placed next to—or in contact with—rainwater reservoirs such as damp concrete or rain-filled masonry block.

Consider the photo shown in Figure 17.2 taken of a building being built in South Florida.[1] The contractor started the vertical masonry wall. Then when the hollow-core concrete floor planks were delivered, it was raining. His idea was to conserve money by using the crane once. He used the crane to take the hollow core planks off the delivery truck and set them in place during the rain storm. The water from the rain saturated the hollow-core planks. The vertical masonry walls kept much of that rain water in the concrete planks.

The first of the two photographs in Figure 17.2 was taken 14 days after the rain event. Water-driven leaching of the concrete has started to become apparent, seen here as the pale straight lines shown in the image. By the time of the second photo, taken 30 days later, the leaching had produced the stalactites seen in a straight line under the hollow cores that stored the rainwater. If the building had been sealed up and its interior finishes applied, mold growth would have been a significant risk.

Another example was a fully-completed 3-story bank building in central Florida.[2] High humidity problems were recorded from the first day of the building's occupancy. Soon, the walls of the bank's steel vault actually rusted, along with many of its safety deposit boxes. Suspecting construction moisture, the investigator took the finished ceiling down, and drilled small holes into the concrete floor planks above the vault. Water began running out of the planks. Within a matter of hours, more than *35 gallons* [133 liters] of water drained out of the concrete. This drainage took place approximately 9 months after the building was completed.

In both of these examples, the problem is not with the material which stored the rain. Concrete is very tolerant of excess moisture.

Hollow concrete floor planks...

... contain rainwater, trapped by walls.

14 days after the building was closed in. Water was still leaking out of the rain-filled hollows of the concrete planks.

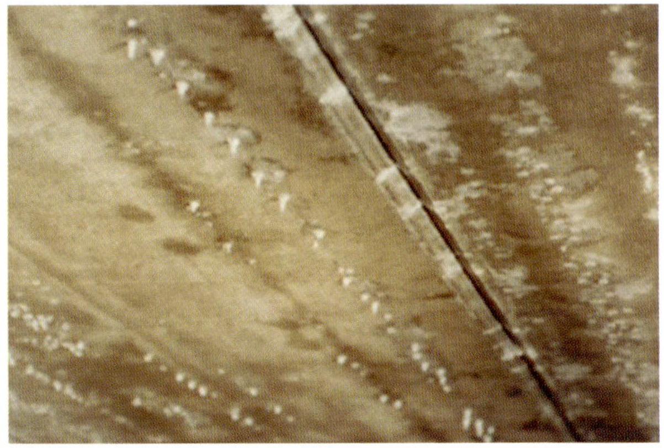

30 days later, so much water still remained that it created stalactites, as it continued to drip slowly out of the concrete, from the rain-filled hollows.

Fig. 17.2 Concrete can be a large moisture reservoir

Concrete does not grow mold. But the rainwater it absorbs during construction can provide the moisture needed to grow mold in nearby organic materials.

On this job site, dozens of gallons of water were trapped in the planks, because they were erected and sealed up during a rainstorm. So much water was trapped that it created stalactites as it drained slowly out—and months later the structure still was not dry enough for interior finish.[1]

But when massive amounts of liquid water are trapped indoors, humidity rises and moisture moves to less water-tolerant materials, where mold will grow. Of course, that problem arises even more quickly when the moisture-sensitive materials themselves are not kept out of the rain.

Cautions for Each Construction Phase

The Associated General Contractors of America (AGC) provides advice to contractors and owners on managing the risk of mold during construction.[3] In addition to providing general background information, that advice divides into suggestions and concerns for three phases of construction, which differ according to their exposure to rain. Several of those suggestions are included below. Note that some of the original AGC suggestions have been modified or expanded to reflect an HVAC perspective.

Exposed phase - Keep fibrous glass insulation dry

While the construction is fully exposed to the elements, it is simply impractical and cost-prohibitive to protect everything from rain from above, and from surface water and groundwater below. Buildings are built outdoors, and rain happens.

On the other hand, some materials might arrive on-site at an early stage, become saturated, and then release that water to nearby moisture-sensitive materials at some later stage. Fibrous glass insulation is one example. It is used in exterior walls, or for lining or wrapping of air ducts.

Fibrous glass is quite durable, and will tolerate excessive moisture for long periods without growing mold—on its own surface. Sometimes contractors, especially those with less experience in humid climates, do not always take the precautions needed to keep this material dry after it arrives at the job site.

If rain saturates fibrous glass insulation, the moisture will be very difficult to remove. That moisture may become the reason for mysterious mold and moisture problems during the early days of building occupancy, as it leaves the insulation and moves to more sensitive materials. So it is wise to keep any material dry which might later be installed indoors.

Partially-enclosed - Allow concrete and fireproofing to dry

After the roof protects the construction from rain, it should be possible to store any moisture-sensitive material, such as pallets of wall

Fig. 17.3 Sprayed-on fireproofing and insulation
Water based spray-on products must be dried out completely before interior finishes such as paper-faced gypsum board are placed over them, or nearby. Wet fireproofing has been responsible for mold growth in nearby materials in all climates.

Fig. 17.4 Get rid of mold
When bare gypsum board grows this much mold during construction, its usually faster and more economical to simply pull it out and replace it rather than trying to remediate the damage. Gypsum board is cheap, and mold remediation is expensive. But after finishes have been applied or woodwork set in place, the economics are not so clear. Drying and cleaning is often the faster and less costly alternative, as long as the owner agrees.

board or lumber, away from rain and raised well above the damp ground. That would be a basic precaution.

But also, it becomes important to shelter (as much as possible) any masonry block walls, fire walls and concrete floor slabs. These need to be drying as quickly as possible, rather than loading up on water before the building is closed in.

Saturated masonry block and saturated concrete are often the cause of major problems with mold for two reasons: They can store so much water, and moisture-sensitive materials like flooring and paper-faced gypsum board are often fastened only fractions of an inch away from their damp surfaces.

Similarly, fire walls and elevator shaft linings are often made of heavy-duty, fire-rated gypsum board. Fire walls must usually be erected before the roof can protect them from rain. Their mass is large, so they can absorb a great deal of rain water. If concrete floor slabs, masonry walls and fire-rated gypsum walls are not fully-dried, the sensitive materials fastened nearby can grow mold in a very short period (days or weeks).

Also at the partially-enclosed stage, fireproofing for steel structural members will be either spray-applied, or will be fabricated of heavy-duty, fire-rated gypsum board. The difficulty with spray-applied fireproofing is that it must release a great deal of water before finish materials can be installed nearby. The difficulty with gypsum board fireproofing for columns is that the gypsum board must be set tightly to the concrete floor and the ceiling. So if either of those concrete surfaces is still saturated, it may take a very long time to dry out after gypsum board has been set in place, eliminating air flow across damp concrete surfaces near the columns.

Begin measurements at this phase, to guide low-cost drying

The proactive owner or contractor can use the partially-enclosed phase of construction to begin monitoring the moisture content of concrete and masonry. Using portable fans to keep warm outdoor air moving through exposed construction is a productive option. As long as the moisture content of concrete or fireproofing is still high, drying is easier and faster than at later stages. The last bits of moisture will be far more difficult to remove.

After the building is closed in, fans alone are not going to be effective. The air will need to be dry, which adds costs. Using the partly-enclosed stage to dry with outdoor air can reduce both the time and the cost of drying later—especially if rain is kept out of the concrete and out of the masonry when the construction is still partly exposed.

If either the basic structure or the fireproofing is taking too long to dry during the partially-enclosed phase, it begins to make economic sense to investigate the drying services described later in this chapter. Drying the structure and fireproofing after the roof is on—but before the finish materials are applied—is much faster, more certain and less costly than waiting until after the finish materials *cover up* the excess moisture in concrete, masonry and fireproofing.

Some additional cautions relate to keeping gypsum wall board dry during the partially-enclosed phase. Certainly, one must not install unprotected wall board over wet concrete or wet concrete block. But also recognize that wall board will absorb moisture from humid air. The amount of absorption increases when the wall board surface is cool and the dew point of the air is close to that surface temperature.

This often happens on construction sites, especially in parts of the building which are below grade or deep inside the structure, such as elevator shafts and plumbing or electrical shafts. It usually happens during early morning hours. The outdoor air dew point rises as morning dew evaporates, but the building surfaces remain cool because of their large mass, and because they have been cooling all night long.

In that situation, it *seems* logical to heat the building or flow more outdoor air through with fans to increase drying. Those are good ideas *only if the air is dry*. One must not use direct-fired heaters, which add more than a gallon of water to the air for every gallon of fuel they burn. And one must not flow humid outdoor air over surfaces which are still cool. Flowing more high dew point air through a cool building only increases the amount of moisture absorption.

Hot air *which has a low dew point* is a useful tool. But damp hot air usually causes more problems than it solves. Portable dehumidifiers which reduce the dew point while adding heat are useful. And indirect-fired heaters or electric radiant heaters are also useful. These don't dry the air like dehumidifiers, but at least the heat they provide comes without raising the air's dew point.

Just make sure the gypsum board is not heated above any temperature limit established by its manufacturer. Heating gypsum board too high for extended periods can strip out the chemically-bound water molecules which actually provide the fire protection. For most types of gypsum board, keeping the product temperature below 95°F [35°C] will avoid degrading its fire-protection properties.

Controlled phase - Watch out for wall board, and for HVAC

After both the roof and the walls have been closed in, finish materials will be arriving on the site. These will usually include large amounts of paper-faced gypsum board. And for cost reasons, the wall board specified for general interior use is likely to be more moisture-sensitive than the gypsum board used for shaft liners. It is very important that wall board and any other absorptive cellulosic materials be kept dry, and that these not be installed near to, nor in contact with, damp concrete or masonry. Further, if paper-faced wall board does get wet, it's important to dry it out quickly.

Gypsum board manufacturers are all quite clear on this point in their guidance to designers and contractors. Even their products impregnated with antifungals and waxes to resist mold and moisture over their lifetimes are not intended to be installed over wet surfaces. Nor should their surfaces be sealed up with wall covering or paint if they have been soaked during construction. Drops and spatters on gypsum board may be acceptable, because these will probably dry out. But installing saturated wall board, or failing to dry it after it has been soaked is a very risky practice in any climate, and even more risky when the climate is hot and humid.

From an economic perspective, the least expensive alternative is to avoid installing wet wall board. If material becomes saturated before installation, either replace it with dry material, or dry it out before it is installed.

If the wall board gets wet *after* installation, drying it out may be the lowest-cost alternative. Drying out unfinished wall board—which is only wet and *not yet moldy*—can be quick and fairly economical. But if unfinished material stays wet for long periods and grows mold, there be no point in drying it. (See Figure 17.4) The cost of drying and mold remediation could be more expensive than tearing it out and replacing it. Unfortunately by the time costly remediation is required, the cost of extending the schedule to replace the already-installed moldy material could be higher still. That's why it's a very risky idea to let it stay wet. Keeping wall board dry helps avoid a common cause of lost profits and busted schedules during construction.

Avoid early HVAC startup

Early startup of the cooling system is a controversial issue. After the building is fully closed-in, many general contractors are eager to start the cooling systems to provide worker comfort and to speed drying of fireproofing and wall board joints. However, ASHRAE Standard 62.1 strongly cautions against this practice, because it often leads to indoor air quality problems.[5]

Beyond the written cautions outlined in ASHRAE standards, in the everyday world the HVAC contractors know that early startup of cooling systems usually causes problems. Early startup nearly always ignores or postpones the comprehensive testing, balancing and commissioning of that system, and therefore often takes years off the life of the equipment and may even void the owner's warranty.

Also, even with construction filters in place, startup during wall board sanding and painting often results in fine-particle gypsum

Fig. 17.5
Std 62.1 cautions against early startup
Section 7 of ASHRAE standard 62.1 clearly cautions against starting up HVAC systems during construction, before the system has been tested, balanced and commissioned. Early startup can lead to indoor air quality problems through mold in both the system and in the building, because HVAC systems are not designed to dry out wet buildings.

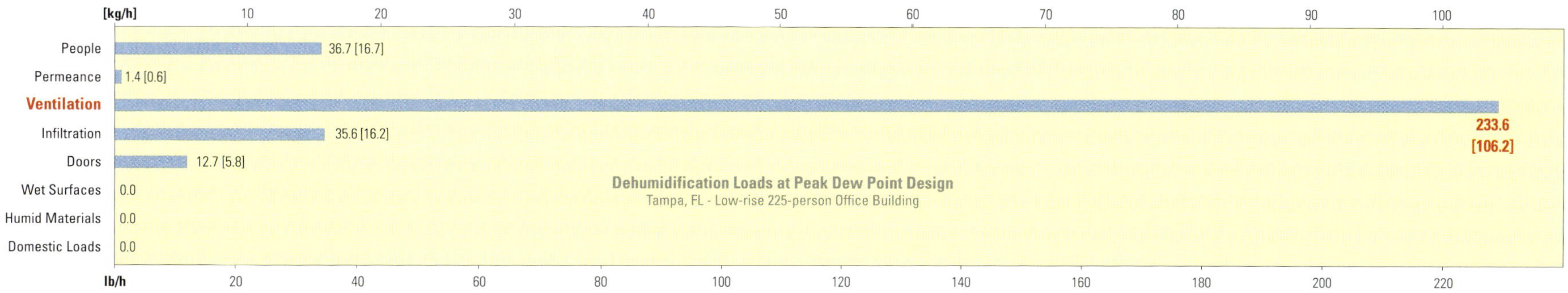

Fig. 17.6 Ventilation humidity load
Ventilation generates the largest HVAC humidity load in hot and humid climates. If the system is started before commissioning, it may not be capable of removing this load, which then floods the building with excessive moisture to support mold growth.

and paint dust on the wetted surfaces of the cooling coils. That fine particulate fills the spaces between the cooling fins. It solidifies like concrete, clogging the equipment and sometimes requiring its replacement.

Another common problem with early startup of HVAC in hot and humid climates is the building's outdoor air ventilation—especially if there is no dedicated ventilation dehumidification system. In hot and humid climates, ventilation air pours a continuously high humidity load into the building, which does not help the building dry out. (See Figure 17.6) Instead of drying the building, humid ventilation air loads the materials with even more moisture.

The ventilation air humidity problem becomes even worse when the contractor attempts to minimize the strain on a cooling system by setting the building's thermostat to a higher-than-usual temperature. At higher set points, the cooling system may not remove any of the ventilation humidity load at all, because it might not run its compressors for a *long enough period* (at least 40 continuous minutes in each hour for conventional constant-volume DX equipment) to actually drop out any significant amount of moisture.[6]

The opposite thermostat strategy is no improvement. In a damp building, lower-than-usual temperature set points are not helpful. The cooling system may run long enough to partly dry the air, but that colder air will also create cold surfaces all over the damp building. Cool surfaces often lead to condensation and then to mold growth on the back side of finished walls installed over any damp structure.

So if the structure has not dried out *before* being closed-in, or if the wall board is not already dry enough to tape and paint, it's wiser to consider temporary drying services than to start the HVAC systems prematurely. That advice immediately poses two questions—how dry is *dry enough* to avoid mold problems in moist materials, and how can that moisture be reliably measured?

How Dry Is Dry Enough To Prevent Mold?

The answer is simple—until one has to consider the realities of the construction schedule, the practical economics of the budget and the uncertainty of the exact location of excess moisture.

A simple answer would be that, in paper-faced gypsum board or any wood-based material, as long as a wood-based moisture meter shows a reading of 13% WME or lower, that material is very unlikely to grow mold. (WME refers to the "Wood Moisture Equivalent" - the moisture content as measured with a typical, low-cost wood-based meter rather with a more expensive and less easily available meter scaled for gypsum board.)

Unfortunately, that simple answer does not cover all job site realities. Some examples include:

- The *exact location* of the moisture is critical. A reading of 11% WME in one location could rise to a reading of 23% only one inch away [within a distance of less than 24mm], as shown in Figure 17.7.
- The moisture content of a moisture-sensitive material may be measured as being adequately dry at one moment in time. But if that material is fastened to a reservoir of moisture (such as wall board over damp masonry) the fact that the initial moisture content of the wall board is under 13% WME may not provide adequate protection. If that 13% wall board is then covered with vinyl wall covering, its moisture content could rise very rapidly. The vinyl will trap water vapor, preventing moisture from the masonry from passing through that porous wall board and into the open air.
- A common specification for framing lumber requires a maximum of 19% moisture content for the wood. It is true that this amount of moisture will not support mold growth on most species. But again, if paper-faced wall board is nailed in place over that damp wood, and if that wall board is then painted with impermeable paint or covered

11% WME, measured on the dry side

15% WME, measured across the visible edge

19% WME, measured on the moldy side

23% WME, measured in the moldy area

Fig. 17.7 The critical micro-geography of moisture measurements

Moisture content can vary widely over a distance of just a few millimeters. So it's important to take many readings to be sure the construction is dry all over, before attaching finish materials.

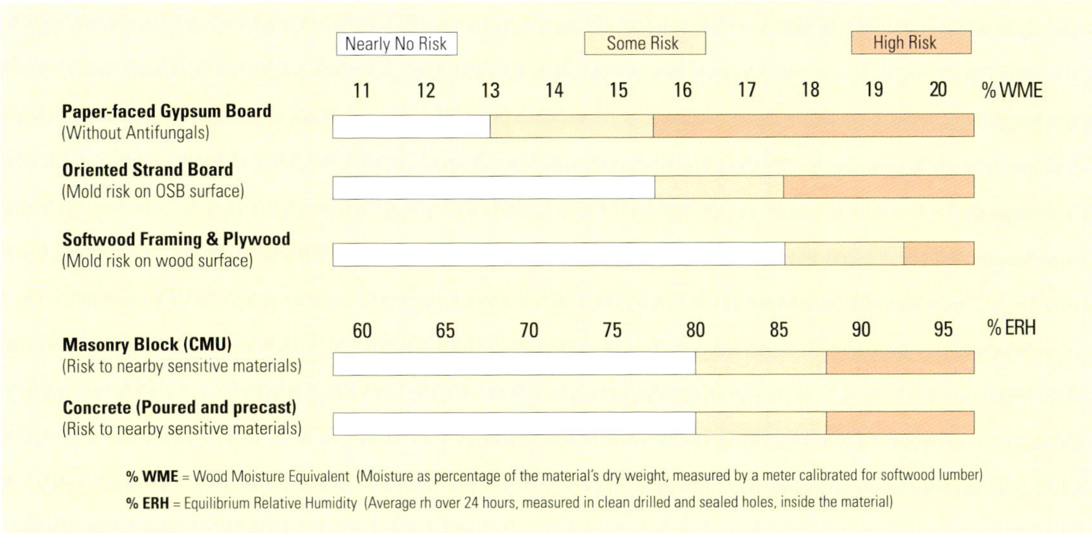

Fig. 17.8 Mold risk increases at higher moisture content

The real world of construction and buildings is complex. Assemblies get wet, and then they dry out. Also, antifungal treatments can delay the onset of mold growth at a given moisture content. So there are no firmly established threshold limits on mold growth versus moisture content.

The rate of mold growth increases with increasing moisture content. It also increases with more time at warm temperatures, and with the presence of organic nutrients on surfaces.

When all other factors are equal, lumber and plywood resist mold growth longer than paper or OSB. That's because paper and engineered wood products which have been "chopped up, broiled and pre-cooked" are easier for mold to colonize and digest.

with vinyl, and if all of the wood framing and the wood exterior sheathing is at 19% moisture content, the wall board is *very* likely to grow mold. Annoyingly, the reverse could also be true. Framing lumber showing a moisture content of 22% may represent no problem, as long as that moisture level is only typical of one or two locations on a few timbers, and as long as the rest of the structure is much drier and can absorb the excess moisture as those few damp sections dry out.

- If the moisture content of paper-faced gypsum wall board stays at 15% WME, it would probably not grow mold, even if it stayed that moist for months or years. However—that level is "teetering on the edge of catastrophe." Some mold may growing, but perhaps at a rate so slow as to be difficult to see with the naked eye. Then, when any condensation occurs (adding bulk moisture to that pre-dampened board) visible mold growth could occur within a few hours or a few days. Conversely, if the wall board has been treated with antifungals, a moisture content of 15% WME or even slightly higher may present little long term risk for mold—until after the material becomes really soaked. For a less risky approach, keep the moisture content of untreated paper-faced gypsum board below 13% WME.

In the flooring industry, perhaps because of repeated failures and litigation, moisture measurements are quite commonly required to support the warranty of the finished floor and its adhesives. Most manufacturers clearly specify both the testing methods and the threshold moisture content values in their product installation guidelines.

Unfortunately for products other than flooring, the author is unaware of any authoritative, peer-reviewed field studies which correlate moisture content readings with mold growth. Also unfortunately, the manufacturers of wall board, acoustic ceiling tile, fibrous glass insulation and wall coatings have not established an *actionable* and *quantitative* definition of their vague requirements to "make sure material is not wet before installation, nor installed over wet materials."

Until comprehensive and conclusive field research has been accomplished, the ranges shown in Figure 17.8 may be helpful in assessing the risk of mold at different material moisture contents in new construction in hot and humid climates.

Please note that these ranges are based only on the author's experience, and on informal input gathered from drying experts, laboratory researchers, material suppliers, insurance adjustors and forensic investigators. The author hopes and expects that at some point in the future, material suppliers will focus their resources (and find the courage) to publish more definitive and better-researched mold risk moisture contents for their products.

All that can be said with certainty at this time is that the values shown in Fugure 17.8 appear to economically achievable in the real world, and that above these ranges the risk will be higher, and below these ranges the risks will be lower. These suggestions are also based on the uncertainty of low-cost moisture meter readings in general, combined with the increased uncertainty of such measurements under construction job site conditions—issues which will be discussed in the next section.

Measuring Moisture

In the laboratory, moisture content measurements are usually made by weighing a sample of the moist material and then drying it in an oven until the material is no longer losing weight. Then, the difference between the initial and final weights of the sample is divided by the final weight of the sample. The resulting value is the percent moisture content of the original sample, on a dry basis.

There are many sources of error in this type of measurement, including the fact that heating the sample may drive off volatile vapors other than water, and that the sample may not be entirely dry at the end of the process. But for most research purposes, weight-loss measurements are accurate enough to be useful.

However, in the field such precise and time-consuming toasting of representative samples, each one torn out of the fabric of a building, is simply not a practical means of measuring moisture content. Beyond the awkwardness of destructive testing, there are the all-important issues of moisture geography and the normal drying of new construction.

As shown in Figure 17.7, one measurement may differ from another by important amounts, over a distance of just a few millimeters. In the field, the need is for fast, approximate measurements. Hundreds or thousands of measurements will be needed in a building. And the need for accuracy is not so great as in the laboratory, because there is great uncertainty about what maximum moisture level is appropriate during construction, given the fact that over time, buildings tend to dry out.

There are four methods typically used for field measurements of construction moisture, when the goal is to confirm the approximate dryness of concrete, masonry, wood products and gypsum board.

FIG. 17.9
Penetrating moisture meters
These meters measure the resistance between two pins inserted into the material, and convert that resistance to a moisture content reading. Usually, this is based on a calibration for soft wood. So measurements taken in gypsum board or other materials should carry the suffix "WME", to alert the reader of any report that the numbers represent the wood moisture equivalent. This type of meter is also called a "pin-type" meter, or a "resistance-type" moisture meter.

1. Electrical Resistance - "Penetrating Meters"

Figure 17.9 shows examples of electrical resistance moisture meters, also known in the trade as pin-type meters or penetrating meters. The inspector pushes the meter's two sharp pins into the material. The electrical resistance between those pins is very high in dry material, but very low in moist material. This large difference can be measured reliably within the range of interest for moisture content vs. mold risk, especially in wood.

In wood, the correlation between different resistances and moisture contents is well-characterized for nearly all wood species in the range between about 10% and 30% of dry weight. Above 30% moisture content, the fibers are saturated, so electrical resistance is quite low. Reliable measurements then become more difficult. When moisture rises above 40% moisture, it becomes very difficult if not impossible to obtain repeatable measurements. A similar difficulty occurs when material is nearly dry. Electrical resistance is extremely high, so reliable measurements are very unlikely when the true moisture content of wood is below 7%.

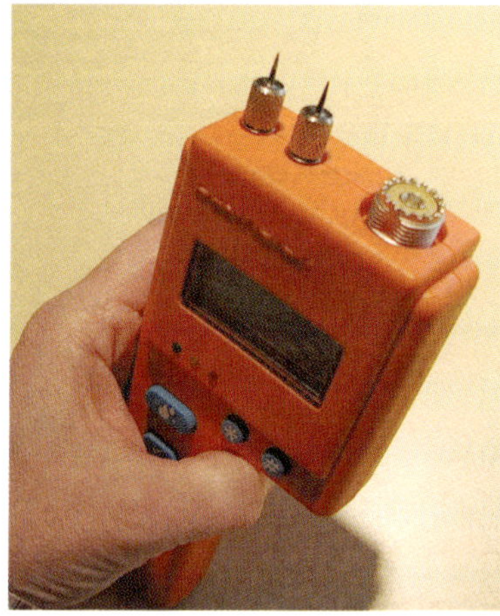

FIG. 17.10
Pins for measuring moisture content
Most penetrating meters have pins on the instrument, and also an attachment point for remote probes. The screw terminal shown here serves as an attachment point for hammer probes—for measuring moisture in hardwood floors, and for long-needle probes, which allow the instrument to read moisture content deep inside insulation or underneath it.

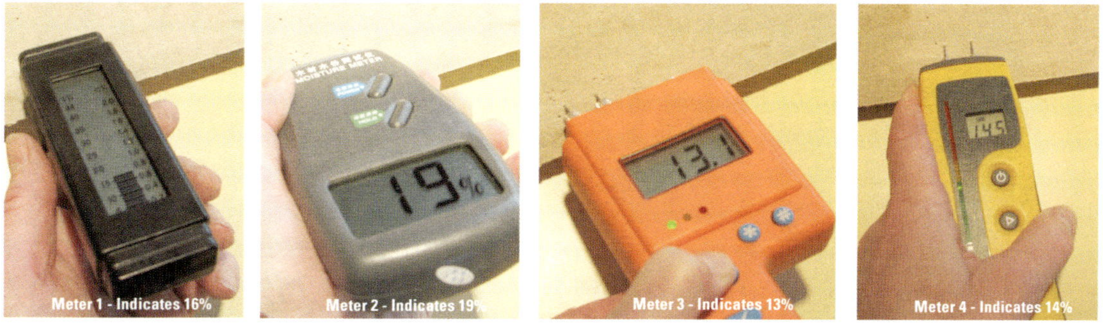

Fig. 17.11 Variation between meters

In precisely the same location at precisely the same time, different moisture meters report strikingly different values. That's one good reason why the make and model number of the meter should accompany any record of moisture readings, and also a reason to distrust any decimal fractions displayed by current state-of-the-art meters.

But for construction purposes and for measuring moisture contents between 10 and 30%, this type of meter is very economical, and is reasonably repeatable between identical models from the same manufacturer. For building construction and mold risk assessment, it's enough to know that resistance-based measurements above 30% indicate the material is "much too wet" and measurements below 10% indicate it's "plenty dry enough."

One unfortunate fact is that the instruments often display values which appear to be precise to within 1/10th of one percent moisture content. In fact the moisture measurement tolerances, even within the range of 10 to 30%, are probably no better than ±2 to 3% of the true moisture content, and then only in the specific wood species for which the instrument has been calibrated. The variation could be even more than ±3% when used in manufactured wood products, which have glues and other materials such as preservatives cooked into their structure.

So on construction job sites, it's wise not to get too excited about decimal fractions shown by moisture meters, except perhaps as an indicator of relative differences in the same material, in the same exact location of the pin holes, when using the exact same model, by the same meter manufacturer.

Wood-based meters are also used for measuring moisture content of gypsum board, which provides another common confusion when reading reports. In gypsum board, the true range of moisture contents will be between 0.4% and 2.0% of dry weight. Above that percentage,

	Softwood Moisture Meter Readings (Moisture as a percent of dry weight)					Surface Temperature °F [°C]
	8%	12%	16%	20%	24%	
Corrected Values for % Moisture in Aspen OSB	10	14	18	22	26	14° [-10°]
	8	13	17	21	25	32° [0°]
	8	12	15	19	23	50° [10°]
	7	11	14	18	22	68° [20°]
	6	10	13	17	20	86° [30°]

Fig. 17.12 Moisture meter corrections for OSB[7]

The electrical resistance of oriented strand board, with all of its glues, waxes and tiny voids, is different from the resistance of the solid softwood it is made from. This table was developed for the Canada Mortgage and Housing Corporation to correct resistance-based moisture meter readings taken in aspen-based OSB.

the facing and backing paper debonds from the gypsum, and nearly all strength is lost—the material crumbles easily.

Unfortunately, in the range of 0.4 to 2% gypsum moisture contents, a wood-based meter will indicate values between 10% and 40% or even higher. That's why it's good practice to state the abbreviation "WME" (Wood Moisture Equivalent) when recording measurements from a wood-based moisture meter used to measure gypsum board, or when measuring engineered wood products, insulation, masonry, concrete or any other material which is not actually softwood lumber.

OSB (Oriented Strand Board) can also be measured with a wood-based meter, but corrections are needed to support accurate assessments of mold risk. That material is very widely used in construction in North America in applications such as exterior sheathing and subflooring. Because it looks like wood and is composed mostly of wood chips, many professionals assume a wood-based meter gives reliable measurements for OSB. But the electrical resistance of OSB is different from softwood framing lumber within the range of interest.

Corrections are useful, because the difference between 18% and 15% represents a significant difference in mold risk, and there is a significant increase in the time and costs needed to dry down below 13% moisture content to reach the lowest range of mold risk. Figure 17.12 shows correction factors developed by the National Research Council of Canada for using a resistance-based moisture meter in oriented strand board made mostly from aspen wood chips.[7]

Another issue with wood-based resistance moisture meters is the variation between instruments made by different manufacturers. Figure 17.12 illustrates the problem. In the exact same holes, in the exact same moldy wall board, measurements taken with different instruments in the same five-minute period differ by 6% moisture content. This fact is useful to keep in mind when reading reports. Measurements taken by different people at different times with different meters are very unlikely to agree, even if the measurements were taken in *precisely* the same location.

Consequently, good practice would include documentation of the make and model number of the instrument, the date and time of the measurement, along with a photo or series of photos showing the exact locations and full context of those measurements. An example of such photos is shown in Figure 17.13. With those two

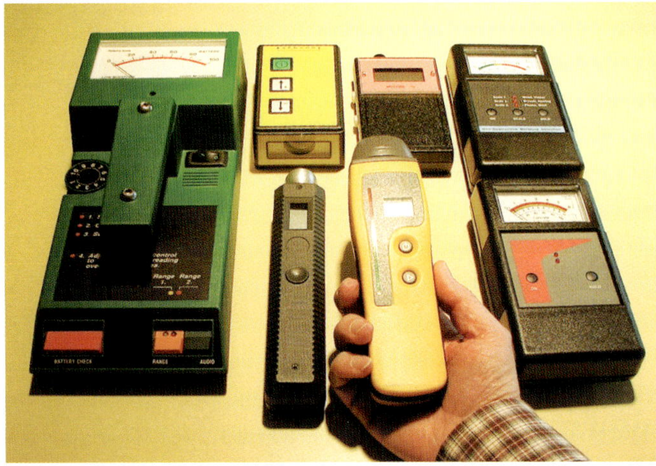

Fig. 17.14 Non-penetrating meters

images, one is quite certain of the context of the measurements, and their exact locations, and the range of values around the probable problem area.[8]

2. Electrical field variation - "Non-Penetrating Meters"

When measuring moisture in materials which must not be punctured, or when measuring moisture in layered assemblies, investigators usually use a "non-penetrating" meter. Figure 17.14 shows several examples of these instruments. Each works on a slightly different principle, but they all set up some form of electrical field, which changes in some way as the instrument is on top of the material in question.

As long as the material is well-characterized, and the meter is measuring only a single material, and provided there are no air gaps between the meter and the material, and there is no metal or other conductive layer inside the material being measured—one can obtain an approximation of its moisture content. These conditions exist in cabinet-making shops, where a craftsman is measuring the moisture

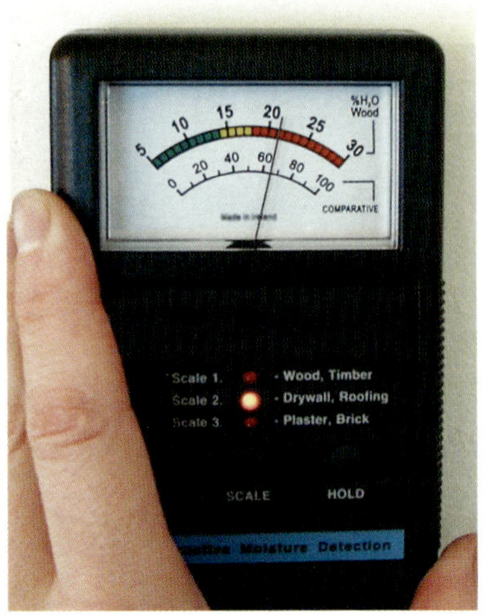

Fig. 17.15
Absolute and relative scales
Most non-penetrating meters show both absolute and relative (comparative) scales. In reports, its important to note which scale was used to record the reading.

Fig. 17.13 Quick and comprehensive documentation using photos
A table of moisture meter readings is not nearly as informative as photos showing the location and context of those readings. These photos show measurements recorded on labels made with black markers on white masking tape. The images provide comprehensive, convincing and very fast documentation of moisture content readings.

Fig. 17.16
Meter for concrete and CMU
These meters are most useful for fast surveys, to locate suspect areas. ERH tests are more accurate, but take days.

content of clean, uniform boards of a well-characterized species of wood. But on construction sites or inside the complexity of an existing building, the substrates and assemblies are not uniform, and their electrical properties are not well-characterized.

With all of those conditions to satisfy before accuracy can be assumed, it's easy to understand why these meters are usually used for quick scans and for relative changes in moisture content rather than for absolute measurements. Results from non-penetrating meters depend on the operator's skill and his or her consistency more than they do on the technology employed by the meter.[8] That's one reason that readings from these meters are usually read from a "relative scale," which has only a loose correlation with moisture content. (See Figure 17.15) Even worse, different instrument manufacturers use different relative scales.

For example, one prominent North American manufacturer uses a relative scale of 0 to 100 as shown in Figure 17.15, numbers which are easily confused with percent moisture content. To avoid *that* confusion, another manufacturer uses a relative scale which extends from 0 to 200, and one German manufacturer uses a scale from 100 to 600. Based on this author's tests, there appears to be no direct correlation between percent moisture content and the relative scales used by non-penetrating meters.

So this is another useful fact to keep in mind when reading reports. A report will be very confusing when it mixes measurements from penetrating and non-penetrating meters, or if it is not clear that the scale used is a "relative indicator" rather than a percentage of dry weight. For example, what would one make of an OSB moisture content reading of "162," if the report did not clarify that this was read from a relative scale using a non-penetrating meter rather than a percentage of dry weight read from a pin-type meter?

This is not to suggest that non-penetrating meters are not useful. In fact, they are nearly essential for quickly scanning an assembly to locate pockets of elevated moisture content. And when there are layers to consider, as in the case of moisture in subflooring under

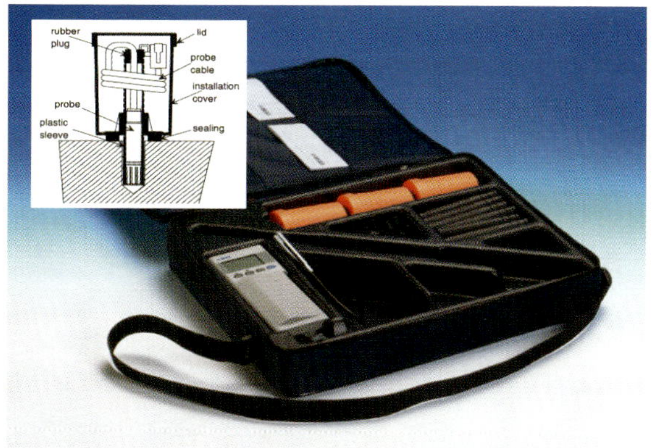

Fig. 17.17 Equilibrium rh test (ERH)
In concrete, the internal rh inside the material is a reliable indicator of its moisture content. But measurements are not reliable until 72 hours after the hole is drilled, cleaned out and sealed up.

tile, or moisture in exterior sheathing covered by rigid board insulation, non-penetrating meters are really the only practical instrument to use. One should simply keep in mind that the readings are best used as *relative indicators* of moisture differences, as opposed to repeatable measurements of percentage moisture content.

3. Equilibrium Relative Humidity (ERH)

In well-characterized solid materials, one can reliably estimate moisture content according to the relative humidity of air trapped in small voids in that material. When air is trapped *completely inside* the material, those small air pockets have a good chance of reaching equilibrium with the material's temperature and the moisture inside its micropores—given enough time.

In contrast, out at the exposed surface, reaching equilibrium between wet material and the air is very unlikely. Drifting air is constantly delivering more moisture to the surface or drying it out, and heating the surface of the material or cooling it. Equilibrium won't happen on surfaces exposed to the open air.

But if one drills a hole *inside* the material and cleans it out, and then seals up a relative humidity sensor inside that air pocket, and allows that micro-environment to stabilize at constant temperature for three days (72 hours after drilling), the relative humidity inside the clean, sealed hole will be a useful and repeatable indicator of the material's moisture content.

Figure 17.17 shows an example of a kit for measuring the moisture content of poured concrete in this way. A hole is drilled in the concrete. Then the drilling dust is carefully removed. A plastic sleeve, closed at the top and open at the bottom containing a relative humidity sensor is inserted into the hole and sealed to the upper surface of the concrete. The air inside the sleeve stabilizes over at least 48 hours. After that period, the relative humidity reading has usually stopped changing. The moisture content can be read from a chart which correlates equilibrium relative humidity with moisture content, for the specific concrete mix in question.

Drilling holes to different depths of the concrete allows an investigator to see a profile—the gradient of different moisture contents at different depths in the slab. With several measurements, one can be more certain of whether moisture is still trapped deep in the material.

This method has been used in Great Britain and in Scandinavia for decades. In 1992, ASTM International established a test standard formalizing this method[9] for use in parts of the world which adopt ASTM standards.

In the ASTM procedure, one test is performed for each 1,000 ft^2 of surface [100 m^2]. Also for slabs on grade or below grade, one test is required within 3 ft. [1 meter] of each exterior wall. The flooring manufacturer and the manufacturer of any adhesives used in the floor will usually have established maximum moisture contents for the slab, expressed as a percent equilibrium relative humidity.

85% ERH in concrete slabs is a typical maximum to prevent moisture reactions with flooring adhesives. But many manufacturers require lower levels for good reasons, such as the formulation of their product, its expected duty and service life, the duration of the required warranty and the uncertainty of moisture measurements made in the field. Maximums as low as 75% ERH are sometimes specified, which presents a significant challenge for keeping construction on schedule in a humid climate, even when using drying services.

Another emerging use for some variation of this test is in masonry block walls. The open cells of concrete blocks often fill with rainwater during construction, before the roof is set in place to keep rain out of the block. If the block stays wet when the building is closed in, the moisture in the block can evaporate slowly, feeding moisture to other more sensitive materials nearby. This problem is common in two locations: the exterior wall and interior fire walls.

In exterior walls, the sun warms the saturated block, driving moisture towards the inside of the finished exterior wall, where it can condense into interior gypsum board cooled by the air conditioning system. In fire walls separating different portions of a building, the trapped moisture comes out more slowly, because it is not forced out by heat from the sun. The slow evaporation of moisture from interior fire walls may support mold growth in the gypsum board which lines those walls, even when there is no other source of water leakage inside the building. Such problems tend to become apparent long after construction is complete. That time delay can obscure the fact that the problem originated from construction-related moisture.

To help avoid these problems, one can use the relative humidity probe on a standard portable thermohygrometer. Drill a hole into the bottom of the block, into the hollow core. Seal the air gap around the probe and wait until the relative humidity reading of the air inside the core stops changing. If the relative humidity is elevated, it may be an indication that the block is still saturated with rain water and needs to be dried before finish is applied.

As of the publication of this book, there are no maximum limits established for what that center-of-block relative humidity should be, either to reduce mold risk or to apply finishes. But certainly at air temperatures above 75°F [24°C] any reading above 75% rh would suggest that a great deal of moisture remains in the block.

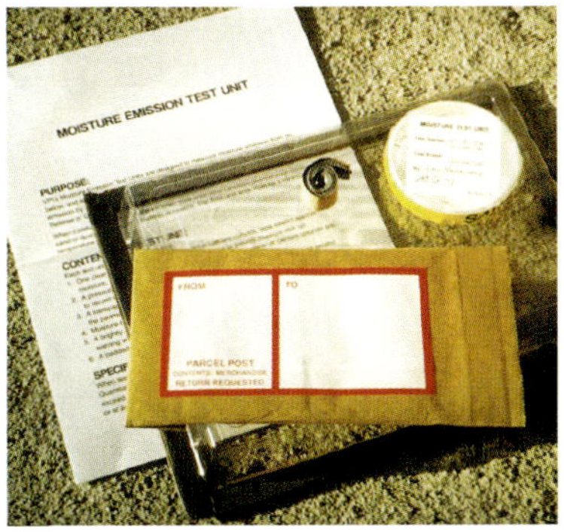

Fig. 17.18 Vapor emission test kit

4. Vapor emission rate - The "Calcium chloride test"

Before the ERH method came into general use in the US, the vapor emission test was the standard method of measuring moisture in concrete for flooring applications.

Figure 17.18 shows such a test kit. A measured weight of a powerful desiccant—calcium chloride powder—is placed in a dish set on the concrete. An air-tight enclosure is placed over the open dish and sealed to the concrete. After a waiting period of 72 hours, the weight gain of the calcium chloride is measured, and normalized to an emission rate, expressed as pounds of water vapor per 1,000 ft^2 per 24 hours. This emission rate is often abbreviated in conversation (and sometimes in specifications) to the inaccurate and rather confusing description of "a maximum of _x_ pounds moisture content." In fact, this method does not measure how much moisture is in the concrete. It only measures the rate at which that moisture is currently leaving the exposed surface, at the exact location of the test kit.

An emission rate of less than 5 lb /1,000 ft^2/24 hrs was once the traditional maximum for the safe installation of flooring adhesives. More recently, flooring adhesive manufacturers have required lower emission rates—4 lbs. and even 3 lbs.—as the maximum for safe application. The suppliers of wooden sports floors have traditionally required these lower limits, and the manufacturers of water-based adhesives often require those lower values as well.

As with any technology, there are cautions when using this method. When there are coatings on the concrete, the vapor emission rate will be misleadingly slow. Typical coatings include "curing compounds" (spray-on coatings applied to keep water in the concrete to complete the hydration reaction), and sealants which keep moisture in the slab from interfering with flooring adhesion.

That slow rate can lead to a false sense of security with respect to the risk of moisture problems at a later date. If moisture continues to evaporate from the slab, it may still moisten the adhesive enough to prevent curing, grow mold or carry alkali salts to the surface to interfere with adhesion—even if the vapor emission rate test showed favorable values.

Also, the emission rate is heavily influenced by the temperature of the concrete. More heat on the surface propels faster emission from that surface. Cooler surfaces slow down the emission rate. So specifications for the test method specify that the test be performed "at the normal service temperature of the floor." Obviously most buildings are cooled in service, while most slabs are exposed outdoors and get quite hot during construction. So this "requirement" is rather impractical and often ignored in hot climates, leading to less reliable conclusions about residual moisture.

Another related problem not exposed by testing by this method is temporary surface drying during construction. A heat-dried surface can provide a temporarily-low vapor emission rate. Heat drives moisture outwards to the air—but also downwards into the cooler layers of the slab. Slowly rebounding moisture can accumulate in adhesives later, after construction-related heat is gone.

These limitations explain why the ERH method is gaining favor in the flooring industry over the vapor emission test. But used thoughtfully and with attention to its limitations, the calcium chloride test remains a reliable standby in many flooring-related situations.

The next obvious question is *what to do* about excessive moisture if the schedule is too short to allow natural drying, or if something happens to saturate the construction after the building has been closed in. The next section provides some suggestions.

Construction Drying

HVAC systems are often designed so they can keep the indoor dew point low enough for comfort. But they are rarely designed to dry out construction-related moisture. The humidity load from wet materials is simply too large, and deep-drying a structure is too complex for systems which are cost-optimized for comfort air conditioning. The better plan is to dry materials quickly with aggressive temporary drying if they get wet through extreme weather or human error.[10]

In soaked materials, moisture will be supplied to the surface mold *from inside the material* rather than through adsorption from the surrounding air. Therefore, after actual wetting the challenge is to remove the excess moisture from *inside* the material as well as from its surface. Structural drying requires a combination of air movement across all wetted surfaces at very low surface humidity over an extended period—usually below 30% surface rh for several days or more than a week.

The tools and techniques for drying out building assemblies were developed over the last 30 years in response to the needs of the insurance industry, which uses restorative drying to limit the cost of water damage claims after floods, fires and disasters.[11]

Equipment For Construction Drying

HVAC designers have become familiar with some of this equipment because a variant has been applied to the problem of humidity control in commercial buildings. Portable mechanical and desiccant dehumidifiers are the tool of choice for much construction drying, since they dry the air to low dew points. The dehumidifiers used for building drying are configured quite differently from those used for permanent installations. Portable dehumidifiers are trailer-tolerant and will withstand the rugged environment of a construction site. Also, they are configured for the highly aggressive drying needed to bring the controlled space down to a low dew point, providing a very low relative humidity at the surface of the moist material.

Fig. 17.19 Portable desiccant dehumidifiers with supplemental indirect-fired heaters for construction drying

Figure 17.19 shows a pair of portable desiccant dehumidifiers being used to keep a new school building dry in Austin, Texas as a precaution to reduce mold risk on a fast-track project. The principal energy requirement of desiccant dehumidifiers is for heat to reactivate the desiccant wheel. Using propane or natural gas for reactivation, as shown in this photo, cuts the power requirement down to the point where supplemental generators can be avoided, saving rental costs.

In hot and humid weather, mechanical dehumidifiers and hybrid desiccant-mechanical dehumidifiers are also a popular alternative for building drying. These use more electrical power than gas-fired all-desiccant units, but they do not require external reactivation heat.

Construction Drying Techniques

There are several advantages to drying out construction moisture using temporary equipment. First, the building's HVAC system does not have to be turned on, which keeps construction dust out of its duct work, cooling coils and filters. Second, the contractor can flow dry air over the moist material directly, rather than hoping it will drift there through duct work designed to cool and heat rooms rather than to dry wet concrete or wallboard. Finally, the full warranty period on HVAC system components will be preserved if that system can remain off until commissioning is complete.

Fig. 17.20
Temporary dry air distribution ducts

Fig. 17.21 "Air movers"

These are used to in combination with temporary duct work to direct the dry air across the moist surfaces.

Fig. 17.22 Contain the dry air

Temporary doors will be needed to keep humid air out of the building. These simple spring-loaded plywood panels do the job well.

Figure 17.20 shows how dry air can be distributed through the construction zone. Note the use of inflatable plastic ducts rather than using the building's own air distribution system.

Figure 17.21 shows the small portable centrifugal blowers called "air movers" in the drying trade. These are moved constantly. Some materials such as uncoated wall board will dry quickly. Other material such as concrete will require a higher air flow rate, drier air and perhaps some heat to dry properly.

To dry in the shortest possible time, a drying contractor periodically measures material moisture content and re-adjusts the position of air movers every day, to make sure all materials dry quickly and completely. Drying technicians speak of "chasing the moisture" throughout the wetted area. This is critical. The building drying industry has discovered, through painful experience and occasional litigation, that simply supplying super-dry air to a wetted building does not ensure that the wet structure will dry quickly and completely.

Dry Air Must Be Contained

Buildings under construction are usually wide open. Even after closing in, exterior doors are usually open for easy access. To dry out moist material, the drying air must be contained. Figure 17.22 shows a typical temporary enclosure and doorway, built to retain the dry air in the part of the building that needs it. Keeping doors closed is never popular with workers, but the irritation can be minimized through springs on lightweight swinging doors. Hermetic air seals are not necessary in building drying, but humid outdoor winds must be kept out of the area being dried.

Drying Poured Concrete

In every yard of concrete, about 50 gallons of water is excess to the curing process. [Each cubic meter of concrete contains roughly 248 liters of water more than what is needed for curing.] [13]

Concrete "cures" through hydration. The cement particles need a certain amount of water to develop strength, which is accomplished through a huge increase in their surface area. For example, a pea-sized piece of cement has a total surface area of about 2,000 cm^2. During hydration (curing) the surface of that granule increases nearly 1000 times to about 2,000,000 cm^2 [(14)]. After an acceptable strength is reached, the excess free water becomes a potential liability—not to the concrete, but to the interior finish.

Most concrete used in commercial construction reaches 70% of its final strength after 28 days. After that, excess moisture must be removed so that flooring and interior gypsum board can be installed. At this point, the job superintendent has three choices:

1. Wait for the concrete to dry through natural convection.
2. Seal up the surface so excess moisture is trapped in the concrete.
3. Use dehumidifiers to dry the material more rapidly than it would through natural convection.

Waiting is often the first choice. When the schedule allows the time, simply waiting a few weeks avoids the costs of additional material or services. And in hot climates, although the dew point may be high there's often enough energy and natural convection to dry the structure quickly enough to keep the project schedule on-track. During rainy or highly humid weather, however, natural drying of concrete may take too long.

Sealing can be effective, especially when there is not much excess moisture left in the concrete—but not enough time left for drying to the level required by flooring manufacturers. In the US, manufacturers' warranties generally specify a water vapor emission rate from the surface of no more than 3 or 4 lb./1000 ft^2/24 hrs.[15] International standards vary widely.[16]

When the problem is " far too much water" still left in the concrete, the third choice—drying with dehumidifiers—is a popular choice. It costs about 80% less than the cost of vapor-sealing wet concrete, but it takes more time. Occasionally, these two techniques

Fig. 17.23
Warped sports floor
Problems like this are a certain indicator of excessive moisture in the concrete, and-or excessive humidity after installation.

are combined. Portable dehumidifiers are brought in to accelerate drying. Then surface sealing is added after most, but not quite all, of the excess water has been removed from the concrete.

Maple Flooring Installation

New schools often contain an engineered wooden sports floor in the gymnasium. To maintain the warranty of such floors, the concrete underslab must be dried to demanding specifications. Then the wood flooring strips must be brought to equilibrium with the center of the humidity range they will encounter over their service life. Figure 7.23 shows what can happen if the humidity is not controlled during construction as well as afterwards, when the school is closed for the summer. The floor warped and buckled when uncontrolled summertime humidity caused the flooring to expand beyond its design limit.

To prevent such problems, the Maple Flooring Manufacturers Association has established environmental specifications during installation as well as during service. The relative humidity in the space where the floor is installed must be held between 35% and 55% rh at all times. And ideally it should be kept at the center of that range during installation to minimize potential damage during subsequent seasonal extremes (MFMA, 1998).

This specification is a challenge because a school usually demands beneficial occupancy by the start of the school year—just when the outdoor humidity is at its peak. Bringing a dehumidifier on site lets the flooring contractor slowly and gently bring the wood to the appropriate equilibrium moisture content before installation. This is useful when wood has been stored under less than ideal circumstances.

Drying Time For Taped Drywall Seams (hrs)

Relative Humidity (% RH)	Air Temperature °F	40	50	60	70	80	90
	°C	4.4	10	15	21	27	32
	90	216	144	102	72	49	36
	80	114	78	54	38	27	19
	70	84	54	38	26	19	14
	60	60	42	29	20	14	10
	50	48	36	24	17	12	9
	40	43	29	20	14	10	7
	30	38	26	18	12	9	6
	20	34	24	16	11	8	6

Fig. 17.24 Approximate time needed to dry out taped seams on gypsum board [18]

Drying Wallboard Joints

Rainy weather creates a problem for installing and finishing gypsum wall board. In a hot and humid climate, some of that moisture inevitably ends up in the wall board as the building is being constructed. The drying rate is a function of both air temperature and relative humidity at the surface of the material.

For example, Figure 17.24 shows that at 60°F [18.3°C] and 85% rh, it will take about 3 days to dry the tape joints of the wallboard sheets.[18] In contrast, the warm dry air from a dehumidifier can easily bring the air to 80°F [26.7°C] and 20% rh. In that environment the joints usually dry in about 6 hours. In other words, the joints dry before each work shift is finished. (Note, however, that gypsum wall and ceiling installers caution against heating the indoor environment above 95°F [35°C].)[4]

Using dry air to dry tape joints also helps expose potential future problems while they can still be eliminated economically. Joints that *fail* to cure quickly in dry air are usually being fed by moisture from some hidden source such as water leaks in the exterior envelope, pipe leaks or overly-wet concrete or masonry block walls.

Specifications To Keep New Construction Dry

The HVAC designer is seldom involved with construction moisture. The owner, architect and general contractor have the real control over moisture in materials during and after construction. When either of these organizations wish to take a proactive approach to speeding construction and reducing mold risk, steps to consider include:

1. Specify that all gypsum wallboard be installed with a fire sealant bead of 3/8" [9mm] between the floor and the bottom edge of the gypsum. This avoids the common problem of water from accidental spills or floor cleaning wicking up into the wall to feed mold behind the cove molding at the base of the wall. The sealant also eliminates the air path that reduces the fire protection and acoustical barrier properties of walls.
2. Specify that the moisture content on the indoor side of the base course of all masonry block walls be measured and documented by the general contractor, and that no gypsum board be hung on those walls until the moisture content of the base course of blocks, measured on their interior surfaces, is the same as an identical block which has been stored away from all rain or other water contact.
3. Specify that the moisture content of the gypsum board walls be measured after fastening in place but before taping, and documented by the general contractor at four locations on each wall: twice at 1 inch [24mm] above the bottom edge of the wall, and twice at half-way between the floor and ceiling. Specify that neither the taping nor the interior finish may be applied until the moisture content of the wall board is below 13% WME (Wood Moisture Equivalent - moisture content as measured by a wood-based meter).
4. Specify that the moisture content of the concrete floor slab shall be measured in accordance with ASTM standard F 2170-02. If the moisture content is excessive compared to the requirements of the manufacturer of the leveling compound, or the flooring or flooring adhesive, the air above the moist concrete shall be held below 30% rh until the slab is dry enough to meet the specification established by all of the manufacturers of materials used in the flooring system.
5. Specify that the general contractor shall provide humidity control to the standard defined by the Maple Flooring Manufacturers Association during the installation of any wooden flooring or any fine wooden cabinetwork.

Summary

Mold problems in new construction are obviously preventable since very few new buildings have such problems, even in hot and humid climates. Owners have a right to assume that a new building will be free of mold growth—assuming they do not force a design process or a construction schedule which ensures trapped moisture. In addition to better specifications from the architect, the moisture measurement and construction drying techniques described by this chapter can be useful tools in that effort.

References

1. Halyard, Paul, P.E., Fellow, ASHRAE. Peninsula Forensic Engineering, Orlando, FL. Personal communication, regarding examples of problems with construction moisture in Florida.
2. Halyard, Paul, P.E., Fellow, ASHRAE. Peninsula Forensic Engineering, Orlando, FL. Personal communication, regarding examples of problems with construction moisture in Florida.
3. Associated General Contractors of America. 2003. "Managing the risk of mold in the construction of buildings." Constructor magazine, May, 2003 pp. 11-31. No-cost pdf at: www.agc.org
4. AWCI International. Technical Manual No. 14: Site conditions for the installation of gypsum board. Association of the Wall & Ceiling Industry, Falls Church, VA. www.awci.org
5. ASHRAE Standard 62.1 User's manual - Section 7: Construction and system startup. ASHRAE, Atlanta, GA. www.ashrae.org
6. Henderson, H. 1998. "The impact of part-load air conditioner operation on dehumidification performance: Validating a latent capacity degradation model." Proceedings of the 1998 ASHRAE Indoor Air Quality Conference, 1998.
7. Forintek Canada, for the Canada Mortgage and Housing Corporation. 2001. "Guidelines for on-site measurement of moisture in wood building materials." (Full report, rather than the "Research Highlight" version) CMHC, Ottawa, Canada. www.cmhc.ca
8. Harriman, Lewis G. III, 2007. "Practical aspects of measuring moisture in buildings." Proceedings of the Bugs, Mold & Rot IV Conference, Minneapolis, MN, June, 2008. National Institute of Building Science, Washington, DC. www.nibs.org.
9. ASTM F 2170-02. Standard test method for determining relative humidity in concrete floor slabs using in situ probes. ASTM International, West Conshohocken, PA, www.astm.org.
10. Harriman, Lewis G. III, 2002. Preventing mold by keeping new construction dry." ASHRAE Journal, May 2002. pp.22-34. ASHRAE, Atlanta, GA. www.ashrare.org.
11. Lee, Mickey. 2000. "Wringing out extra costs from water damage claims." Claims magazine. August 2000. National Underwriter Company, Publishing Division. Erlanger, KY, USA
13. Harriman, Lewis G. III, 1995. "Drying Concrete." Construction Specifier magazine. March 1995, Construction Specification Institute, Arlington, VA. USA.
14. Hansen, Torben, 1989. "Physical structure of hardened cement paste, a classical approach." Materiaux et Constructions, Essais et Recherches # 19.
15. Kanare, Howard M. 2005. Engineering Bulletin 119.01: Concrete floors and moisture Portland Cement Association, Skokie, IL. www.cement.org
16. Hedenblad, Göran. 1997. Drying of construction water in concrete - Drying times and moisture measurement. Byggforskninggsgrådet (The Swedish Council for Building Research) Stockholm, Sweden. ISBN 91-540-5785-X
17. MFMA. 1998. Humidity Control During Installation. Maple Flooring Manufacturers Association. Northbrook, IL USA
18. NWCB. 2001. Field Technical Bulletin 301 - Recommended application of gypsum board, and Field Technical Bulletin 303 - Gypsum wallboard and winter weather. Northwest Wall & Ceiling Bureau, Seattle, WA www.nwcb.org.

Image Credits

17.2 - Paul Halyard, Peninsula Forensic Engineering

17.3 - Krezgroup.com

17.4 - Daniel Friedman, www.inspect-ny.com

17.7 - Mason-Grant Consulting, www.masongrant.com

17.8 - Mason-Grant Consulting

17.13 - Mason-Grant Consulting

17.17 - Vaisala Inc, www.vaisala.com

17.19 - Munters Moisture Control Services, www.munters.com

17.20 - Munters Moisture Control Services

17.21 - Munters Moisture Control Services

17.22 - Munters Moisture Control Services

Appendix

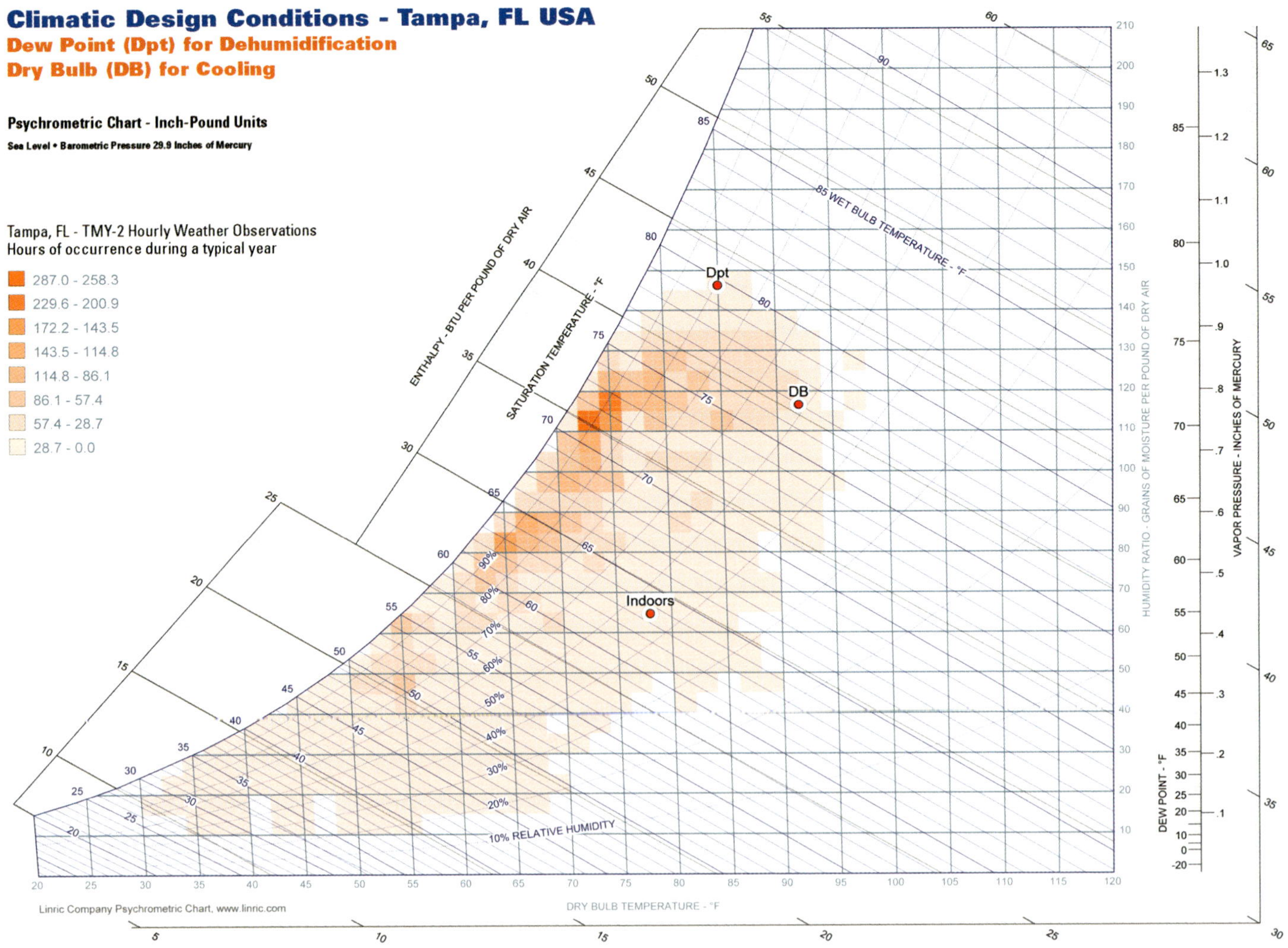

Fig. A.1 (I-P)
Comparing Design Conditions

This chart illustrates how much more humidity is in the air at the **peak dew point** *condition compared to the much lower humidity level at the familiar dry bulb design condition.*

The peak dry bulb is appropriate for calculating sensible cooling loads—but not for calculating the dehumidification loads. When designing the dehumidification components of the system, it's important to use the peak outdoor dew point as the design extreme. These data can all be found in the ASHRAE Handbook—Fundamentals, begining with the 1997 edition. (Earlier editions only had the peak dry bulb values).

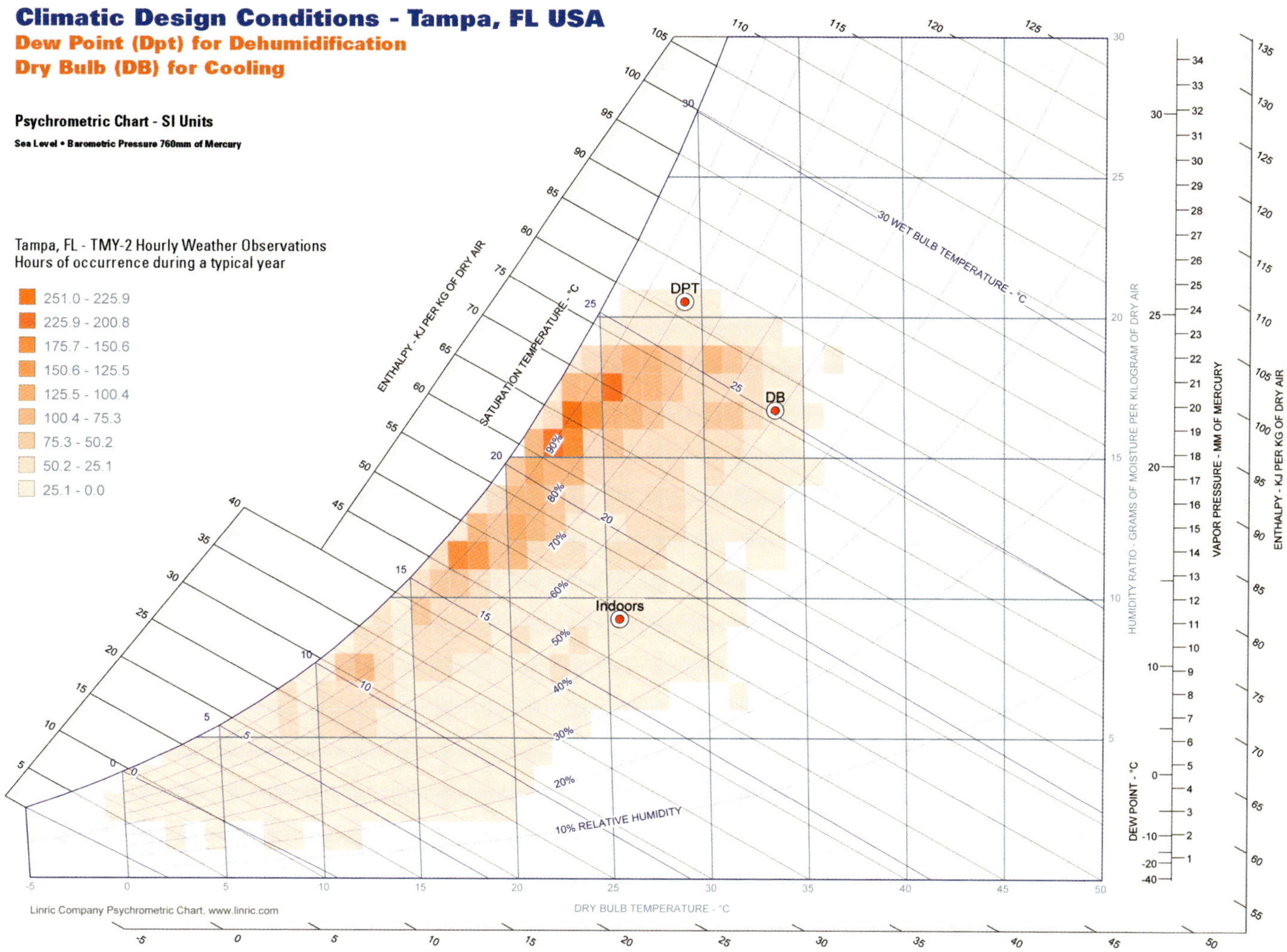

Fig. A.1 (SI)

Comparing Design Conditions

This chart illustrates how much more humidity is in the air at the ***peak dew point*** *condition compared to the much lower humidity level at the familiar dry bulb design condition.*

The peak dry bulb is appropriate for calculating sensible cooling loads—but not for calculating the dehumidification loads. When designing the dehumidification components of the system, it's important to use the peak outdoor dew point as the design extreme. These data can all be found in the ASHRAE Handbook—Fundamentals, begining with the 1997 edition. (Earlier editions only had the peak dry bulb values).

Fig. A.2
Equations for Dehumidification Design

Equation 1
Dehumidification Capacity

What is the dehumidification capacity of the supply air? Said another way; how much moisture can a known amount of air with a known humidity ratio remove from the conditioned space every hour?

Q (supply air flow rate - scfm or l/s)
W_1 (humidity ratio of air in the controlled space - gr/lb or g/kg)
W_2 (humidity ratio of the supply air - gr/lb or g/kg)

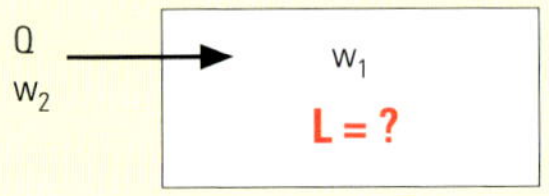

$$\text{lbs/h} = \frac{Q \times (W_1 - W_2) \times 4.5}{7000} \qquad \text{kg/h} = \frac{Q \times (W_1 - W_2) \times 4.32}{1000}$$

Equation 2
Required Supply Air Humidity Ratio

If the supply air flow rate and the indoor humidity ratio are fixed, how dry does the supply air have to be in order to remove a known internal moisture load?

Q (supply air flow rate - scfm or l/s)
W_1 (humidity ratio of air in the controlled space - gr/lb or g/kg)
L (moisture load in the controlled space - lb/h or kg/h)

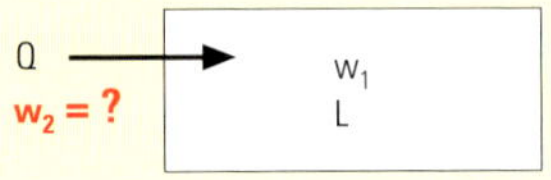

$$\text{gr/lb} = W_1 - \left(\frac{7000 \times L}{Q \times 4.5}\right) \qquad \text{g/kg} = W_1 - \left(\frac{1000 \times L}{Q \times 4.32}\right)$$

Equation 3
Required Supply Air Flow Rate

If the supply air humidity ratio and the indoor humidity ratio are both fixed, how much supply air is needed to remove a known internal moisture load?

W_1 (humidity ratio of air in the controlled space - gr/lb or g/kg)
W_2 (humidity ratio of the supply air - gr/lb or g/kg)
L (moisture load in the controlled space - lb/h or kg/h)

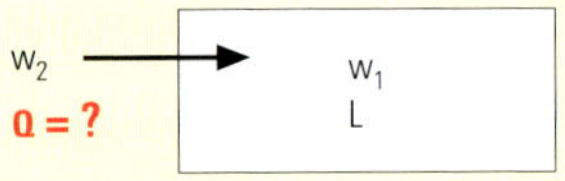

$$\text{scfm} = \frac{7000 \times L}{(W_1 - W_2) \times 4.5} \qquad \text{l/s} = \frac{1000 \times L}{(W_1 - W_2) \times 4.32}$$

Equation 4
Mixed-air Humidity Ratio

When two air flows with different humidity ratios are blended, what will be the humidity ratio of the resulting mixture?

Q_1 (Air flow rate of the first air stream - scfm or l/s)
Q_2 (Air flow rate of the second air stream - scfm or l/s)
W_1 (humidity ratio of air in the controlled space - gr/lb or g/kg)
W_2 (humidity ratio of the supply air - gr/lb or g/kg)

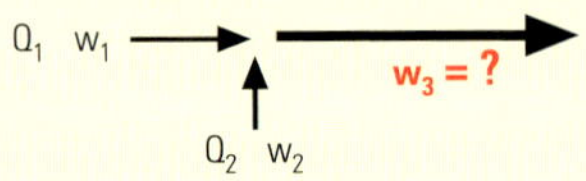

$$\text{gr/lb} = \frac{(Q_1 \times W_1) + (Q_2 \times W_2)}{Q_1 + Q_2} \qquad \text{g/kg} = \frac{(Q_1 \times W_1) + (Q_2 \times W_2)}{Q_1 + Q_2}$$

Equation 5
Dehumidified Air vs. Bypass Air

*Usually, a dehumidifier only needs to dry **part** of the supply air rather than its entire volume. How much of the supply air should be dried by the dehumidifier in order to bring the entire supply air flow to the required humidity ratio?*

Q_3 (Supply air flow rate - scfm or l/s)
W_1 (Humidity ratio of air leaving the dehumidifier - gr/lb or g/kg)
W_3 (Required humidity ratio of the supply air - gr/lb or g/kg)

$$\text{scfm} = \frac{(Q_3 \times W_2) - (Q_3 \times W_3)}{W_2 - W_1}$$

$$\text{l/s} = \frac{(Q_3 \times W_2) - (Q_3 \times W_3)}{W_2 - W_1}$$

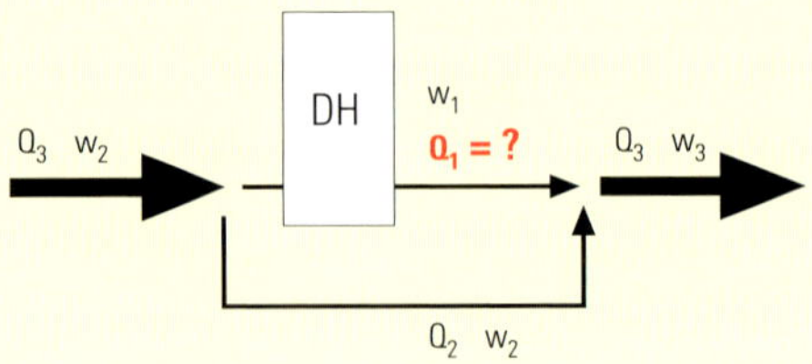

Fig. A.3 I-P to SI Conversion Factors

To Convert	Into	Multiply By
Btu	joules	1055
Btu/h	watts	0.293
Btu/lb	kJ/kg	2.326
celsius	fahrenheit	(°C x 1.8) + 32
cubic feet/min (cfm)	litres/sec	0.4720
fahrenheit	celsius	(°F -32) ÷ 1.8
feet	metres	0.3048
feet/min (fpm)	metres/sec (m/s)	5.08×10^{-3}
gallons	pounds	8.346
gallons	ft^3	0.1337
gallons	litres	3.785
gallons/hr	millitres/sec	1.05
gallons/min	litres/sec	0.06308
grains/lb	grams/kg	0.143
grains	grams	15.43
grams	pounds	2.2046×10^{-3}
inches	millimetres	25.4
inches of water column	inches of mercury	0.07355
inches of water column	pascals	249
joules	Btu	9.48×10^{-4}

To Convert	Into	Multiply By
kilograms	pounds	2.205
kilowatts	Btu/h	3414
litres/sec	cfm	2.19
metres	feet	3.281
metres/sec	miles/hr	2.237
metres/sec	ft/min	88
millimetres	inches	0.03937
pascals	inches of water	4.016×10^{-3}
pascals	inches of mercury	2.959×10^{-4}
pints	pounds	1
pints/day	pounds/hr	0.0416
pounds (mass)	grains	7000
pounds (mass)	kilograms	0.4536
pounds of water	ft^3	0.01603
pounds of water	gallons	0.1199
pounds	pints	1
pounds/hr	pints/day	24
tons, refrigeration	Btu/h	12000
tons, refrigeration	kW	3.52
watts	Btu/h	3.414

°F	gr/lb	in.hg.	°C	g/kg	kPa
160	2096	9.665	71.11	299.70	32.67
159	2023	9.436	70.56	289.34	31.89
158	1954	9.213	70.00	279.40	31.14
157	1887	8.994	69.44	269.87	30.40
156	1823	8.779	68.89	260.73	29.67
155	1762	8.569	68.33	251.95	28.96
154	1703	8.362	67.78	243.52	28.27
153	1646	8.161	67.22	235.42	27.58
152	1592	7.963	66.67	227.62	26.91
151	1539	7.769	66.11	220.13	26.26
150	1489	7.580	65.56	212.91	25.62
149	1440	7.394	65.00	205.97	24.99
148	1394	7.212	64.44	199.27	24.38
147	1348	7.034	63.89	192.83	23.78
146	1305	6.860	63.33	186.61	23.19
145	1263	6.689	62.78	180.62	22.61
144	1223	6.522	62.22	174.84	22.05
143	1184	6.359	61.67	169.27	21.49
142	1146	6.199	61.11	163.89	20.95
141	1110	6.042	60.56	158.69	20.42
140	1075	5.889	60.00	153.67	19.91
139	1041	5.740	59.44	148.82	19.40
138	1008	5.593	58.89	144.14	18.90
137	976	5.450	58.33	139.61	18.42
136	946	5.310	57.78	135.24	17.95
135	916	5.172	57.22	131.01	17.48
134	888	5.038	56.67	126.91	17.03
133	860	4.907	56.11	122.96	16.59
132	833	4.779	55.56	119.13	16.15
131	807	4.654	55.00	115.42	15.73
130	782	4.531	54.44	111.83	15.32
129	758	4.412	53.89	108.36	14.91
128	734	4.295	53.33	105.00	14.52
127	711	4.180	52.78	101.74	14.13
126	689	4.069	52.22	98.59	13.75
125	668	3.959	51.67	95.54	13.38
124	647	3.853	51.11	92.58	13.02
123	627	3.749	50.56	89.71	12.67

°F	gr/lb	in.hg.	°C	g/kg	kPa
122	608	3.647	50.00	86.93	12.33
121	589	3.548	49.44	84.24	11.99
120	571	3.451	48.89	81.63	11.66
119	553	3.356	48.33	79.10	11.34
118	536	3.263	47.78	76.65	11.03
117	519	3.173	47.22	74.27	10.72
116	503	3.085	46.67	71.97	10.43
115	488	2.999	46.11	69.74	10.14
114	473	2.915	45.56	67.57	9.85
113	458	2.833	45.00	65.47	9.58
112	444	2.753	44.44	63.43	9.30
111	430	2.675	43.89	61.45	9.04
110	416	2.599	43.33	59.54	8.78
109	403	2.525	42.78	57.68	8.53
108	391	2.452	42.22	55.87	8.29
107	378	2.382	41.67	54.12	8.05
106	367	2.313	41.11	52.43	7.82
105	355	2.246	40.56	50.78	7.59
104	344	2.180	40.00	49.18	7.37
103	333	2.117	39.44	47.63	7.15
102	323	2.054	38.89	46.13	6.94
101	312	1.994	38.33	44.67	6.74
100	302	1.935	37.78	43.26	6.54
99	293	1.877	37.22	41.88	6.35
98	284	1.821	36.67	40.55	6.16
97	275	1.767	36.11	39.26	5.97
96	266	1.714	35.56	38.00	5.79
95	257	1.662	35.00	36.79	5.62
94	249	1.612	34.44	35.61	5.45
93	241	1.562	33.89	34.46	5.28
92	233	1.515	33.33	33.35	5.12
91	226	1.468	32.78	32.27	4.96
90	218	1.423	32.22	31.23	4.81
89	211	1.379	31.67	30.22	4.66
88	204	1.336	31.11	29.23	4.52
87	198	1.294	30.56	28.28	4.38
86	191	1.254	30.00	27.35	4.24
85	185	1.214	29.44	26.46	4.10

Fig. A.4a

Dew points, Humidity Ratios and Vapor Pressures at Saturation (Upper range)

Fig. A.4b
Dew points, Humidity Ratios and Vapor Pressures at Saturation (Middle range)

°F	gr/lb	in.hg.	°C	g/kg	kPa	°F	gr/lb	in.hg.	°C	g/kg	kPa
84	179	1.176	28.89	25.58	3.98	46	46	0.312	7.78	6.59	1.05
83	173	1.139	28.33	24.74	3.85	45	44	0.300	7.22	6.34	1.02
82	167	1.103	27.78	23.92	3.73	44	43	0.289	6.67	6.10	0.98
81	162	1.067	27.22	23.13	3.61	43	41	0.278	6.11	5.87	0.94
80	156	1.033	26.67	22.36	3.49	42	39	0.268	5.56	5.64	0.91
79	151	1.000	26.11	21.61	3.38	41	38	0.258	5.00	5.43	0.87
78	146	0.967	25.56	20.89	3.27	40	37	0.248	4.44	5.22	0.84
77	141	0.936	25.00	20.19	3.16	39	35	0.238	3.89	5.02	0.81
76	136	0.905	24.44	19.51	3.06	38	34	0.229	3.33	4.82	0.77
75	132	0.876	23.89	18.85	2.96	37	32	0.220	2.78	4.64	0.74
74	127	0.847	23.33	18.21	2.86	36	31	0.212	2.22	4.46	0.72
73	123	0.819	22.78	17.59	2.77	35	30	0.204	1.67	4.28	0.69
72	119	0.792	22.22	16.99	2.68	34	29	0.196	1.11	4.11	0.66
71	115	0.765	21.67	16.41	2.59	33	28	0.188	0.56	3.95	0.64
70	111	0.740	21.11	15.85	2.50	32	26.5	0.1805	0.00	3.79	0.610
69	107	0.715	20.56	15.30	2.42	31	25.3	0.1724	-0.56	3.62	0.583
68	103	0.691	20.00	14.77	2.33	30	24.2	0.1646	-1.11	3.46	0.556
67	100	0.667	19.44	14.26	2.26	29	23.1	0.1572	-1.67	3.30	0.531
66	96	0.645	18.89	13.76	2.18	28	22.0	0.1501	-2.22	3.15	0.507
65	93	0.622	18.33	13.28	2.10	27	21.0	0.1432	-2.78	3.01	0.484
64	90	0.601	17.78	12.82	2.03	26	20.1	0.1367	-3.33	2.87	0.462
63	86	0.580	17.22	12.37	1.96	25	19.1	0.1304	-3.89	2.74	0.441
62	83	0.560	16.67	11.93	1.89	24	18.3	0.1244	-4.44	2.61	0.420
61	80	0.541	16.11	11.51	1.83	23	17.4	0.1186	-5.00	2.49	0.401
60	78	0.522	15.56	11.10	1.76	22	16.6	0.1131	-5.56	2.37	0.382
59	75	0.504	15.00	10.70	1.70	21	15.8	0.1078	-6.11	2.26	0.365
58	72	0.486	14.44	10.32	1.64	20	15.1	0.1028	-6.67	2.16	0.347
57	70	0.469	13.89	9.95	1.58	19	14.4	0.0980	-7.22	2.05	0.331
56	67	0.452	13.33	9.59	1.53	18	13.7	0.0933	-7.78	1.96	0.315
55	65	0.436	12.78	9.24	1.47	17	13.0	0.0889	-8.33	1.86	0.300
54	62	0.420	12.22	8.90	1.42	16	12.4	0.0847	-8.89	1.77	0.286
53	60	0.405	11.67	8.58	1.37	15	11.8	0.0806	-9.44	1.69	0.272
52	58	0.391	11.11	8.27	1.32	14	11.2	0.0767	-10.00	1.61	0.259
51	56	0.376	10.56	7.96	1.27	13	10.7	0.0730	-10.56	1.53	0.247
50	54	0.363	10.00	7.67	1.23	12	10.2	0.0695	-11.11	1.46	0.235
49	52	0.349	9.44	7.38	1.18	11	9.7	0.0661	-11.67	1.38	0.224
48	50	0.337	8.89	7.11	1.14	10	9.2	0.0629	-12.22	1.32	0.213
47	48	0.324	8.33	6.84	1.10	9	8.8	0.0598	-12.78	1.25	0.202

ASHRAE

Guide for Buildings in Hot & Humid Climates

Psychrometric Chart Inch-Pound Units

Sea Level • Barometric Pressure 29.9 Inches of Mercury

Chart generated by PsychPro software
www.linric.com

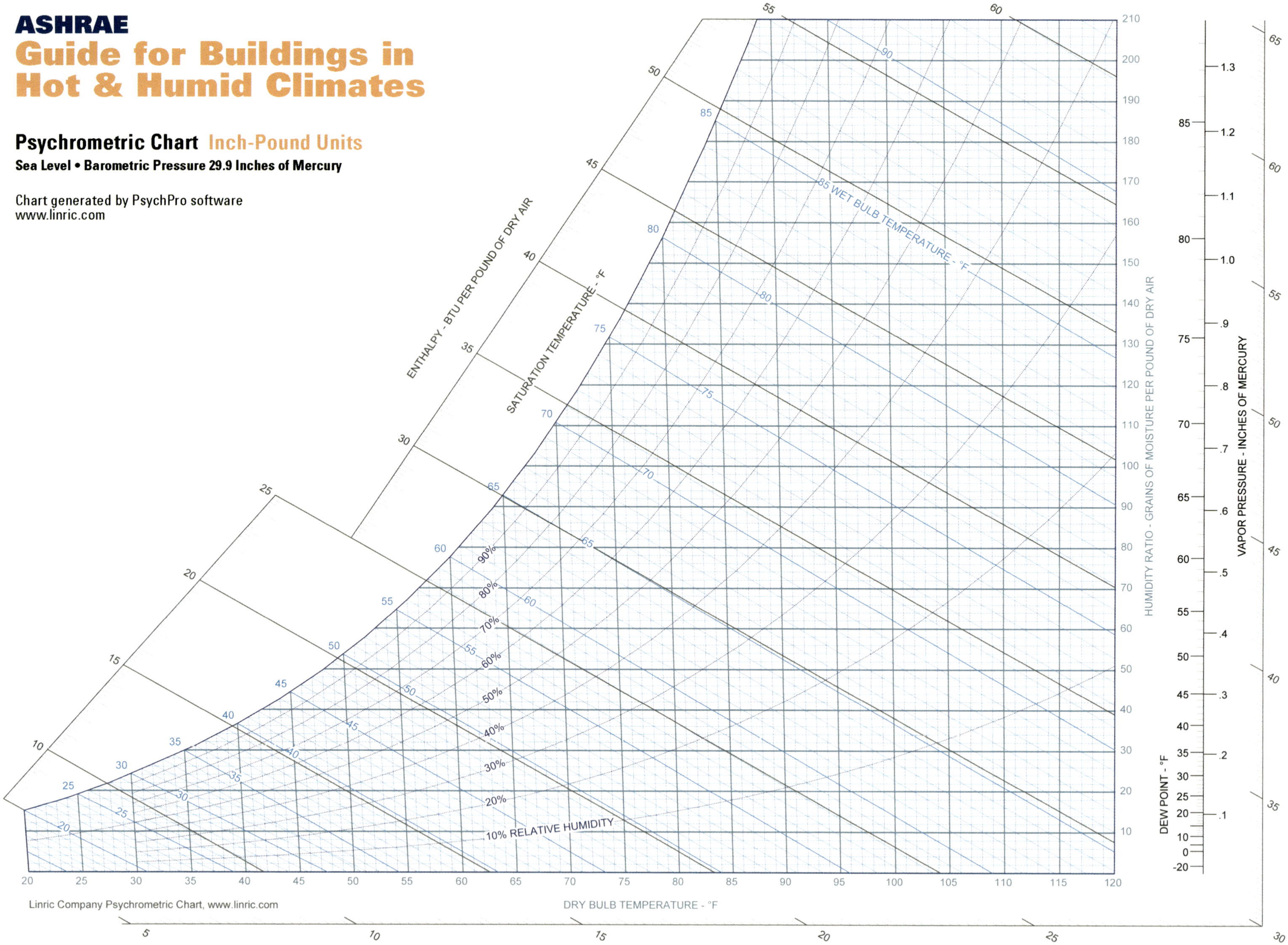

ASHRAE
Guide for Buildings in Hot & Humid Climates

Psychrometric Chart SI Units

Sea Level • Barometric Pressure 760 mm of Mercury

Chart generated by PsychPro software
www.linric.com

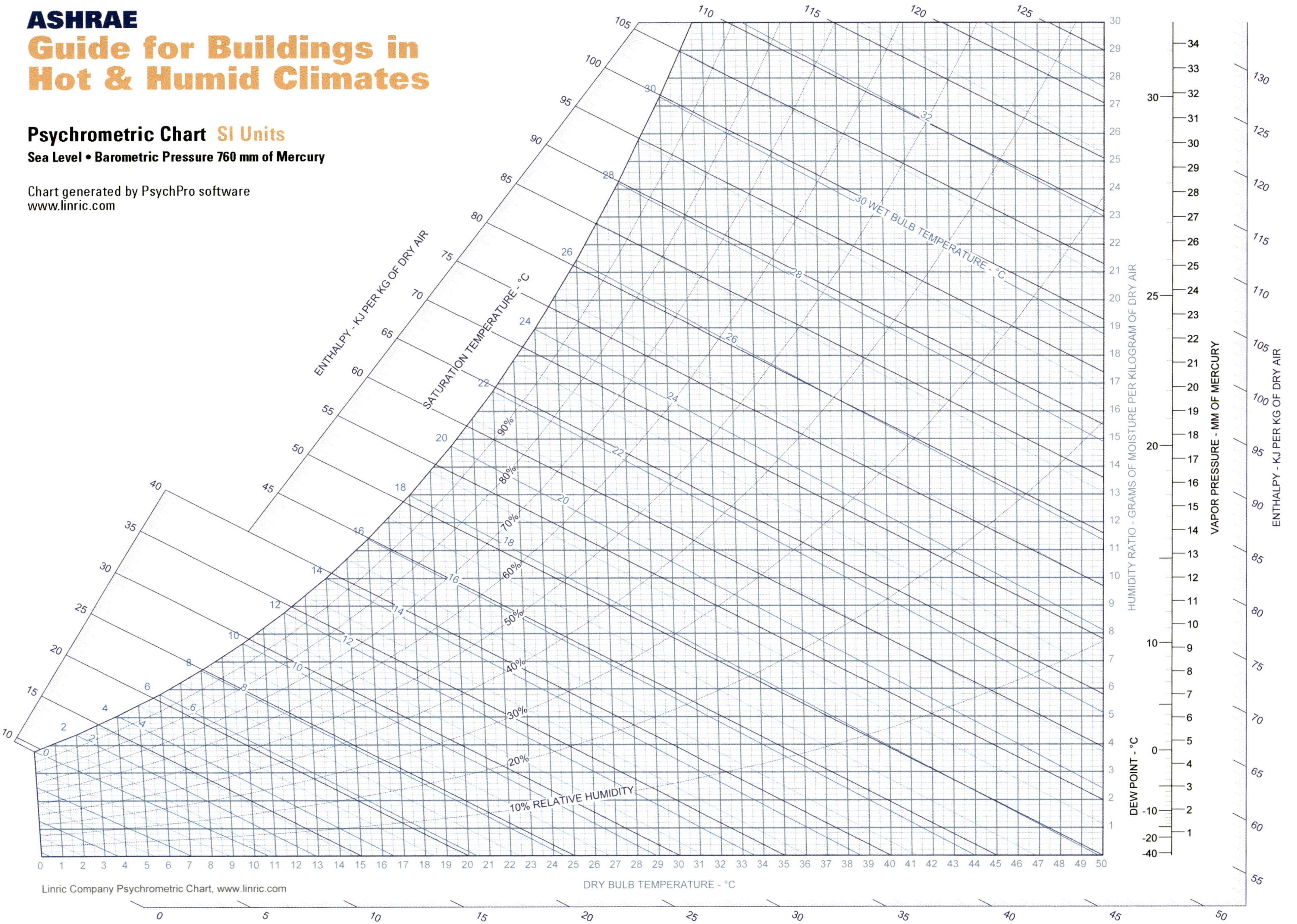

Production notes for the expanded 2nd edition

This book was written, illustrated and designed by Lew Harriman of Mason-Grant Consulting, who also prepared the print-ready electronic flies. Hardware used for the project included an Apple Inc. 15" MacBookPro computer, running Mac OSX 10.5.5. with an Apple 20" external display monitor. Software to write the text and to produce the graphics, layout and calculations included Adobe InDesign CS3, Macromedia Freehand MX, iPhoto, Photoshop CS3, SketchUp Pro, Linric PsychPro and MSoft EXCEL. The body text font is ITC Garamond book condensed. The font for section heads and for the labels in illustrations is ITC Univers condensed.